T0271263

SCATTERING METHODS IN COMPLEX FLUIDS

Summarising recent research on the physics of complex liquids, this in-depth analysis examines the topic of complex liquids from a modern perspective, addressing experimental, computational and theoretical aspects of the field.

Selecting only the most interesting contemporary developments in this rich field of research, the authors present multiple examples including aggregation, gel formation and glass transition, in systems undergoing percolation, at criticality, or in supercooled states. Connecting experiments and simulation with key theoretical principles, and covering numerous systems including micelles, microemulsions, biological systems and cement pastes, this unique text is an invaluable resource for graduate students and researchers looking to explore and understand the expanding field of complex fluids.

SOW-HSIN CHEN is Professor Emeritus of Applied Radiation Physics, Department of Nuclear Science and Engineering, Massachusetts Institute of Technology, specialising in the dynamics of soft condensed matter using thermal neutron, synchrotron X-ray and laser light spectroscopy methods.

PIERO TARTAGLIA is a retired Professor of Physics, Department of Physics, University of Rome 'La Sapienza' and a former member of the Scientific Committee and the Physical Committee of the Italian National Research Council. His research focuses on the dynamic properties of complex liquids and liquid systems.

SCATTERING METHODS IN COMPLEX FLUIDS

SOW-HSIN CHEN

Department of Nuclear Science and Engineering
Massachusetts Institute of Technology

PIERO TARTAGLIA

Dipartimento di Fisica
Università di Roma 'La Sapienza'

CAMBRIDGE
UNIVERSITY PRESS

CAMBRIDGE
UNIVERSITY PRESS

University Printing House, Cambridge CB2 8BS, United Kingdom

One Liberty Plaza, 20th Floor, New York, NY 10006, USA

477 Williamstown Road, Port Melbourne, VIC 3207, Australia

314-321, 3rd Floor, Plot 3, Splendor Forum, Jasola District Centre, New Delhi - 110025, India

79 Anson Road, #06-04/06, Singapore 079906

Cambridge University Press is part of the University of Cambridge.

It furthers the University's mission by disseminating knowledge in the pursuit of education, learning and research at the highest international levels of excellence.

www.cambridge.org
Information on this title: www.cambridge.org/9780521883801

© Cambridge University Press 2015

First published 2015

A catalogue record for this publication is available from the British Library

Library of Congress Cataloging in Publication data
Chen, Sow-Hsin, 1935– author.
Scattering methods in complex fluids : selected topics / Sow-Hsin Chen
(Department of Nuclear Science and Engineering, Massachusetts Institute of Technology),
Piero Tartaglia (Dipartimento di Fisica, Università degli Studi di Roma 'La Sapienza', Italy).
    pages   cm
Includes bibliographical references.
ISBN 978-0-521-88380-1
1. Complex fluids.   2. Scattering (Physics)   I. Tartaglia, Piero,
1942- author.   II. Title.
QD549.2.C66C44   2015
530.4'2–dc23   2014035058

ISBN 978-0-521-88380-1 Hardback

Contents

Illustrations

List of illustrations

Preface

The central theme of this book is 'slow dynamics in supercooled, glassy liquids and dense colloidal systems' which has been an intense area of current research for some time. Although it can be well described by the mode-coupling theory of dense liquids, controversial viewpoints persist. Thus, the authors have written about the exciting modern aspects of the physics of liquids by selecting only the most interesting contemporary development in this rich field of research in the last decades.

This book presents and summarises a wide variety of recent research on the physics of complex liquids and suggests that the use of established techniques, essentially neutron, X-ray and light scattering together with theoretical and computer molecular dynamic simulation approaches, can be fruitfully applied to solve many new phenomena. These techniques are also central to investigating new interesting findings in liquid water such as liquid–liquid transition and its associated low-temperature critical point.

Although many materials found in nature can be classed as complex fluids, the authors have chosen to focus on water and colloids in this book for the following reasons:

- Water is the most important liquid for life on Earth. It covers 71% of the Earth's surface and is probably the most ubiquitous, as well as the most essential, molecule on Earth. It is a vital element controlling not only all aspects of life itself but also the environmental factors that make life enjoyable. Water is a simple molecule yet possesses unique and anomalous properties at low temperatures that have fascinated scientists for many years. Thus in selecting the categories of complex liquids to include in this book, water is the obvious top choice.
- Colloids are another class of complex liquids characterised by the slowing down of the dynamics. They are becoming increasingly studied for their potential applications and the availability of degrees of freedom that are relatively simple

to vary experimentally through physical and chemical control parameters, giving rise to a much larger variety of phenomena compared to simple liquids. Initially a few relevant and classical examples of clustering and percolation in supramolecular colloidal aggregates are treated. Then various aspects of the physics of complex liquids are considered, focusing in particular on glass transition in colloidal systems, emphasising the role of the mode-coupling theory of the kinetic glass transition. The theory predicted and allowed us to study in detail many interesting new phenomena in colloids, such as re-entrant transitions and higher-order singularities in systems where short-range attraction is added to the usual short-distance repulsion between particles.

In order to provide an in-depth analysis examining the topics of complex liquids from a modern perspective, addressing the experimental, computational and theoretical aspect of the field, the book consists of ten chapters divided into three parts:

- Part I with three chapters deals with 'scattering and liquids'.
- Part II with three chapters deals with 'structural arrest' phenomena.
- Part III with four chapters deals mainly with 'water'.

Setting a good foundation for the rest of the book, the first two chapters cover elements of scattering techniques and theories commonly used in studying the structure and dynamics of liquid state matter. They are the outgrowth of parts of SHC's lecture notes used in two of his graduate courses at MIT for many years – 'Photon and neutron scattering spectroscopy and its applications in condensed matter' and 'Statistical thermodynamics of complex liquids'. In some of the chapters, certain sections are prefaced 'Module' to show that the topics they cover are significant, although they may not be in strict sequential order within the chapter.

Both authors, SHC and PT, have spent a large portion of their lives studying complex liquids, specifically water and colloidal systems, and collectively they have published several hundred research papers on these topics. Furthermore, they have been collaborating on these subjects for over 40 years, which has resulted in more than 50 joint scientific papers. Thus, it is natural for them to want to complement their mutual research interests and summarise their respective research on these topics throughout these long and productive years. A selection of arguments is made in the book, collecting what they consider relevant to the modern physics of liquids, in order to share their knowledge and insights with their readers. The research coverage is very up-to-date to June 2014.

This unique book should be of interest to all scientists who are interested in the dynamical properties of glassy liquids. It will also be an invaluable resource for science and engineering graduate students and researchers looking to explore and understand the advancing field of complex fluids.

The authors want to thank their colleagues, former Ph.D. students and post-doctoral associates with whom they have shared many research topics reported in this book. They acknowledge in particular the long and fruitful collaboration with, amongst others, Professors Chung-Yuan Mou, Francesco Mallamace, Piero Baglioni and Paola Gallo (for SHC), and Francesco Sciortino (for both PT and SHC). For over a decade, all Chen's research projects have been funded by the Office of Basic Energy Sciences of the US Department of Energy. Their support is gratefully acknowledged. SHC also wants to thank Dr. P. Thiyagarajan for his encouragement and discussions.

During the course of preparing and writing this book, for SHC, due to his physical limitation in efficient typing, particularly the equations, he wishes to gratefully acknowledge the persistent help from his Ph.D student, Zhe Wang, for preparing some initial background information and typing revised paragraphs and equations of a few initial chapters during 2011–2013. He appreciates his postdoc Kao-Hsiang Liu for similar help in a few months around late 2012. He also wishes to thank his Ph.D. student, Peisi Le, who has redrawn numerous required graphics from April to August 2014. To his co-author, PT, he has enjoyed greatly the extensive discussions and interactions with PT, and appreciates greatly the additional typesetting work his co-author, PT, had to take on his behalf. Finally, SHC is particularly thankful to his wife, Ching-chih, for her indispensible and time-consuming work in reformatting all his chapters and references, converting and processing many colour graphics to publishable form, as well as performing the final editing, proofreading and indexing of his part of the book, without whose efforts this work could never have been completed.

Sow-Hsin Chen at MIT, Cambridge, MA
Piero Tartaglia at University of Rome "La Sapienza", Rome
January 1, 2015

Part I

Scattering and liquids

1

Scattering techniques for the liquid state

1.1 Introduction

1.1.1 Radiation scattering used for condensed matter spectroscopy

Spectroscopy measures the structure and dynamics of the ground or low-lying excited states of condensed matter. Radiation is useful as a tool for spectroscopy if it couples weakly (in a sense to be discussed later) to the many-body system. In this case the double differential cross-section (per unit solid angle, per unit energy transfer) for the radiation scattering can be written schematically as

$$\frac{d^2\sigma}{d\Omega d\omega} \approx \left(\frac{d\sigma}{d\Omega}\right)_0 \sum_{i,f} P_i \left| \langle f | \sum_{l=1}^{n} e^{i(\mathbf{k_1}-\mathbf{k_2})\cdot\mathbf{r}_l} | i \rangle \right|^2 \delta(\hbar\omega - E_i + E_f). \tag{1.1}$$

In Eq. (1.1) the first factor $\left(\frac{d\sigma}{d\Omega}\right)_0$ refers to the differential scattering cross-section from the basic unit of scattering medium in the system and the second factor, usually called the dynamic structure factor, represents the time-dependent structure of the system as seen by the radiation. This clear separation of the basic scattering problem, as represented in the first factor, from the dynamic structure factor of the many-body system itself, is only possible when the radiation couples weakly to the system, and therefore the use of Born approximation in deriving Eq. (1.1) is valid. Both thermal neutrons and photons with energy up to the X-ray region satisfy this criterion and thus are useful as probes for condensed matter time-dependent structures. The dynamic structure factor contains two parameters related to the energy and momentum of the probe, namely, the momentum transfer (Eq. (1.2a)) and the energy transfer (Eq. (1.2b)) to the system in the scattering process:

$$\hbar\mathbf{Q} = \hbar(\mathbf{k_1}-\mathbf{k_2}), \tag{1.2a}$$

$$\hbar\omega = \hbar(\omega_1 - \omega_2). \tag{1.2b}$$

A schematic diagram of indicating the incident and scattered radiation with respect to the scattering medium is shown in Figure 1.1. In general the double differential cross-section depends on all four parameters: $\vec{k}_1, \omega_1, \vec{k}_2, \omega_2$, namely,

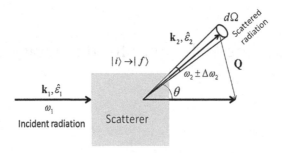

Figure 1.1 Schematic diagram of scattering geometry.

both the incident and the scattered wave vectors and frequencies. But this complexity is reduced greatly when the Born approximation is applicable. In this case the dynamic structure factors depend, apart from the state of the system, on only two external parameters, Q and ω. This has an immediate significant experimental implication. We shall explain in Section 1.1.6 that, qualitatively speaking, when the radiation imparts momentum $\hbar \vec{Q}$ and energy $\hbar \omega$ to the system it effectively probes the structure and dynamics of the system with a spatial resolution of $R = 2\pi Q^{-1}$ and time scale of $\tau = 2\pi \omega^{-1}$. For the purpose of this introductory chapter, we are mostly interested in situations where $nk_1 \simeq nk_2 \simeq k$ and n is the refractive index of the medium. In this case the magnitude of the wave vector transfer Q, which we call the Bragg wave number, can be expressed in terms of the wave number k and scattering angle θ as

$$Q = 2k \, \sin \frac{\theta}{2}. \tag{1.3}$$

In order to probe the spatial structure of the system at different levels, one would like to change Q accordingly. For instance, to detect a periodic structure of spacing $d \approx 10\,\text{Å}$, one needs to have $Q \simeq 2\pi/d \simeq 0.628\,\text{Å}^{-1}$. One can use cold neutrons of wavelength $\lambda \approx 4\,\text{Å}$ ($k = 2\pi/\lambda = 1.57\,\text{Å}^{-1}$) and work at a scattering angle around $\theta \approx 23°$ so that $2k \sin \frac{\theta}{2} = 2 \times 1.57 \sin \left(\frac{23°}{2}\right) \simeq 0.628\,\text{Å}^{-1}$. Alternatively one can use X-rays of wavelength $\lambda = 0.62\,\text{Å}$ and work at an angle around $\theta \approx 3.6°$. Or one can even use γ-rays of $\lambda = 0.03\,\text{Å}$ (412 keV γ-rays from gold Au[198]) and work at an angle around $\theta \approx 0.17°$. Similar consideration applies to the frequency or the time scale. Since the relevant variable is the difference between incident and scattered energies, one can tune the dynamic range of the probe by varying the accuracy of the energy difference measurement. A good example is a neutron scattering study of polymer chain dynamics by Higgins *et al.* (1981), using the so-called spin-echo technique. Using 8 Å neutrons ($\hbar \omega_1 = 1279\,\mu\text{eV}$) they were able to measure energy difference of $E = \hbar \omega$ in the scattered neutrons and thus were able to probe the chain dynamics at a time scale τ of $0.4\,\mu\text{s}$ over the

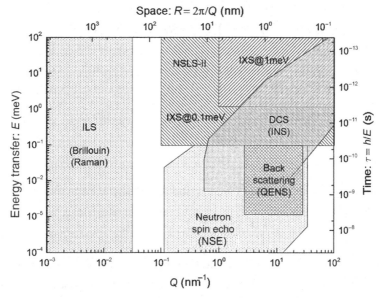

Figure 1.2 The $(Q, E = \hbar\omega)$ ranges covered in the scattering experiment and the corresponding space-time, (R, τ), ranges probed in the material system for each of the three scattering techniques. Due to the recent advances in instrumentation, the overlaps between neutron and X-ray scattering spectroscopy are clearly visible. The disk chopper spectrometer (DCS, used for INS), quasi-elastic backscattering spectrometer (QENS) and neutron spin echo spectrometer (NSE) are presently available in both NIST Center For Neutron Research (NCNR) and Spallation Neutron Source (SNS) in Oak Ridge National Laboratory in USA.

distances scale of up to $2\pi/Q = 200\,\text{Å}$ $(Q = 0.03\,\text{Å}^{-1})$. Figure 1.2 summarises the present status of neutron, X-ray and light-scattering spectroscopy relevant to the discussions here in terms of its (Q, E) and its corresponding (R, τ) space-time coverage. The important point to notice is the regions of overlap of the three scattering methods due to the recent advances in techniques and instrumentations of neutron and X-ray spectroscopies.

The primary purpose of this chapter is to derive the important relation given in Eq. (1.1) and to identify in each of the three cases – thermal neutron, X-ray and light scattering – the corresponding basic scattering unit that contributes to the factor $\left(\frac{d\sigma}{d\Omega}\right)_0$. We shall comment on the meanings of the dynamic structure factor using some examples in Section 1.1.6.

1.1.2 Thermal neutron scattering

Consider a combined system of a material medium, described by a Hamiltonian H_s, and a neutron with a kinetic energy $\frac{p^2}{2m}$ and their mutual interaction V. The time-independent Schrödinger equation for the system is

$$\left(\frac{p^2}{2m} + H_s + V\right)\Phi(\mathbf{r}, \{R\}) = E\Phi(\mathbf{r}, \{R\}). \tag{1.4}$$

We denote by $\mathbf{r}$ position of the neutron and R the collection of coordinates of all particles in the system. The material system has a set of stationary states $\{|n\rangle\}$ defined by

$$H_s|n\rangle = E_n|n\rangle, \tag{1.5a}$$

$$\langle n|n'\rangle = \delta_{nn'}. \tag{1.5b}$$

Using this set of stationary states we expand the total wave function as

$$\Phi(\mathbf{r}, \{R\}) = \sum_n \langle \psi_n(\mathbf{r})|n\rangle. \tag{1.6}$$

Substituting Eq. (1.6) into Eq. (1.4) and taking a scalar product with $\langle n|$ on both sides, we obtain a wave equation for the neutron in the presence of the material system:

$$(\nabla^2 + K_n^2)\psi_n(\mathbf{r}) = \frac{2M}{\hbar^2} \sum_{n'} \langle n|V|n'\rangle \psi_{n'}(\mathbf{r}), \tag{1.7}$$

where $\frac{\hbar^2 K_n^2}{2M} = E - E_n$ is the kinetic energy of the neutron in the system.

The neutron wave (Eq. (1.7)) can be used in two ways. The first application is to consider the propagation of neutrons in the material medium. For this application we take the system to be in the ground state, i.e. $|n\rangle = |n'\rangle = |0\rangle$, and introduce an optical potential $U(r) = \langle 0|V|0\rangle$. Then putting $\psi_0 \equiv \psi$ and $K_0 \equiv K$, Eq. (1.7) reduces to a wave equation:

$$(\nabla^2 + K^2)\psi(\mathbf{r}) = \frac{2M}{\hbar^2}U(\mathbf{r})\psi(\mathbf{r}). \tag{1.8a}$$

One normally takes the optical potential to be a pseudo-potential (Sears, 1978),

$$U(\mathbf{r}) = \frac{2\pi\hbar^2}{M} \sum_{l=1}^{n} b_l\delta(\mathbf{r} - \mathbf{R_l}). \tag{1.8b}$$

A simple example is to take a homo-nuclear system with identical bound atom scattering length $b_j = b$ (Koester et al., 1981),[1] and work out the index of refraction of neutrons in the medium.

---

[1] Bound scattering length refers to scattering length of a nucleus that is fixed in space. Measurement of scattering length is normally made in a situation where the nucleus is free to recoil. The measurement gives the free scattering length a, which is related to b by $b = \frac{A+1}{A}a$, where A is atomic weight of the nucleus. See Koester et al. (1981) for tabulation of values of b.

Write

$$U(\mathbf{r}) = \frac{2\pi\hbar^2 b}{M} \sum_{l=1}^{n} \delta(\mathbf{r} - R_l) = \frac{2\pi\hbar^2 b}{M} n(\mathbf{r}), \tag{1.8c}$$

where $n(r)$ is the local number density of nuclei. Putting Eq. (1.8c) into Eq. (1.8a) we have

$$(\nabla^2 + K^2)\psi(\mathbf{r}) = 4\pi bn(\mathbf{r})\psi(\mathbf{r}). \tag{1.9a}$$

Take a plane wave $\psi(\mathbf{r}) = \exp(i\mathbf{K}' \cdot \mathbf{r})$ propagating in an optically homogeneous medium where we can set $n(r) = n$ (this is valid when no Bragg condition is satisfied). We have from Eq. (1.9a) a relation

$$-K'^2 + K^2 = 4\pi bn. \tag{1.9b}$$

From Eq. (1.9b) the index of refraction follows as

$$n \equiv \frac{K'}{K} = \sqrt{1 - \frac{4\pi bn}{K^2}} = 1 - \frac{bn\lambda^2}{2\pi}. \tag{1.10}$$

Take a typical case of nickel for which $n = 9.13 \times 10^{22}$ cm^{-3} and $b = 1.03 \times 10^{-12}$ cm. We have for a 4 Å neutron, $\frac{bn\lambda^2}{2\pi} \sim 2.4 \times 10^{-5}$. The refractive index of a material with $b > 0$ is therefore slightly optically rare with respect to neutron wave. Notice the λ^2 dependence in Eq. (1.10). This means for ultra cold neutrons of $\lambda = 800$ Å (speed ≈ 5 m/sec), the index of refraction is as large as $n = 0.04$ and the total reflection from the surface of Ni film of up to $\theta_c = 87.7°$ is possible. If one takes into account the periodic variation of $n(r)$ in a crystalline solid then one can proceed to work out the dynamic theory of neutron diffraction in a perfect crystal (Rauch and Petrascheck, 1978). The second application is to solve Eq. (1.5) for the scattered wave function of neutrons. Scattering of neutrons in general induces excitation or de-excitation of the medium, so the matrix element $\langle n|V|n'\rangle$ of the interaction potential has to be evaluated between the initial and final states of the scattering medium. To calculate the scattering cross-section, we are interested in the asymptotic wave function at distances large compared to the sample size. First, solve for Green's function:

$$(\nabla^2 + K_n^2)G_n(|\mathbf{r} - \mathbf{r}'|) = -\delta(\mathbf{r} - \mathbf{r}'), \tag{1.11}$$

which gives

$$G_n(|\mathbf{r} - \mathbf{r}'|) = \frac{e^{iK_n|\mathbf{r}-\mathbf{r}'|}}{4\pi|\mathbf{r} - \mathbf{r}'|} \xrightarrow[r\gg r']{} \frac{e^{iK_n r}}{4\pi r} e^{iK_n \hat{\mathbf{r}} \cdot \mathbf{r}'}. \tag{1.12}$$

Constructing an inhomogeneous solution from Eq. (1.12) and adding to it the homogeneous solution which represents the incident plane wave, we get

$$\psi_n(\mathbf{r}) = \delta_{nn_0}e^{i\mathbf{K}\cdot\mathbf{r}} - \frac{e^{iK_n r}}{r}\left(\frac{M}{2\pi\hbar^2}\right)\sum_{n'}\int_V d^3r' e^{-i\mathbf{K}_n\cdot\mathbf{r}'}V_{nn'}(\mathbf{r}')\psi_{n'}(\mathbf{r}'), \quad (1.13)$$

where $\mathbf{K} \equiv \mathbf{k}_1$ denotes the incident wave vector, $\mathbf{K}_n \equiv \mathbf{k}_2 = K_n\hat{r}$ denotes the scattered wave vector and the integration is over the volume the system.

To go further from here one makes the Fermi approximation, which essentially consists of two parts: make the Born approximation by taking

$$\psi_{n'}(\mathbf{r}') \simeq \delta_{n'n_0}e^{i\mathbf{K}\cdot\mathbf{r}'}, \quad (1.14a)$$

and simultaneously use the pseudo-potential

$$V(\mathbf{r}) = \frac{2\pi\hbar^2}{M}\sum_{l=1}^{n}b_l\delta(\mathbf{r}-\mathbf{R}_l), \quad (1.14b)$$

to get

$$\psi_n(\mathbf{r}) = \delta_{nn_0}e^{i\mathbf{K}\cdot\mathbf{r}} - \frac{e^{iK_n r}}{r}\left(\frac{M}{2\pi\hbar^2}\right)\sum_{n'}\int_V d^3r' e^{i(\mathbf{K}-\mathbf{K}_n)\cdot\mathbf{r}'}V_{nn_0}(\mathbf{r}'). \quad (1.15)$$

There are a number of discussions in the literature with regard to the Fermi approximation, such as Sachs (1953), so we shall not go into it here. The accuracy of this approximation is estimated to be of the order of 0.1% in the case of a hydrogen atom bound in a molecule. Taking a typical neutron–nuclear potential of depth $-V_0 \approx 36\,\text{MeV}$ and width $r_0 \approx 2 \times 10^{-13}$ cm, the validity of the Fermi approximation rests on the following considerations:

(i) $Kr_0 \ll 1$ (in fact $\approx 10^{-4}$) – a consideration for validity of low-energy scattering from a bound nucleus where a scattering length b is sufficient to characterise the cross-section.
(ii) For a square-well potential $b \propto V_0 r_0^3$ and one can redefine fictitious potential parameters $\bar{V}_0$, $\bar{r}_0$ such that not only the scattering length is preserved, i.e. $\bar{V}_0\bar{r}_0^3 = V_0r_0^3$, but at the same time the condition for validity of the Born approximation is also satisfied, i.e. $\frac{m\bar{V}_0\bar{r}_0^3}{\hbar^2} \ll 1$.
(iii) The fictitious range of nuclear force $\bar{r}_0$ chosen above can still be much smaller than amplitude of the zero point vibration A of the bound nucleus in the molecule.

Physically speaking, even though neutron–nuclear interaction is so strong that the Born approximation is not applicable, a fortunate situation arises because the interaction is over such a short range that one can smear out the interaction range considerably, so as to decrease its strength in such a way that the Born approximation can be used. In practice, the amplitude of the zero point vibration A is 10^{-9} cm and $\bar{r}_0$ can be taken to be $100\,r_0 = 2 \times 10^{-11}$ cm. We now define the scattering amplitude $f_{nn_0}(\theta)$ by writing Eq. (1.15) in asymptotic form:

$$\psi_n(\mathbf{r}) = \delta_{nn_0} e^{i\mathbf{K}\cdot\mathbf{r}} + f_{nn_0}(\theta)\frac{e^{iK_n r}}{r}, \tag{1.16a}$$

and identify

$$f_{nn_0}(\theta) = -\sum_{l=1}^{N} b_l \langle n | e^{i(\mathbf{k}_1 - \mathbf{k}_2)\cdot\mathbf{R}_l} | n_0 \rangle. \tag{1.16b}$$

The first term in Eq. (1.16a) represents the incident wave ψ_{inc} while the second term represents the scattered wave ψ_{sc} in the region $\theta \neq 0$. The differential cross-section for elastic scattering $d\sigma$ can then be obtained by calculating the ratio of the number of neutrons elastically scattered into $d\Omega$ in the direction of $\mathbf{k}_2$ per second to the incident neutron flux,

$$d\sigma = \frac{k_2 |\psi_{sc}|^2}{k_1 |\psi_{inc}|^2} r^2 d\Omega, \tag{1.17a}$$

or

$$\frac{d\sigma}{d\Omega} = \frac{k_2}{k_1} \sum_{n_0} P_{n_0} |f_{n_0 n_0}(\theta)|^2, \tag{1.17b}$$

where Eq. (1.17b) is obtained from Eq. (1.17a) by averaging over the initial distribution P_{n_0} of the initial states $|n_0\rangle$ of the system.

The inelastic scattering cross-section can likewise be obtained from the scattering amplitude by

$$\frac{d^2\sigma}{d\Omega d\omega_2} = \frac{k_2}{k_1} \sum_{n,n_0} P_{n_0} |f_{n,n_0}(\theta)|^2 \delta\left(\omega_1 - \omega_2 + \frac{E_{n_0} - E_n}{\hbar}\right), \tag{1.18a}$$

where an additional average over the unknown final states $|n\rangle$ is made together with a delta function factor, ensuring the energy conservation

$$\hbar\omega = \hbar\omega_1 - \hbar\omega_2 = E_n - E_{n_0}. \tag{1.18b}$$

Using the expression for the scattering amplitude (Eq. (1.17a)) in Eq. (1.18a) we then get (by writing $|i\rangle \equiv |n_0\rangle$, $|f\rangle \equiv |n\rangle$)

$$\frac{d^2\sigma}{d\Omega d\omega_2} = \frac{k_2}{k_1} \sum_{i,f} P_i |\langle f | \sum_{l=1}^{N} b_l e^{i\mathbf{Q}\cdot\mathbf{R}_l} | i \rangle|^2 \delta\left(\omega + \frac{E_i - E_f}{\hbar}\right). \tag{1.18c}$$

In expression similar to Eq. (1.1) is recovered except for a trivial kinematic factor k_2/k_1. When all nuclei are identical, i.e. $b_j = b$,

$$\frac{d^2\sigma}{d\Omega d\omega_2} = b^2 \frac{k_2}{k_1} \sum_{i,f} P_i |\langle f| \sum_{l=1}^{N} e^{i\mathbf{Q}\cdot\mathbf{R}_l}|i\rangle|^2 \delta\left(\omega + \frac{E_i - E_f}{\hbar}\right). \tag{1.18d}$$

We note that in this case $\left(\frac{d\sigma}{d\Omega}\right)_0 = b^2$ is the differential cross-section for a single bound state nucleus.

Coherence and incoherence

The cross-sections in Eqs. (1.17b) and (1.18c) contain a coherent superposition of phase factors from each nuclei $\exp(i\mathbf{Q} \cdot \mathbf{R}_l)$ only when all nuclei look identical to incoming neutrons. In practice there are two sources of incoherence. First, even a chemically homogeneous system contains isotopes of the same element. Second, neutron–nuclear interaction is spin dependent, namely the scattering length is dependent on mutual orientations of neutron spin relative to nuclear spin. Therefore, even for an unpolarised incident neutron beam, Eq. (1.18c) has to be averaged over all possible nuclear isotopic and spin states. We denote this average by a bar. Consider then the factor

$$\overline{\sum_{i,f} P_i |\langle f| \sum_{l=1}^{N} b_l e^{i\mathbf{Q}\cdot\mathbf{R}_l}|i\rangle|^2 \delta\left(\omega + \frac{E_i - E_f}{\hbar}\right)}, \tag{1.19a}$$

which, by using an integral representation of the delta function,

$$\delta\left(\omega + \frac{E_i - E_f}{\hbar}\right) = \frac{1}{2\pi} \int_{-\infty}^{\infty} dt e^{-i\omega t} e^{\frac{i}{\hbar}(E_f - E_i)t}, \tag{1.19b}$$

can easily be transformed into the time-dependent form:

$$\frac{1}{2\pi} \int_{-\infty}^{\infty} dt e^{-i\omega t} \overline{\sum_i P_i |\langle i| \sum_{l,l'}^{N} b_l b_{l'} e^{-i\mathbf{Q}\cdot\mathbf{R}_l(0)} e^{i\mathbf{Q}\cdot\mathbf{R}_l(t)}|i\rangle|}. \tag{1.19c}$$

If one now assumes that the isotopic states and spin states of each nucleus are completely uncorrelated with its position one can rewrite the above expression as

$$\frac{1}{2\pi} \int_{-\infty}^{\infty} dt e^{-i\omega t} \sum_i P_i |\langle i| \sum_{l,l'}^{N} \overline{b_l b_{l'}} e^{-i\mathbf{Q}\cdot\mathbf{R}_l(0)} e^{i\mathbf{Q}\cdot\mathbf{R}_l(t)}|i\rangle|. \tag{1.19d}$$

Consider now the average

$$\overline{b_l b_{l'}} = \overline{b_l^2} = \overline{b^2}, \quad \text{when } l = l',$$
$$\overline{b_l b_{l'}} = (\overline{b})^2, \quad \text{when } l \neq l',$$

and combine both equations into one equation,

$$\overline{b_l b_{l'}} = [\overline{b^2} - \bar{b}^2]\delta_{ll'} + (\bar{b})^2 \equiv b_{\text{inc}}^2 \delta_{ll'} + b_{\text{coh}}^2. \tag{1.20}$$

Substituting Eq. (1.20) back into Eq. (1.19d), we see that the dynamic structure factor (Eq. (1.19d)) can always be decomposed into a single-particle term (incoherent scattering term):

$$Nb_{\text{inc}}^2 S_s(\mathbf{Q}, \omega) = Nb_{\text{inc}}^2 \int_{-\infty}^{\infty} \frac{dt}{2\pi} e^{-i\omega t} \sum_i P_i \langle i | e^{-i\mathbf{Q}\cdot\mathbf{R}(0)} e^{i\mathbf{Q}\cdot\mathbf{R}(t)} | i \rangle, \tag{1.21a}$$

and a pair-of-particles term (coherent scattering term):

$$Nb_{\text{coh}}^2 S(\mathbf{Q}, \omega) = Nb_{\text{coh}}^2 \int_{-\infty}^{\infty} \frac{dt}{2\pi} e^{-i\omega t} \sum_i P_i \langle i | \frac{1}{N} \sum_{l,l'} e^{-i\mathbf{Q}\cdot\mathbf{R}_l(0)} e^{i\mathbf{Q}\cdot\mathbf{R}_{l'}(t)} | i \rangle. \tag{1.21b}$$

It is of interest to work out an example of spin incoherence for two isotopes of hydrogen. For a given nuclear spin I there are two possible spin states of the neutron–nuclear system, as shown in the following:

Combined spin states	Multiplicity	Scattering length
$I + 1/2$	$2I + 2$	a_+
$I - 1/2$	$2I$	a_-

Therefore

$$\bar{b} = \frac{2I + 2}{4I + 2} a_+ + \frac{2I}{4I + 2} a_-, \tag{1.22a}$$

$$\overline{b^2} = \frac{2I + 2}{4I + 2} |a_+|^2 + \frac{2I}{4I + 2} |a_+|^2. \tag{1.22b}$$

For light hydrogen (H) and heavy hydrogen (deuterium, D) we have the values reported in Table 1.1.

It should be noted that, because $a_- < 0$, in H^1 the coherent cross-section is much smaller than the incoherent cross-section. Since the average scattering length $\bar{a}$ of H^1 is negative, it is possible for example to mix a proportion of light water and

Table 1.1 *Light and heavy hydrogen*

	I	a_+ $(10^{-12}\,cm)$	a_- $(10^{-12}\,cm)$	$\bar{a}$ $(10^{-12}\,cm)$	σ_{coh} (barn)	σ_{inc} (barn)
H^1	1/2	1.085	−4.750	−0.3742	1.77	79.9
D^2	1	0.950	0.098	0.6671	5.59	2.2

heavy water so that the average scattering amplitude of the mixture changes continuously from a negative value to positive value: $\bar{b}$ for O^{16} is 0.5804×10^{-12} cm; therefore, $\bar{b}(H_2O) = -0.168 \times 10^{-12}$ cm, and $\bar{b}(D_2O) = 1.915 \times 10^{-12}$ cm. The mixture has an average coherent length per unit volume (in units of 10^{-12} cm Å^{-3}) $\langle b \rangle = [-0.00562 + 0.0697 P_{D_2O}]$ where P_{D_2O} is the percentage of D_2O in water. This possibility is the basis for the so-called contrast variation method.

1.1.3 X-ray scattering

The scattering of X-rays is usually treated classically in standard textbooks, starting from Maxwell's equations. This is usually satisfactory for understanding X-ray diffraction phenomena and also for the propagation of X-rays in perfect crystals. However, besides these coherent elastic phenomena, there are other incoherent inelastic phenomena such as Compton scattering and X-ray Raman scattering. In order to include these latter phenomena in a coherent way, a quantum mechanical treatment is more suitable. We shall therefore use a non-relativistic quantum electrodynamic treatment for X-ray scattering (Landau and Lifshitz, 1960). The primary interaction of X-rays with a material system is through photon–electron coupling. The standard Hamiltonian of an N-electron system in the presence of an electromagnetic field is

$$H = H_s + H_R + \sum_{j=1}^{N} -\frac{e}{mc}\overrightarrow{p}_j \cdot \overrightarrow{A}_j + \sum_{j=1}^{N} \frac{e^2}{2mc^2}A_j^2, \tag{1.23a}$$

where the first term is the Hamiltonian of the electronic system in the absence of the electromagnetic (EM) field, the second term is the Hamiltonian of a pure radiation field, and the third (V_1) and fourth terms (V_2) are interaction terms; $\overrightarrow{p}_j$ is the momentum operator of the j-th electron (m is the electron mass) and $\overrightarrow{A}_j$ is the vector potential of the EM field at the position of the j-th electron, given by

$$\overrightarrow{A}_j = \sum_{k,\lambda} \sqrt{\frac{2\pi\hbar c}{kV}}\hat{\varepsilon}_{k\lambda}[a_{k\lambda}e^{i\mathbf{k}\cdot\mathbf{r_j}} + a_{k\lambda}^+ e^{-i\mathbf{k}\cdot\mathbf{r_j}}]. \tag{1.23b}$$

In Eq. (1.23b), $\hat{\varepsilon}_k$ is the polarisation vector of photon state $|k, \lambda\rangle$, $a_{k\lambda}$, $a_{k\lambda}^+$ are the annihilation and creation operators of the photons, and V is the volume of the box enclosing the EM radiation. In terms of the creation and annihilation operators, the Hamiltonian of the radiation field can be written as

$$H_R = \sum_{k,\lambda} \hbar\omega_k \left(a_{k\lambda}^+ a_{k\lambda} + \frac{1}{2}\right). \tag{1.23c}$$

Single free electron initially at rest

Since the coupling of an EM wave to an electron is weak (coupling constant $\approx 1/137$), we can use first-order time-dependent perturbation theory. In the free electron case, $\vec{p} = 0$, and we have only to consider V_2 interaction. We want to calculate the transition rate from an initial state

$$|i\rangle = |A; 1_{k\,\lambda}, 0_{k'\,\lambda'}\rangle, \tag{1.24a}$$

to a final state

$$|f\rangle = |B; 0_{k\lambda}, 1_{k'\lambda'}\rangle, \tag{1.24b}$$

where A and B denote the initial and final states of the electron. Using the golden rule

$$W_{fi} = \frac{2\pi}{\hbar^2}|V_{fi}|^2 \rho(E_f), \tag{1.25a}$$

and noting that

$$
\begin{aligned}
V_{fi} &= \langle B; 0_{k\lambda}, 1_{k'\lambda'}|\frac{e^2}{2mc^2}A^2|A; 1_{k\lambda}, 0_{k'\lambda'}\rangle \\
&= \frac{e^2}{mc^2}\frac{2\pi\hbar c}{V}\frac{1}{\sqrt{kk'}}(\hat{\varepsilon}_{k\lambda'}, \hat{\varepsilon}_{k'\lambda'})e^{i(\mathbf{k_1}-\mathbf{k_2})\cdot\mathbf{r}}\delta_{AB},
\end{aligned} \tag{1.25b}
$$

and

$$\rho(E_f) = \frac{V}{(2\pi)^3}\frac{\omega_2^2}{hc^3}\left(\frac{\omega_2}{\omega_1}\right)d\Omega, \tag{1.25c}$$

for the final density of states of a freely recoiling electron, we can express the differential cross-section as (write $\omega_1 = ck_1 = ck$, $\omega_2 = ck_2 = ck'$)

$$d\sigma = \frac{W_{fi}}{c/V} = \left(\frac{e^2}{mc^2}\right)(\hat{\varepsilon}_{k_1} \cdot \hat{\varepsilon}_{k_2})^2\left(\frac{\omega_2}{\omega_1}\right)^2 d\Omega, \tag{1.26a}$$

or

$$\left(\frac{d\sigma}{d\Omega}\right)_{pol} = r_0^2(\hat{\varepsilon}_{k_1} \cdot \hat{\varepsilon}_{k_2})^2\left(\frac{\omega_2}{\omega_1}\right)^2, \tag{1.26b}$$

where r_0 is the classical radius of an electron $\left(\frac{e^2}{mc^2}\right)$. Equation (1.26b) is the non-relativistic Thomson cross-section for an unbound electron. For unpolarised incident X-rays one averages Eq. (1.26a) over all possible initial polarisations to get

$$\left(\frac{d\sigma}{d\Omega}\right)_{unpol} = r_0^2\left(\frac{1+\cos^2\theta}{2}\right)\left(\frac{\omega_2}{\omega_1}\right)^2, \tag{1.26c}$$

where θ is the scattering angle.

Bound electrons

For a bound electron $\vec{p}_j \neq 0$ and, in principle, both V_1 and V_2 have to be included in the calculation of transition rate. However, it was shown by Eisenberger and Platzman (1970) that when $\hbar\omega_1$ and $\hbar\omega_2$ are much larger than $E_f - E_i$, where E_i and E_f are the initial and final state energies of the atom, the contribution to the transition rate from V_1 is small enough to be neglected. Typically, $\hbar\omega_1 \sim \hbar\omega_2 \sim$ 20 keV for the Mo K_α line, $\lambda = 0.62$ Å, while $E_f - E_i$ varies from 20 to 100 eV for most light elements. Thus, the above-mentioned condition is met and one has only to retain V_2. A straightforward calculation such as the free electron case then gives

$$\left(\frac{d\sigma}{d\Omega}\right)_{pol} = r_0^2(\hat{\varepsilon}_{k_1} \cdot \hat{\varepsilon}_{k_2})^2 \left(\frac{\omega_2}{\omega_1}\right) \left| \langle f| \sum_{j=1}^{N} e^{i\mathbf{Q}\cdot\mathbf{r_j}}|i\rangle \right|^2. \tag{1.27a}$$

We have only the (ω_2/ω_1) factor because in this case of a bound electron one should not include the recoil factor in the density of the state factor. One uses instead

$$\rho(E_f) = \frac{V}{(2\pi)^3} \frac{\omega_2^2}{\hbar c^3} d\Omega. \tag{1.27b}$$

The double differential cross-section is likewise put into a form similar to Eq. (1.1):

$$\frac{d^2\sigma}{d\Omega d\omega_2} = r_0^2(\hat{\varepsilon}_{k_1} \cdot \hat{\varepsilon}_{k_2})^2 \left(\frac{\omega_2}{\omega_1}\right) \sum_f \left| \langle f| \sum_{j=1}^{N} e^{i\mathbf{Q}\cdot\mathbf{r_j}}|i\rangle \right|^2 \delta(\omega + \frac{E_i - E_f}{\hbar})$$

$$= \left(\frac{d\sigma}{d\Omega}\right)_{Thomson} \left(\frac{\omega_2}{\omega_1}\right) \sum_f \left| \langle f| \sum_{j=1}^{N} e^{i\mathbf{Q}\cdot\mathbf{r_j}}|i\rangle \right|^2 \delta(\omega + \frac{E_i - E_f}{\hbar}). \tag{1.27c}$$

Note that the kinematic factor in this case is $\frac{\omega_1}{\omega_2} = \frac{k_1}{k_2}$, which is the inverse of that in the neutron cross-section (e.g. Eq. (1.18c)).

Referring to the dynamic structure in Eq. (1.27c), we can distinguish between three cases for an atomic system:

(i) Take $|i\rangle = |0\rangle$ to be the ground state of the atom. Then if $|f\rangle = |n\rangle$ is a discrete excited electronic state, the scattering is called *electronic Raman scattering* and is an incoherent scattering.

(ii) If $|f\rangle$ is in the continuum states of the atom then the scattering is called *Compton scattering*, which is also an incoherent scattering.

(iii) If $|i\rangle = |f\rangle = |0\rangle$, that is, when the scattering does not involve the excitation of the electronic system (therefore it is elastic), then it is called *Rayleigh scattering*, which is a coherent scattering. We are interested only in the Rayleigh scattering in the case of the X-ray diffraction experiment.

Eisenberger and Platzman (1970) calculated the relative magnitudes of the contributions from each of the three processes for a hydrogen atom as a function of the parameter Qa (where a is the Bohr radius). They showed that for small Qa (< 0.5) Rayleigh scattering is dominant, but for large Qa (> 2.0) Compton scattering is the dominant factor. Raman scattering is always small and only appreciable around $Qa \sim 1.0$.

For unpolarised X-rays, the Rayleigh scattering cross-section is then given by

$$\left(\frac{d\sigma}{d\Omega}\right)_R = r_0^2 \left(\frac{1 + \cos^2\theta}{2}\right) \left| \langle 0| \sum_{j=1}^{N} e^{i\mathbf{Q}\cdot\mathbf{r}_j}|0\rangle \right|^2, \tag{1.28a}$$

where $|0\rangle$ is the ground state of the electronic system. Consider the so-called scattering factor

$$f(\mathbf{Q}) = \langle 0| \sum_{j=1}^{N} e^{i\mathbf{Q}\cdot\mathbf{r}_j}|0\rangle, \tag{1.28b}$$

and introducing the average charge density at $\mathbf{r}$ by

$$\rho(\mathbf{r}) = \langle 0| \sum_{j=1}^{N} \delta(\mathbf{r} - \mathbf{r}_j)|0\rangle, \tag{1.28c}$$

we can rewrite the scattering factor as

$$f(\mathbf{Q}) = \int d^3r \rho(\mathbf{r}) e^{i\mathbf{Q}\cdot\mathbf{r}}, \tag{1.28d}$$

which is seen to be a Fourier transform of the charge density distribution in the system. In condensed phases where electrons can be regarded as tightly bound to atoms with centre position at $\mathbf{R}_l$, we can further write

$$\begin{aligned}
f(\mathbf{Q}) &= \langle 0| \sum_{l} \sum_{j=1}^{Z} e^{i\mathbf{Q}\cdot(\mathbf{R}_l + \mathbf{r}_{l_j})}|0\rangle \\
&= \sum_{l} e^{i\mathbf{Q}\cdot\mathbf{R}_l} \langle 0| e^{i\mathbf{Q}\cdot\mathbf{r}_{l_j}} |0\rangle \\
&= \sum_{l} f_l(\mathbf{Q}) e^{i\mathbf{Q}\cdot\mathbf{R}_l},
\end{aligned} \tag{1.28e}$$

where $f_l(\mathbf{Q})$ is the atomic ground state form factor given by

$$f_l(\mathbf{Q}) = \int d\tau \psi_0^* \sum_{j=1}^{Z} e^{i\mathbf{Q}\cdot\mathbf{r}_j} \psi_0, \tag{1.28f}$$

and ψ_0 is the ground state function. We notice $\lim_{Q \to 0} f_l(Q) = Z$ (atomic number) so that the X-ray scattering amplitude is very weak for light elements. In terms of the atomic form factor, Eq. (1.28a), the differential cross-section can be rewritten as

$$\left(\frac{d\sigma}{d\Omega}\right)_R = r_0^2 \left(\frac{1 + \cos^2\theta}{2}\right) \left|\sum_l f_l(Q)e^{iQ\cdot R_l}\right|^2. \tag{1.29a}$$

For a monatomic single crystal, $f_l(Q) = f(Q)$, and the diffraction condition is satisfied only at reciprocal lattice points, i.e. only when $Q = G$. Hence, an absolute atomic structure factor measurement at various G can be used to determine

$$|f(G)|^2 = \left| \int d\tau \psi_0^* \sum_{j=1}^{Z} e^{iG\cdot r_j} \psi_0 \right|^2, \tag{1.29b}$$

which can provide information on the ground state electronic wave function.

1.1.4 Light scattering

Quantum mechanical formulation of the scattering of electromagnetic waves in the optical frequencies can be based on the same Hamiltonian, Eq. (1.23a), which we used before for the X-ray case,

$$H = H_s + H_R + V_1 + V_2. \tag{1.30a}$$

Since for optical frequencies $\hbar\omega_1 \sim \hbar\omega_2 \sim 2\,\text{eV}$ are of the same order of magnitude as the atomic energy levels, both interaction terms V_1 and V_2 contribute to the matrix element in the transition rate. In general it can be shown (Power and Thirunamachandran, 1978) that the sum of the interaction energies V_1 and V_2 can be transformed into an equivalent series of multipole interactions. This is especially convenient at low frequencies when $ka \ll 1$ or, in other words, when the wavelength of light is large compared to the atomic size. In this case we have only to retain the leading term in the multipole expansion, which is just the electric dipole term, i.e.

$$V_1 + V \leftrightarrow -\sum_l \mu_l \cdot \vec{E}_l, \tag{1.30b}$$

where $\vec{\mu}_l$ is the dipole moment of the l-th atom in the system, and the electric field vector at the l-th atomic site is given by

$$\vec{E}_l = -\frac{1}{c}\frac{\partial \vec{A}_l}{\partial t} = \sum_{k,\lambda} i\sqrt{\frac{2\pi\hbar\omega_k}{V}}\hat{\varepsilon}_{k\lambda}\left[a_{k\lambda}e^{ik\cdot R_l} + a_{k\lambda}^+ e^{-ik\cdot R_l}\right]. \tag{1.31}$$

Since the interaction V is a linear combination of $a_{k\lambda}$ and $a_{k\lambda}^+$, the transition matrix element between the initial and final states of the form

$$|i\rangle = |A; 1_{k\lambda}, 0_{k'\lambda'}\rangle, \qquad E_i = E_A + \hbar\omega_k,$$
$$|f\rangle = |B; 0_{k\lambda}, 1_{k'\lambda'}\rangle, \qquad E_f = E_B + \hbar\omega_{k'},$$

is of the second-order type, i.e.

$$\sum_I \frac{\langle f|V|I\rangle\langle I|V|i\rangle}{E_i - E_I}.$$

Furthermore, there are two possible intermediate states:

$$|I\rangle_1 = |I; 0_{k\lambda}, 0_{k'\lambda'}\rangle, \qquad E_{I_1} = E_I,$$
$$|I\rangle_2 = |I; 1_{k\lambda}, 1_{k'\lambda'}\rangle, \qquad E_{I_2} = E_I + \hbar\omega_k + \hbar.\omega_{k'}.$$

Taking into account the energy conservation relation, $\hbar\omega_k + E_A = \hbar\omega_{k'} + E_B$, the transition matrix element works out to be a sum of the two terms for each atom:

$$M_l = -2\pi\hbar\sqrt{\omega_k\omega_{k'}}e^{i\mathbf{Q}\cdot\mathbf{R}_l}\sum_I \left\{ \frac{\langle B|\mu_{k'}^l|I\rangle\langle I|\mu_k^l|A\rangle}{E_A - E_I + \hbar\omega_k} + \frac{\langle B|\mu_k^l|I\rangle\langle I|\mu_{k'}^l|A\rangle}{E_A - E_I - \hbar\omega_{k'}} \right\},$$
(1.32)

where the abbreviations $\mu_k^l \equiv \hat{\varepsilon}_{k\lambda} \cdot \vec{\mu}_l$, $\mu_{k'}^l \equiv \hat{\varepsilon}_{k'\lambda'} \cdot \vec{\mu}_l$ are made.

Scattering cross-section for an atom

Using the *golden rule* we work out the differential cross-section to be

$$\left(\frac{d\sigma}{d\Omega}\right)_{Raman} = kk'^3\left|\sum_I \left\{ \frac{\langle B|\mu_{k'}|I\rangle\langle I|\mu_k|A\rangle}{E_A - E_I + \hbar\omega_k} + \frac{\langle B|\mu_k|I\rangle\langle I|\mu_{k'}|A\rangle}{E_A - E_I - \hbar\omega_{k'}} \right\}\right|^2. \quad (1.33)$$

In the special case of elastic scattering (Rayleigh scattering) we set $\omega_k = \omega_{k'} = \omega$, $E_A = E_B = E_0$, $|A\rangle = |B\rangle = |0\rangle$, $|I\rangle = |n\rangle$ to get

$$\left(\frac{d\sigma}{d\Omega}\right)_R = k^4\left|\frac{1}{\hbar}\sum_n \left\{ \frac{\langle 0|\mu_{k'}|n\rangle\langle n|\mu_k|0\rangle}{\omega_{n0} - \omega} + \frac{\langle 0|\mu_k|n\rangle\langle n|\mu_{k'}|0\rangle}{\omega_{n0} + \omega} \right\}\right|^2$$

$$= k^4\left|\sum_{\alpha,\beta} \varepsilon'_\alpha P_{\alpha\beta}(\omega)\varepsilon_\beta\right|^2. \quad (1.34a)$$

In Eq. (1.34a) the quantity $P_{\alpha\beta}(\omega)$, given by

$$P_{\alpha\beta}(\omega) = \frac{1}{\hbar}\sum_n \left\{ \frac{\langle 0|\mu_\alpha|n\rangle\langle n|\mu_\beta|0\rangle}{\omega_{n0} - \omega - i\eta} + \frac{\langle 0|\mu_\beta|n\rangle\langle n|\mu_\alpha|0\rangle}{\omega_{n0} + \omega + i\eta} \right\}, \quad (1.34b)$$

is the atomic polarisability tensor. To show this latter point we consider an atomic dipole subjected to a time-dependent EM field with interaction energy of the form

$$\hat{A}F(t) = -\hat{\mu} \cdot \hat{\varepsilon}E_0 e^{-i\omega t + \eta t}, \tag{1.35a}$$

and compute the response of a system property $\hat{B}$ via a standard linear response formula (Nozieres, 1964),

$$\langle \hat{B}(t) \rangle - B_0 = \chi_{BA}(\omega)e^{-i\omega t + \eta t}, \tag{1.35b}$$

where the *admittance* function is given by

$$\chi_{BA}(\omega) = \frac{1}{i\hbar} \int\limits_0^\infty d\tau \langle 0|[\hat{B}_I(\tau), \hat{A}]|0\rangle e^{-i\omega \tau + \eta \tau}. \tag{1.35c}$$

We are interested in the case where $\hat{A} \equiv -\vec{\mu} \cdot \hat{\varepsilon}E_0$ and $\hat{B} = \mu_\alpha$. Since in the absence of the field the atomic dipole is randomly oriented with its equilibrium value $B_0 = 0$, we have

$$\langle \hat{\mu}_\alpha(t) \rangle = \chi_{\alpha\beta}(\omega)e^{-i\omega t + \eta t},$$

where $\chi_{\alpha\beta}(\omega)$ may be worked out from Eq. (1.35c) to be

$$\chi_{\alpha\beta}(\omega) = \frac{1}{\hbar} \sum_n \left\{ \frac{\langle 0|\mu_\alpha|n\rangle \langle n|\mu_\beta|0\rangle}{\omega_{n0} - \omega - i\eta} + \frac{\langle 0|\mu_\beta|n\rangle \langle n|\mu_\alpha|0\rangle}{\omega_{n0} + \omega + i\eta} \right\}$$

$$= P_{\alpha\beta}(\omega)E_0. \tag{1.35d}$$

Notice that, if the atomic charge distribution is spherically symmetric, $P_{\alpha\beta}(\omega)$ reduces to a scalar polarisability $\alpha(\omega)$, i.e.

$$P_{\alpha\beta}(\omega) = \alpha(\omega)\delta_{\alpha\beta}, \tag{1.35e}$$

and the Rayleigh cross-section is simply

$$\left(\frac{d\sigma}{d\Omega}\right)_R = k^4 |\hat{\varepsilon}_{k_1} \cdot \hat{\varepsilon}_{k_2}|^2 |\alpha(\omega)|^2. \tag{1.35f}$$

Now, consider a system of N optically isotropic atoms in thermal equilibrium. Remember from Eq. (1.32) that μ_l contains a phase factor $e^{i\mathbf{Q}\cdot\mathbf{R}_l}$. We obtain the differential cross-section by summing the transition matrix element over l before squaring:

$$\left(\frac{d\sigma}{d\Omega}\right)_R = k^4 |\hat{\varepsilon}_{k_1} \cdot \hat{\varepsilon}_{k_2}|^2 |\sum_l \alpha_l(\omega)e^{i\mathbf{Q}\cdot\mathbf{R}_l}|^2. \tag{1.36}$$

The double differential cross-section is likewise obtained by averaging Eq. (1.36) over the initial external motional states of the thermal equilibrium system and then summing over the final motional states with an energy conserving factor:

$$\frac{d^2\sigma}{d\omega d\Omega} = k^4 |\hat{\varepsilon}_{k_1} \cdot \hat{\varepsilon}_{k_2}|^2 \sum_{i,f} P_i \left| \langle f| \sum_l \alpha_l(\omega) e^{i\mathbf{Q}\cdot\mathbf{R}_l} |i\rangle \right|^2 \delta(\omega + \frac{E_i - E_f}{\hbar})$$

$$= k^4 |\alpha|^2 |\hat{\varepsilon}_{k_1} \cdot \hat{\varepsilon}_{k_2}|^2 S(Q, \omega). \tag{1.37}$$

We see that, for a monatomic system, the basic cross-section $(d\sigma/d\Omega)_0$ is the Rayleigh scattering cross-section for an isolated atom, and here $\alpha(\omega)$ plays the role of the scattering length in the case of thermal neutron scattering.

1.1.5 Classical versus quantum mechanical descriptions of scattering

It is instructive at this point to ask a question: What is the relative merit of using quantum mechanical versus classical descriptions of the scattering of an electromagnetic wave by a material medium? Since Maxwell's equations are a valid description of EM phenomena almost down to the atomic scale, in many ways it is nicer to start from the classical wave equation.

Consider Maxwell's equation in a non-magnetic medium without free charge and current density. Setting the time dependence of all fields to be $e^{-i\omega t}$, we have

$$\nabla \times \mathbf{E} - \frac{i\omega}{c}\mathbf{B} = 0, \quad \text{(first Maxwell equation)}, \tag{1.38a}$$

$$\nabla \times \mathbf{B} + \frac{in^2\omega}{c}\mathbf{E} = 0, \quad \text{(third Maxwell equation)}. \tag{1.38b}$$

Eliminating $\mathbf{B}$ from Eq. (1.38a) and Eq. (1.38b) we get

$$\nabla \times (\nabla \times \mathbf{E}) - \frac{n^2\omega^2}{c^2}\mathbf{E} = 0,$$

or

$$\nabla \times (\nabla \times \mathbf{E}) - n^2 k_0^2 \mathbf{E} = 0, \tag{1.38c}$$

where n is the refractive index of the medium and k_0 is the wave number in a vacuum. Equation (1.38c) with a subsidiary condition $\nabla \cdot \mathbf{E} = 0$ (which follows from Maxwell's fourth equation $\nabla \cdot \mathbf{D} = 0$), supports only the propagation of a plane wave with a modified wave number $k = nk_0$.

In order to have scattering one must have an inhomogeneity of the refractive index. Let the dielectric inhomogeneity be given by

$$n^2 \equiv \varepsilon = \varepsilon_0 + \Delta\varepsilon(\mathbf{r}), \tag{1.38d}$$

where the fluctuation $\Delta\varepsilon(\mathbf{r}) \ll \varepsilon_0$. Then Eq. (1.38c) can be rewritten as

$$\nabla \times (\nabla \times \mathbf{E}) = \nabla(\nabla \cdot \mathbf{E}) - \nabla^2\mathbf{E} = -\nabla^2\mathbf{E},$$

to get

$$\nabla^2\mathbf{E} + k^2\mathbf{E} = -k_0^2\Delta\varepsilon(\mathbf{r})\mathbf{E}, \tag{1.38e}$$

where $k = n_0 k_0$ is the propagation wave number in the medium in the absence of fluctuations. This is a vector analogue of the scalar Schrödinger wave equation which we solved for neutron scattering. It describes the propagation of EM waves in a material with local fluctuations of the dielectric constant. Quantum theory is only needed to calculate the material property in terms of atomic properties. Once this is done, the solution of the scattering problem is entirely classical. The standard method, analogous to what we did in Section 1.1.2, can be used to establish a scattering solution:

$$\mathbf{E}_s(\mathbf{r}, t) = -\frac{e^{ikr}}{r}\frac{k^2}{4\pi}(\overleftrightarrow{I} - \hat{r}\hat{r}) \cdot \mathbf{E}_0 e^{-i\omega t} \int_V d^3r' \Delta\varepsilon(\mathbf{r}')e^{i\mathbf{Q}\cdot\mathbf{r}'}, \tag{1.39a}$$

where $\overleftrightarrow{I}$ is the unit dyad, $\hat{r}$ the unit vector in the scattered direction, and $\mathbf{E}_0$ the incident field vector.

In the case of Rayleigh scattering of X-rays $k = k_0$ (in vacuum), the dielectric constant fluctuation at the X-ray frequency is given by a free electron expression (Kittel, 1963),

$$\Delta\varepsilon(\mathbf{r}) = -\frac{4\pi e^2 n(\mathbf{r})}{m\omega^2}, \tag{1.39b}$$

where $n(\mathbf{r})$ is the number density of electrons at $\mathbf{r}$. Therefore

$$\mathbf{E}_s(\mathbf{r}, t) = \frac{e^{ikr}}{r}(\overleftrightarrow{I} - \hat{r}\hat{r}) \cdot \mathbf{E}_0 \frac{e^2}{mc^2} \int_V d^3r' n(\mathbf{r}')e^{i\mathbf{Q}\cdot\mathbf{r}'}, \tag{1.39c}$$

and we observe that

$$\left|(\overleftrightarrow{I} - \hat{r}\hat{r}) \cdot \mathbf{E}_0\right|^2 = \sin^{-2}\gamma, \tag{1.39d}$$

where γ is the angle between $\mathbf{E}_0$ and $\hat{r}$. The differential cross-section is just the scattering amplitude squared, with a unit amplitude incident wave. Thus

$$\left(\frac{d\sigma}{d\Omega}\right)_R = \frac{e^2}{mc^2}\sin^2\gamma\,|n(\mathbf{Q})|^2, \tag{1.40a}$$

which is exactly the same as the quantum mechanical result obtained from Eq. (1.27a) by setting $|i\rangle = |f\rangle = |0\rangle$, i.e.

$$\left(\frac{d\sigma}{d\Omega}\right)_R = \left(\frac{e^2}{mc^2}\right)^2 (\hat{\varepsilon}_{k_1} \cdot \hat{\varepsilon}_{k_2}) \left| \langle 0| \sum_{j=1}^{N} e^{i\mathbf{Q}\cdot\mathbf{r}_j} |0\rangle \right|^2, \tag{1.40b}$$

noting $\sin^2\gamma = (\hat{\varepsilon}_{k_1} \cdot \hat{\varepsilon}_{k_2})^2$. The Fourier transform of the electron density in the classical expression can be understood quantum mechanically as

$$n(\mathbf{Q}) = \langle 0| \sum_{j=1}^{N} e^{i\mathbf{Q}\cdot\mathbf{r}_j} |0\rangle. \tag{1.40c}$$

In the case of light scattering in the commonly used VV geometry, the factor $(\overset{\leftrightarrow}{I} - \hat{r}\hat{r}) \cdot \mathbf{E}_0$ equals $\mathbf{E}_0$. In condensed phases the dielectric constant fluctuation may contain a slow time dependence caused by the slowly varying density fluctuation of wavelengths of the same order of magnitude as the wavelength of light. We therefore put

$$\Delta\varepsilon(\mathbf{Q}, t) = \left(\frac{\partial\varepsilon}{\partial n}\right)_T n(\mathbf{Q}, t), \tag{1.41}$$

in Eq. (1.39a) to get

$$\mathbf{E}_s(\mathbf{r}, t) = -\frac{e^{ikr}}{r} \mathbf{E}_0 e^{-i\omega t} \frac{k^2}{4\pi} \left(\frac{\partial\varepsilon}{\partial n}\right)_T n(\mathbf{Q}, t). \tag{1.42}$$

Equation (1.42) can then be used to obtain the commonly used expression for the double differential cross-section (Chen, 1971)[2]

$$\frac{d^2\sigma}{d\omega_2 d\Omega} = \frac{k^4}{16\pi^2} \left(\frac{\partial\varepsilon}{\partial n}\right)^2 \frac{1}{2\pi} \int_{-\infty}^{\infty} dt e^{-i\omega t} \langle n(-\mathbf{Q}, 0)n(\mathbf{Q}, t)\rangle. \tag{1.43}$$

1.1.6 Physical interpretation of the dynamical structure factors

In previous sections we have shown that all three scattering experiments, involving thermal neutrons, X-rays and light lead to double differential cross-sections expressible in terms of a dynamic structure $S(\mathbf{Q}, \omega)$ or its self part $S_s(\mathbf{Q}, \omega)$. It is desirable to give some physical interpretation on the relation between dynamics as measured in $(\mathbf{Q}, \omega)$ space and dynamics as pictured in real space-time $(\mathbf{r}, t)$.

---

[2] For the derivation of Eq. (1.43) from Eq. (1.42), see Chen (1971).

Define a thermal average of a physical quantity by $\hat{O}$ $\langle\hat{O}\rangle = \sum_i P_i\langle i|\hat{O}|i\rangle$. In terms of this average the dynamic structure factors can be written as

$$S(\mathbf{Q}, \omega) = \frac{1}{2\pi} \int_{-\infty}^{\infty} dt\, e^{-i\omega t}\frac{1}{N}\sum_{l,l'} \langle e^{-i\mathbf{Q}\cdot\mathbf{R}_l(0)}e^{i\mathbf{Q}\cdot\mathbf{R}_{l'}(t)}\rangle,$$

$$S_S(\mathbf{Q}, \omega) = \frac{1}{2\pi} \int_{-\infty}^{\infty} dt\, e^{-i\omega t}\langle e^{-i\mathbf{Q}\cdot\mathbf{R}(0)}e^{i\mathbf{Q}\cdot\mathbf{R}(t)}\rangle. \tag{1.44}$$

One often calls the factor in the integrand the intermediate scattering functions:

$$F(\mathbf{Q}, t) = \frac{1}{N}\sum_{l,l'} \langle e^{-i\mathbf{Q}\cdot\mathbf{R}_l(0)}e^{i\mathbf{Q}\cdot\mathbf{R}_{l'}(t)}\rangle,$$

$$F_S(\mathbf{Q}, t) = \langle e^{-i\mathbf{Q}\cdot\mathbf{R}(0)}e^{i\mathbf{Q}\cdot\mathbf{R}(t)}\rangle. \tag{1.45}$$

A frequency domain experiment measures the dynamic structure factor, whereas a time domain experiment measures the intermediate scattering function. Among frequency domain techniques, we have, for neutrons, crystal spectrometry and time-of-flight measurements, and for photons, optical mixing spectroscopy used in light-scattering experiments. Time domain techniques include the neutron *spin-echo* spectrometer, and the extensively used photon correlation spectroscopy. In order to discuss dynamics in $(\mathbf{Q}, \omega)$ space versus $(\mathbf{r}, t)$ space it is most instructive to use examples. Consider, first, incoherent scatterings.

These situations arise in incoherent scattering of neutrons from hydrogen containing substances or in light scattering from a system of independently moving particles. One is observing, in this case, the single particle motion in the system. The simplest model of the single particle motion is a random walk model. A particle moves in an arbitrary direction for a step length l with a mean speed v_0 (for a mean duration $\tau = l/v_0$), then turns around to a randomly selected direction and repeats the same motion. This is a Gauss–Markov process, and one can show that it has an exponentially decaying velocity autocorrelation function (Uhlenbech and Ornstein, 1930)

$$\langle V(0)V(t)\rangle = V_0^2 e^{-t/\tau}. \tag{1.46a}$$

One can calculate the intermediate scattering function by using Eq. (1.46a) to obtain

$$F_S(Q, t) = \exp\left[-Q^2l^2\left(\frac{t}{\tau} - 1 + e^{-t/\tau}\right)\right], \tag{1.46b}$$

where $F_S(Q, t)$ depends only on $|\mathbf{Q}|$ because the motion is isotropic. It is instructive to consider two limiting cases:

(i) $Ql \ll 1$ (or $Q^{-1} \gg l$)

For small values of Ql, the half-width of $F_S(Q,t)$ occurs at large time $(t/\tau \gg 1)$ where we may neglect the factors $e^{-t/\tau}$ and -1 compared to t/τ in the second bracket of Eq. (1.46b). Hence

$$F_S(Q,t) \xrightarrow[Ql\ll1]{} e^{-Q^2(l^2/\tau)t} = e^{-Q^2 Dt}. \tag{1.47a}$$

We can interpret this result by noting that since in this case $R \equiv Q^{-1} \gg l$, the spatial resolution R of the probe is much larger than the particle mean free path l. Therefore the probe is sampling the particle motion with such a coarse resolution that it sees many changes of directions characterised by a particle undergoing diffusive motion. This corresponds to an observation of the long time limit of $F_S(Q,t)$ which is an exponential decay with a diffusion constant given by $D = l^2/\tau$. Fourier transform of Eq. (1.47a) then gives a Lorentzian spectral density function

$$S_S(Q,\omega) \xrightarrow[Ql\ll1]{} \frac{1}{\pi} \frac{DQ^2}{\omega^2 + (DQ^2)^2}. \tag{1.47b}$$

(ii) $Ql \gg 1$ (or $Q^{-1} \ll l$)

In this opposite limit the half-width of the $F_S(Q,t)$ occurs at $t/\tau \ll 1$ so that we are allowed to expand the exponential factor to $e^{-t/\tau}$ of Eq. (1.46b) to obtain the result:

$$F_S(Q,t) \xrightarrow[Ql\gg1]{} e^{-\frac{1}{2}Q^2 l^2 (\frac{t}{\tau})^2} = e^{-\frac{1}{2}Q^2 v_0^2 t^2}. \tag{1.47c}$$

We can interpret this limit by saying that the spatial resolution of the probe is so high that it samples the motions corresponding to small segments of the free moving path length ℓ. This limit is characterised by a Gaussian time dependence of $F_S(Q,t)$ with a decay time $t_{1/2} = (Qv_0)^{-1}$ where $v_0 = \ell/\tau$. The corresponding dynamic structure factor

$$S_S(Q,\omega) = \frac{1}{\sqrt{2\pi}Qv_0} e^{-(\omega^2/2Q^2 v_0^2)}, \tag{1.47d}$$

similarly shows a well-known Doppler broadened Gaussian line shape with a half-width $\omega_{1/2} = \sqrt{2\ln 2}Qv_0$, which is again inversely proportional to $t_{1/2}$. Figure 1.3 gives plots of the reduced decay constant $t_{1/2}Qv_0$ and the reduced spectral width $\omega_{1/2}/Qv_0$ as a function of the observational parameter Ql. An inverse relationship

$$(t_{1/2}Qv_0)^{-1} \propto \frac{\omega_{1/2}}{Qv_0}, \tag{1.47e}$$

follows from the fact that $F_S(Q,t)$ and $S_S(Q,\omega)$ are a Fourier transform pair.

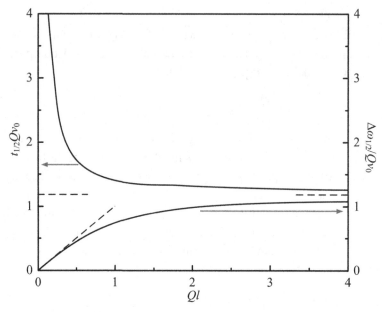

Figure 1.3 Plots of the reduced half-width, $t_{1/2}Qv_0$, of $F_S(Q,t)$ given by Eq. (1.46b) and the reduced half-width, $\omega_{1/2}/Qv_0$, of the corresponding $S_S(Q,\omega)$ as function of the parameter Ql. These curves illustrate nicely a continuous transition from the collision (i.e. change of direction) dominated diffusion behaviour at small Ql to the free particle-like behaviour at large Ql for particles undergoing independent random walks. The dashed line in the lower part of the figure refers to a different kind of random walk, namely, particles undergoing binary collisions as in the case of low-density gases.

Figure 1.4 shows an example of $S_S(Q,\omega)$ as measured by an incoherent neutron quasi-elastic scattering from compressed hydrogen at 85 K (Chen *et al.*, 1973). Here Y is a parameter essentially equal to $(Ql)^{-1}$. This parameter can be changed by changing the scattering angle (thus changing Q) or by changing gas pressure to vary l. Looking from the top to bottom curves in Figure 1.4 one can clearly see a gradual transition from a Doppler broadened Gaussian line at the top to a collision dominated Lorentzian line at the bottom. A plot of reduced width $\omega_{1/2}/Qv_0$ versus Ql is given by a dashed line in the lower part of Figure 1.3. The experimental line widths do not quite agree with the random walk model (given by the solid line) because in a gas a molecule collides with another molecule, leaving with some degree of persistence in the original direction of the incidence. It is therefore not accurate to model the collision with a random change of direction as in the random walk model.

    Figure 1.5 shows an example of the time domain measurement (Holz and Chen, 1978). It gives the reduced width $t_{1/2}Qv_0$ of $F_S(Q,t)$ measured from a collection

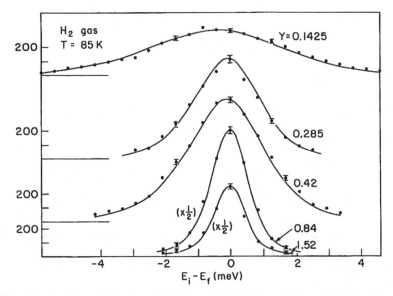

Figure 1.4 A series of the frequency domain function $S_S(Q, \omega)$ as measured by an incoherent neutron quasi-elastic scattering from compressed hydrogen gas at 85 K. The abscissa is $\hbar\omega$ in units of meV and the ordinate is proportional to $S_S(Q, \omega)$. The parameter $Y \equiv (Ql)^{-1}$. As the gas is compressed, the mean free path between collisions decreases, or equivalently the Y parameter increases. Notice that as $Y = 0.14$ the line shape is more like a Gaussian but as Y increases to 1.5 the line shape changes to more like a Lorentzian (Chen *et al.*, 1973). Copyright 1973 by The American Physical Society.

of independent, randomly moving, *E. Coli* bacteria in a quasi-elastic light scattering. The half-width is plotted against the scattering angle θ, which is related to Q by $Q = 2k \sin\theta/2$. The plot resembles the top curve of Figure 1.3 because, qualitatively speaking, bacteria moving in a *mobility buffer* undergo a kind of random walk as a result of chemotaxis.

Information contained in the full (or coherent) dynamic structure factor is of a different nature as compared to the single particle (or self) dynamic structure factor we discussed above. Note that the local density function, defined by

$$n(\mathbf{r}, t) = \sum_l \delta(\mathbf{r} - \mathbf{R}_l(t)), \qquad (1.48a)$$

has a Fourier transform

$$n(\mathbf{q}, t) = \sum_l e^{i\mathbf{q}\cdot\mathbf{R}_l(t)}, \qquad (1.48b)$$

where $n(\mathbf{q}, t)$ denotes a collective density oscillation of a wave vector $\mathbf{q}$ which exists in any condensed phase at finite temperatures. In an equilibrium system,

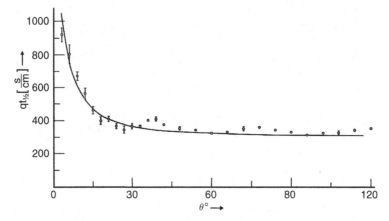

Figure 1.5 An example of a time-domain experiment completed with the photon-correlation technique. This is a plot of reduced half-widths $t_{1/2}Q$ ($Q = q$ in the figure) vs. scattering angle θ, of $F_S(Q, t)$ measured from a collection of independently moving *E. Coli* bacteria. The bacteria are chemotactic to oxygen and are known to undergo *run-twiddle* motions. The twiddle can be regarded as a series of small step motions with $l \sim 0.1$–0.3 µm. This is essentially an experimental realisation of the upper curve in Figure 1.3. At $\theta \leq 10°$ the condition of $Ql < 1$ is satisfied and consequently the reduced half-widths start to rise rapidly. Reprinted from Holz and Chen (1978). Copyright 1978, with permission from Elsevier.

$n(\mathbf{q}, t)$ is a stationary process so it is more natural to talk about the density correlation function

$$\frac{1}{N} \langle n(-\mathbf{q}, 0)n(\mathbf{q}, t)\rangle. \tag{1.48c}$$

From Eq. (1.45) one immediately realises that a coherent scattering experiment measures, directly, such a density correlation function at a particular wave vector imposed by the probe, i.e.

$$\mathbf{q} = \mathbf{Q}. \tag{1.48d}$$

Thus the half-width of $F(Q, t)$ indicates the *decay time* of the density fluctuation of a wave vector $\mathbf{q} = \mathbf{Q}$. Similarly, the half-width of $S(Q, \omega)$ reflects the corresponding *decay rates* of the density oscillation.

An interesting situation occurs when the local order of a fluid is characterised by a correlation length ξ. A class of well-known examples is a one-component fluid near its liquid–gas critical point or a two-component fluid near its demixing critical point. In the limit of $Q\xi \ll 1$ the radiation is probing the system at a spatial length scale $R \approx 1/Q$ much larger than the correlation length ξ, and one is effectively seeing many correlated regions each diffusing independently, like a

Brownian particle having a radius ξ. The density or the concentration fluctuation spectrum has a Lorentzian shape with a line width Γ_Q given by

$$\Gamma_Q \sim DQ^2 \approx \frac{k_B T}{\eta \xi} Q^2 \sim \frac{Q^3}{Q\xi} \quad \text{for} \quad Q\xi \to 0, \tag{1.49a}$$

taking into account the Stokes–Einstein relation $D \approx \frac{k_B T}{\eta \xi}$.

On the other hand, in the limit $Q\xi > 1$, the spatial length scale of the probe is so short that it no longer sees the entire correlation range ξ and one then gets the correlation length independent line width (Kawasaki, 1986)

$$\Gamma_Q \sim Q^3 \quad \text{for} \quad Q\xi > 0. \tag{1.49b}$$

The theory of critical slowing down in fluids is by now well known (Kawasaki, 1986). Thus no attempt is made here to elaborate it except to show one particular example (Figure 1.6) from a two-component fluid of a 50–50 mixture of

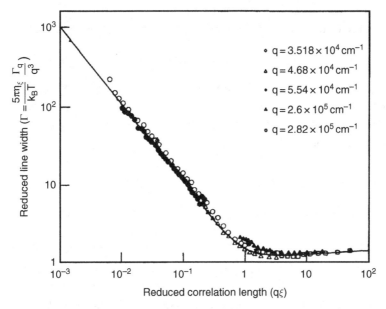

Figure 1.6 A reduced line width of quasi-elastic light-scattering spectra from a binary liquid mixture *n*-hexane and nitrobenzene near the critical point, plotted as a function of a parameter $Q\xi$ ($Q \equiv q$ in the figure) where ξ is the correlation length of the critical fluctuations. The symbols are the decay rates measured by photon correlation spectroscopy (Lai, 1972; Chen *et al.*, 1982). The photon correlation function is closely approximated by an exponential function with a decay rate Γ_q. The solid line represents the fitted theoretical result assuming a background contribution and calculation based on the theory of Kawasaki and Lo (1972); Lo and Kawasaki (1973). From Chen *et al.* (1982). Copyright 1982 by The American Physical Society.

n-hexane and nitrobenzene near room temperature (critical consolute temperature $T_c = 20.29\,^\circ\text{C}$). This work of Lai (1972) clearly shows that the scaled line width Γ_Q/Q^3 versus $Q\xi$ plot has a region inversely proportional to the reduced magnitude of wave vector $Q\xi$ when $Q\xi < 1.0$ and approaching a constant when $Q\xi > 1.0$.

1.2 Module – Small angle neutron scattering

1.2.1 The structure factor

We shall show first that a diffraction experiment with an atomic liquid gives rise naturally to an interference function, called the structure factor $S(Q)$.

Consider a neutron with an incident wave vector $\vec{k}_i$ ($k_i = 2\pi/\lambda$) travelling in the z-direction represented by a plane wave of unit amplitude:

$$\Psi_i(z, t) = \frac{1}{\sqrt{V}} e^{i(k_i z - \omega_i t)}, \quad \text{take } V = 1. \tag{1.50}$$

The incident neutron flux is given in terms of the speed of the neutron v_i as

$$J_i = v_i |\psi_i|^2 = v_i \ \left(\text{cm}^{-2}\text{sec}^{-1}\right). \tag{1.51}$$

The typical speed of thermal neutrons ($E = 25\,\text{meV}$) is $2200\,\text{m s}^{-1}$.

The scattered neutron wave at large distance (a distance much larger than the wavelength of a neutron) for a nucleus fixed at the origin has the form, according to the quantum theory of scattering,

$$\psi_s(\underline{r}, t) = f(\theta, k_i) \frac{e^{i(k_i r - \omega_i t)}}{r}, \tag{1.52a}$$

where r is the distance from the origin to the detector situated at a scattering angle θ and the factor $f(\theta, k_i)$ is called the scattering amplitude.

For thermal neutrons, due to the fact that their wavelengths are much larger than the typical nuclear size, the scattering amplitude is to a very good approximation a constant independent of the scattering angle and the incident wave number of the neutron. This constant is called the *bound scattering length b of the nucleus*, and the magnitude of it is dependent on the range and depth of the neutron–nuclear interaction potential. In practice, b for every isotope is measured experimentally and tabulated (Sears, 1989)

$$f(\theta, k_i) = -b. \tag{1.52b}$$

Next, let the neutron be scattered by a stationary target nucleus having a bound scattering length b_ℓ situated at $\underline{R}_\ell$ (Figure 1.7). The scattered wave is an outgoing spherical wave centred around the scattering nucleus multiplied by a

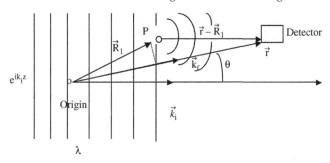

Figure 1.7 Neutron diffraction geometry.

phase factor, $e^{i\underline{k}_i \cdot \underline{R}_l}$, taking into account the additional phase the incident wave needs to accumulate to go from the origin to the location of the nucleus,

$$\psi_s^l(\underline{r}, t) = -b_l e^{i\underline{k}_i \cdot \underline{R}_l} \frac{e^{ik_i |\underline{r}-\underline{R}_l| - i\omega_i t}}{|\underline{r}-\underline{R}_l|}$$

$$\cong -b_l e^{i\underline{k}_i \cdot \underline{R}_l} e^{-ik_i \hat{r} \cdot \underline{R}_l} \frac{e^{i(k_i r - \omega_i t)}}{r} \qquad (1.53)$$

$$= -b_l e^{i\underline{Q} \cdot \underline{R}_l} \frac{e^{i(k_i r - \omega_i t)}}{r},$$

where we used the approximation $k_i |\underline{r}-\underline{R}_l| \cong k_i r - k_i \hat{r} \cdot \underline{R}_l$ and defined the scattering vector,

$$\underline{Q} = \underline{k}_i - k_i \hat{r} = \underline{k}_i - \underline{k}_f.$$

In Eq. (1.52b) the minus sign is a convention such as to make the scattering length positive for most nuclei.

Now consider a target system consisting of N nuclei, we have then

$$\Psi_s(\underline{r}, t) = f(\underline{Q}) \frac{1}{r} e^{i(k_i r - \omega_i t)}, \qquad (1.54a)$$

where

$$f(\underline{Q}) = -\sum_{l=1}^{N} b_l e^{i\underline{Q} \cdot \underline{R}_l}. \qquad (1.54b)$$

We obtain the *differential cross-section dσ per unit solid angle* by defining it as the ratio of the number of neutrons scattered into the unit solid angle in the direction $\hat{r}$ per unit time,

$$\frac{d\sigma}{d\Omega} = \frac{\upsilon_f |\Psi_s|^2 r^2}{\upsilon_i |\Psi_i|^2} = \frac{k_f}{k_i} f^*(Q) f(Q).$$

For a diffraction experiment, where scattering is elastic, $k_i = k_f$, so

$$\frac{d\sigma}{d\Omega} = |f(Q)|^2 = \sum_{i=1}^{N} \sum_{i'=1}^{N} b_i b_{i'} e^{i\vec{Q}\cdot(\vec{R}_i - \vec{R}_{i'})}. \tag{1.55a}$$

In practice, when the system of atoms is in thermal equilibrium at temperature T, we have to take an ensemble average of the above expression to compare with the measured quantity:

$$\frac{d\sigma}{d\Omega} = \left\langle \sum_{l,l'} \overline{b_l b_{l'}} e^{i\underline{Q}\cdot(\underline{R}_l - \underline{R}_{l'})} \right\rangle. \tag{1.55b}$$

The ensemble average should consist of two parts: first, an average over the distribution of nuclei over various spin and isotopic states; second, over the distribution of positions of particles in the system in all its possible configurations consistent with a given temperature T.

Average over spin and isotopic states

Assuming the distribution over spin and isotopic states is uncorrelated with positions of atoms (this is reasonable because energies involved in different orientations of spins are much smaller than the average thermal energy $k_B T$), then

$$\overline{b_l b_{l'}} = \overline{b_l^2} = \overline{b^2}, \quad l = l',$$
$$= \overline{b_l b_{l'}} = \overline{b}^2, \quad l \neq l'.$$

Combining the two possibilities together in one expression, we can write

$$\overline{b_l b_{l'}} = \left[\overline{b^2} - \overline{b}^2 \right] \delta_{ll'} + \overline{b}^2 = b_{inc}^2 \delta_{ll'} + b_{coh}^2. \tag{1.56}$$

Substituting Eq. (1.56) into Eq. (1.55b) one gets

$$\frac{d\sigma}{d\Omega} = N b_{inc}^2 + b_{coh}^2 \left\langle |f(Q)|^2 \right\rangle. \tag{1.57a}$$

Returning to the case of coherent scattering (diffraction of waves) from N nuclei, we have

$$\frac{d\sigma_c}{d\Omega} = N b_{coh}^2 \frac{1}{N} \left\langle |f(Q)|^2 \right\rangle = N b_{coh}^2 S(Q). \tag{1.57b}$$

Define the inter-atomic structure factor by

$$S\left(\underline{Q}\right) = \frac{1}{N}\langle|f\left(Q\right)|^2\rangle = \frac{1}{N}\left\langle\sum_{l,l'}^{N}e^{i\underline{Q}\cdot\left(\underline{R}_l-\underline{R}_{l'}\right)}\right\rangle$$

$$= \int d\underline{r}\, e^{i\underline{Q}\cdot\underline{r}}\frac{1}{N}\left\langle\sum_{l,l'}\delta\left(\underline{r}-\underline{R}_l+\underline{R}_{l'}\right)\right\rangle \quad (\underline{Q}\neq 0). \tag{1.58a}$$

In a fluid every atom is equivalent, so we can choose $l' = 0$ and reduce the double sum to a single sum in the above expression. Furthermore, in the resultant expression we can put $\frac{1}{N}\sum_{l'=0}^{N-1} = 1$. Thus

$$S\left(\underline{Q}\right) = \int d\underline{r}\, e^{i\underline{Q}\cdot\underline{r}}\left\langle\sum_{l=0}^{N-1}\delta\left(\underline{r}-\underline{R}_l\right)\right\rangle$$

$$= 1 + \int d\underline{r}\, e^{i\underline{Q}\cdot\underline{r}}\left\langle\sum_{l=1}^{N-1}\delta\left(\underline{r}-\underline{R}_l\right)\right\rangle. \tag{1.58b}$$

Define a pair correlation function, or a radial distribution function $g(r)$, by equating $\rho = N/V$:

$$\rho g\left(\mathrm{r}\right) = \left\langle\sum_{l=1}^{N-1}\delta\left(\underline{r}-\underline{R}_l\right)\right\rangle. \tag{1.58c}$$

The physical meaning of $g(r)$ can be inferred from the definition as $\rho g\left(\mathrm{r}\right)4\pi r^2 dr =$ average number of atoms situated in a spherical ring of $(r, r+dr)$, given that there is an atom at the origin $r = 0$.

Density autocorrelation function

The number density operator at a point $\underline{r}$ in space is defined as

$$\rho\left(\underline{r}\right) = \sum_{i=1}^{N}\delta\left(\underline{r}-\underline{r}_i\right), \tag{1.59a}$$

where $\underline{r}_i$ is the position vector of the i-th particle in the system. For a homogeneous liquid at equilibrium

$$\langle\rho\left(\underline{r}\right)\rangle = \sum_{l=1}^{N}\langle\delta(\underline{r}-\underline{R}_1)\rangle = \sum_{l=1}^{N}\frac{1}{V} = \frac{N}{V} = \rho. \tag{1.59b}$$

The equilibrium pair correlation function between the densities at two points separated by $\underline{r}$ in a homogeneous fluid is called $G\left(\underline{r}\right)$. Define a density autocorrelation function with the dimension ρ:

$$G\left(\underline{r}\right) = \frac{1}{N} \int d\underline{r}' \left\langle \rho\left(\underline{r}+\underline{r}'\right)\rho\left(\underline{r}'\right)\right\rangle$$

$$= \frac{1}{N} \left\langle \int d\underline{r}' \sum_{i=1}^{N}\sum_{j=1}^{N} \delta\left(\underline{r}'+\underline{r}-\underline{r}_i\right)\delta\left(\underline{r}'-\underline{r}_j\right)\right\rangle \qquad (1.60a)$$

$$= \frac{1}{N} \left\langle \sum_{i=1}^{N}\sum_{j=1}^{N} \delta\left(\underline{r}+\underline{r}_j-\underline{r}_i\right)\right\rangle.$$

The above expression can be reduced to the following sum:

$$G\left(\underline{r}\right) = \frac{1}{N}\left\langle \sum_{i=1}^{N}\delta\left(\underline{r}\right)\right\rangle + \frac{1}{N}\left\langle \sum_{i\neq j}^{N}\delta\left(\underline{r}+\underline{r}_j-\underline{r}_i\right)\right\rangle,$$

and taking $r_j = 0$,

$$G\left(\underline{r}\right) = \delta\left(\underline{r}\right) + \frac{1}{N}\sum_{j=1}^{N}\left\langle \sum_{i\neq j}^{N}\delta\left(\underline{r}-\underline{r}_i\right)\right\rangle$$

$$= \delta\left(\underline{r}\right) + \left\langle \sum_{i\neq j}^{N}\delta\left(\underline{r}-\underline{r}_i\right)\right\rangle = \delta\left(\underline{r}\right) + \rho g\left(\underline{r}\right). \qquad (1.60b)$$

From now on, we shall use k and Q interchangeably, for convenience. Let

$$\rho_k = \int d^3 r \rho\left(r\right)e^{i\underline{k}\cdot\underline{r}} = \sum_{i=1}^{N} e^{i\underline{k}\cdot\underline{r}_i}. \qquad (1.61a)$$

The *static structure factor* $S(k)$ can then be introduced as

$$S\left(\underline{k}\right) = \frac{1}{N}\left\langle \rho_{-k}\,\rho_k\right\rangle. \qquad (1.61b)$$

Explicitly writing out the Fourier transform of the densities

$$S\left(\underline{k}\right) = \frac{1}{N}\left\langle \sum_{i=1}^{N} e^{-i\underline{k}\cdot\underline{r}_i}\sum_{j=1}^{N} e^{-i\underline{k}\cdot\underline{r}_j}\right\rangle$$

$$= \frac{1}{N}\left\langle \sum_{i,j}^{N} e^{-i\underline{k}\cdot(\underline{r}_i-\underline{r}_j)}\right\rangle \qquad (1.61c)$$

$$= 1 + \frac{1}{N}\left\langle \sum_{i\neq j}^{N} e^{-i\underline{k}\cdot(\underline{r}_i-\underline{r}_j)}\right\rangle,$$

which is equal to (as shown in Eq. (1.59a))

$$S(k) = 1 + \rho \int d^3r \, e^{-i\underline{k}\cdot\underline{r}} g(r). \tag{1.61d}$$

An inverse Fourier transform gives

$$\rho g(r) = \frac{1}{(2\pi)^3} \int d^3k \, e^{i\underline{k}\cdot\underline{r}} \left[S\left(\underline{k}\right) - 1 \right]. \tag{1.61e}$$

For an isotropic system, we have

$$S\left(\underline{k}\right) = S(k) = 1 + 4\pi\rho \int_0^\infty dr \, r^2 g(r) \, \frac{\sin kr}{kr}. \tag{1.62a}$$

Let $g(r) = 1 + h(r)$, where $h\left(\underline{k}\right) = \int d^3r \, e^{-i\underline{k}\cdot\underline{r}} h\left(\underline{r}\right)$. Then except for $k = 0$,

$$S\left(\underline{k}\right) = 1 + \rho h\left(\underline{k}\right). \tag{1.62b}$$

1.2.2 Small angle scattering from large particles in solution

Define a scattering intensity function as the scattering cross-section per unit volume as follows:

$$\frac{d\Sigma}{d\Omega} = \frac{1}{V} \frac{d\sigma}{d\Omega}. \tag{1.63a}$$

Then,

$$\frac{d\Sigma}{d\Omega} = \frac{d\Sigma_{coh}}{d\Omega} + \frac{d\Sigma_{inc}}{d\Omega},$$
$$\frac{d\Sigma_{inc}}{d\Omega} = \frac{\Sigma_{inc}}{4\pi} = \frac{N\,\sigma_{inc}}{V\,4\pi} = n\frac{\sigma_{inc}}{4\pi}, \tag{1.63b}$$
$$\frac{d\Sigma_{coh}}{d\Omega} = \frac{1}{V} \left\langle \left| \sum_{l=1}^N b_l e^{i\underline{Q}\cdot\underline{r}_l} \right|^2 \right\rangle.$$

Divide the sample into N_p cells, each cell centred at a colloidal particle. The position vector of the j-th nucleus in the i-th cell can be represented as $\underline{R}_i + \underline{x}_j$, where $\underline{R}_i$ is the position vector of the centre of mass of the particle i, and $\underline{x}_j$ is the position vector relative to the centre of mass of the particle i. Thus

$$\frac{d\Sigma_{coh}}{d\Omega}(\underline{Q}) = \frac{1}{V} \left\langle \left| \sum_{i=1}^{N_p} e^{i\underline{Q}\cdot\underline{R}_i} \sum_{cell\,i}^j b_{i,j} e^{i\underline{Q}\cdot\underline{x}_j} \right|^2 \right\rangle. \tag{1.64}$$

Define the form factor of each cell by

$$F_i\left(\underline{Q}\right) = \sum_{cell\,i}^j b_{i,j} e^{i\underline{Q}\cdot\underline{x}_j}, \tag{1.65a}$$

and introduce the local scattering length density of the i-th cell as

$$\rho_i\left(\underline{r}\right) = \sum_j b_{i,j}\delta\left(\underline{r} - \underline{x}_j\right). \tag{1.65b}$$

We can then write

$$F_i\left(\underline{Q}\right) = \int d^3 r e^{i\underline{Q}\cdot\underline{r}}\rho_i\left(\underline{r}\right)$$

$$= \int_{\text{particle } i} d^3 r e^{i\underline{Q}\cdot\underline{r}}\left[\rho_i\left(\underline{r}\right) - \rho_s\right] + \int_{\text{all } i} d^3 r \rho_s e^{i\underline{Q}\cdot\underline{r}}, \tag{1.65c}$$

where ρ_s is the scattering length density of the solvent which is a constant (i.e. a continuum assumption). So for $\underline{Q} \neq 0$, we can just put

$$F_i\left(\underline{Q}\right) = \int_{\text{particle } i} d^3 r e^{i\underline{Q}\cdot\underline{r}}\left[\rho_i\left(\underline{r}\right) - \rho_s\right]. \tag{1.65d}$$

Using Eq. (1.65d), the cross-section Eq. (1.64) can be written as

$$\frac{d\Sigma_{coh}}{d\Omega} = \frac{1}{V}\left\langle\left|\sum_{i=1}^{N_p} F_i\left(\underline{Q}\right) e^{i\underline{Q}\cdot\underline{R}_i}\right|^2\right\rangle$$

$$= \frac{1}{V}\left\langle\sum_{i,i'}^{N_p} F_i\left(\underline{Q}\right) F_{i'}\left(\underline{Q}\right) e^{i\underline{Q}\cdot\left(\underline{R}_i - \underline{R}_{i'}\right)}\right\rangle. \tag{1.66}$$

For a monodisperse colloidal system the form factor is identical for each particle, so one can write

$$\frac{d\Sigma_{coh}}{d\Omega} = \frac{N_p}{V}\left|F\left(\underline{Q}\right)\right|^2\frac{1}{N_p}\left\langle\sum_{i,i'}^{N_p} e^{i\underline{Q}\cdot\left(\underline{R}_i - \underline{R}_{i'}\right)}\right\rangle$$

$$= n\left|F\left(\underline{Q}\right)\right|^2 S\left(\underline{Q}\right). \tag{1.67c}$$

We call $P\left(\underline{Q}\right) = \left|F\left(\underline{Q}\right)\right|^2$ the particle structure factor and take $n = N_p/V$. With these notations, we have a simple formula for the monodisperse spherical particle system:

$$\frac{d\Sigma_{coh}}{d\Omega} = nP\left(\underline{Q}\right) S\left(\underline{Q}\right), \tag{1.67b}$$

i.e. the small angle scattering intensity is a product of the number density of the particle, the particle structure factor and the inter-particle structure factor.

Example: For a spherical protein, such as Cytochrome C, where $\bar{\rho}$ is the average scattering length density (sld) of the protein and $\Delta\rho = \bar{\rho} - \rho_s$ is called the contrast factor, where ρ_s is the sld of the solvent, we can calculate the form factor $F(Q)$ in the following way:

$$
\begin{aligned}
F(\underline{Q}) = \int_{\text{sphere}} (\bar{\rho} - \rho_s) e^{i\underline{Q}\cdot\underline{r}} &= \Delta\rho \int_0^R dr 4\pi r^2 \frac{\sin Qr}{Qr} \\
&= V_p \Delta\rho \left[\frac{3\,(\sin QR - QR\cos QR)}{Q^3 R^3} \right] \\
&= V_p \Delta\rho \left[\frac{3 j_1\,(QR)}{QR} \right]^2 .
\end{aligned}
\tag{1.68a}
$$

Therefore the particle structure factor is given by

$$
P(\underline{Q}) = (\Delta\rho)^2 V_p^2 \left[\frac{3 j_1\,(QR)}{QR} \right]^2 .
\tag{1.68b}
$$

One often defines a normalised particle structure factor by

$$
\bar{P}(\underline{Q}) = \left[\frac{3 j_1\,(QR)}{QR} \right]^2 .
\tag{1.68c}
$$

Summarising, one writes the differential cross-section as

$$
\begin{aligned}
\frac{d\Sigma_{coh}}{d\Omega} &= n V_p{}^2 (\Delta\rho)^2 \left[\frac{3 j_1\,(QR)}{QR} \right]^2 S(Q) \\
&= n(\bar{\rho} - \rho_s)^2 V_p{}^2 \left[\frac{3 j_1\,(QR)}{QR} \right]^2 S(Q) \\
&= n \left(\sum_i b_i - V_p \rho_s \right)^2 \bar{P}(Q) S(Q) \\
&= C N_A \left(\sum_i b_i - V_p \rho_s \right)^2 \bar{P}(Q) S(Q) ,
\end{aligned}
\tag{1.69}
$$

where V_p is the volume of a protein, C the molar concentration of the protein solution and N_A the Avogadro number.

Non-spherical particles

For a randomly oriented homogeneous cylindrical particle of cross-sectional radius R and length L, the particle structure factor can also be calculated as

$$
\bar{P}(Q) = \int_0^1 d\mu \left[\frac{\sin(QL\mu/2)}{QL\mu/2} \right]^2 \left[\frac{2 J_1(QR\sqrt{1 - \mu^2})}{QR\sqrt{1 - \mu^2}} \right]^2 ,
\tag{1.70a}
$$

which has limiting forms when the cylinder is long compared to its radius as follows:

$$\bar{P}(Q) \underset{QL>2\pi}{\rightarrow} \frac{\pi}{QL} \left[\frac{2J_1(QR)}{QR} \right]^2 \underset{QR<1}{\rightarrow} \frac{\pi}{QL} e^{-\frac{1}{4}Q^2R^2}. \tag{1.70b}$$

On the other hand, when the particle is a flat disc, one has an asymptotic form,

$$\bar{P}(Q) \underset{QR>>1}{\rightarrow} \frac{2}{Q^2R^2} \left[\frac{\sin QL/2}{QL/2} \right]^2 \underset{QL<1}{\rightarrow} \frac{2}{Q^2R^2} e^{-\frac{1}{12}Q^2L^2}. \tag{1.70c}$$

Contrast variation technique

Change ρ_s by mixing H_2O and D_2O so that $\sum b_i - V_p\rho_s = 0$. Then the scattering intensity will go to zero. From the value of this zero contrast solvent scattering length density ρ_s, one can calculate the volume of the particle by $V_p = \sum b_i/\rho_s$. In the case of proteins, one can also obtain the amount of hydration (a layer of water on the surface of protein) in g water / g protein (Chen and Lin, 1986)

$$\rho_s = \beta\rho_{D_2O} + (1 - \beta)\rho_{H_2O} = [6.404\beta - 0.562(1 - \beta)] \times 10^{10}\,\text{cm}^{-2}.$$

Extension to particles of non-spherical shape or particles with a size distribution (Kotlarchyk and Chen, 1983)

For a system of monodisperse particles with non-spherical shape, such as protein molecules having oblate or prolate spherical shapes, one may assume that the particle orientations are uncorrelated with the particle positions. One can then perform the orientational average independently from the centre-to-centre average. Thus

$$\frac{d\Sigma_{coh}}{d\Omega} = \frac{1}{V} \left\langle \sum_{i,j} \langle\langle F_i\left(\underline{Q}\right) F_j\left(\underline{Q}\right)\rangle\rangle e^{i\underline{Q}\cdot(\underline{R}_i-\underline{R}_j)} \right\rangle. \tag{1.71a}$$

The orientational average $\langle\langle F_i\left(\underline{Q}\right) F_j\left(\underline{Q}\right)\rangle\rangle$ can be decomposed into two terms:

$$\langle\langle F_i\left(\underline{Q}\right) F_j\left(\underline{Q}\right)\rangle\rangle = \left\{ \langle|F\left(\underline{Q}\right)|^2\rangle - |\langle F\left(\underline{Q}\right)\rangle|^2 \right\} \delta_{ij} + |\langle F\left(\underline{Q}\right)\rangle|^2. \tag{1.71b}$$

By defining

$$\beta\left(Q\right) = \frac{|\langle F\left(\underline{Q}\right)\rangle|^2}{\langle|F\left(\underline{Q}\right)|^2\rangle}, \tag{1.71c}$$

and

$$\bar{S}\left(\underline{Q}\right) = \frac{1}{N} \left\langle \sum_{i=1}^{N} \sum_{j=1}^{N} e^{i\underline{Q}\cdot(\underline{R}_i-\underline{R}_j)} \right\rangle, \tag{1.71d}$$

Eq. (1.71a) can be written in the form

$$\frac{d\Sigma_{coh}}{d\Omega} = \frac{N}{V}\overline{P}(Q)\left[1 + \beta(Q)(\overline{S}(Q) - 1)\right],\tag{1.72}$$

where $\beta(Q)$ is usually less than unity for some Q for an ellipsoid and $\overline{P}(Q) = \left\langle|F(Q)|^2\right\rangle$. For spherical particles, $\beta(Q) = 1$, Eq. (1.72) reduces to Eq. (1.67b).

For ellipsoidal proteins,

$$\langle F(Q)\rangle = \frac{1}{2}\int_{-1}^{1} d\mu\left(\frac{3j_1(u)}{u}\right), \langle|F(Q)|^2\rangle = \frac{1}{2}\int_{-1}^{1} d\mu\left(\frac{3j_1(u)}{u}\right)^2,$$

$$u = Q\sqrt{a^2\mu^2 + b^2(1 - \mu^2)}.\tag{1.73}$$

For spherical particles with a size distribution, Eq. (1.72) can also be used provided we interpret the average form factors by the following definitions:

$$\left\langle|F(Q)|^2\right\rangle = \int_0^\infty dR f(R)|F(QR)|^2,$$

$$\langle F(Q)\rangle = \int_0^\infty dR f(R) F(QR).\tag{1.74a}$$

A serious approximation is that $\overline{S}(Q)$, called the decoupling approximation in the literature (Kotlarchyk and Chen, 1983), has to be computed by assuming an effective size $\overline{R}$ defined by

$$\overline{R}^3 = \int_0^\infty R^3 f(R)\,dR.\tag{1.74b}$$

At this point, let us briefly summarise the usefulness of the small angle neutron scattering (SANS) technique for colloid and macromolecular solutions. With SANS, it is easy to make an absolute intensity measurement. Therefore, analysis of zero angle intensity gives directly the weight-averaged molecular weight for macromolecules or the mean aggregation number for surfactant micelles. It is possible to do contrast variation measurements by D/H ratio adjustment. Thus SANS technique is able to resolve structure at the 5 Å level. It is sensitive to macromolecular interactions. Thus a proper analysis of SANS intensity will give the excluded volume dimension and charge on proteins or polyelectrolytes.

SANS data correction and normalisation

(i) The transmission T of a plate sample S having a thickness of t and a macroscopic total cross-section Σ is defined as

$$T = \frac{I_t}{I_0} = e^{-\Sigma t},\tag{1.75}$$

where I_0 and I_t are the incident and transmitted beams of neutrons respectively.

(ii) The raw two-dimensional detector counts were first corrected for the room and electronics background (as measured by the blocked beam intensity $I_{b,i}$) and then normalised to a unit monitor counts, M. The corrected intensity at the detector pixel i is given by the following equation:

$$I_s^{0,i} = \left(\frac{I_{s,i}}{M_s} - \frac{I_{b,i}}{M_b} \right) - \frac{T_s}{T_e} \left(\frac{I_{e,i}}{M_e} - \frac{I_{b,i}}{M_b} \right). \tag{1.76a}$$

(iii) The signal intensity (assuming only single scattering) at the detector pixel i which subtends a solid angle $\Delta\Omega$ from the centre of the sample is then given by the following equation:

$$I_s^{0,i} = \Phi_M T_s^0 t_s \left(\frac{d\Sigma_s}{d\Omega} \right)_i \Delta\Omega_i \varepsilon_i, \tag{1.76b}$$

where Φ_M is the incident neutron beam intensity (as measured by the monitor) and Σ is the macroscopic scattering cross-section of the sample and ε_i is the detection efficiency of the pixel i.

(iv) We then construct a normalised intensity $N_{s,i}^s = I_{s,i}^0 \big/ (T_s^0 t_s)$ where T_s^0 and t_s are, respectively, the transmission and the thickness of the sample. In order to further correct for the non-uniformity of the pixel efficiency and to put the intensity finally into an absolute scale, we divide $I_{s,i}^0$ by the scattering from a standard sample (water) processed in the same way. The criterion of the standard sample should be that it is an isotropic incoherent scatterer with a well-known cross-section and it should be thin enough to avoid multiple scattering. In practice, a 1 mm water (H_2O) sample at room temperature approximately satisfies this criterion owing to its large known incoherent cross-section of hydrogen atoms. A water sample, 1 mm in thickness, however, gives rise to an appreciable amount of multiple scattering that the signal needs to be corrected for the effect of multiple scattering by an empirical factor f. We thus performed a separate water sample calibration run to obtain $N_i^w = I_{w,i}^0 \big/ (T_w^0 t_w)$ in obvious notation.

(v) The water intensity is given by

$$I_{w,i}^0 = f \Phi_M \left[\frac{1 - T_w^0}{4\pi} \right] \Delta\Omega_i \varepsilon_i. \tag{1.76c}$$

(vi) Then the normalised intensity ratio can be calculated as

$$N_{s,i} = \frac{I_{s,i}^0}{T_s^0 t_s} \bigg/ \frac{I_{w,i}^0}{T_w^0 t_w} = 4\pi t_w \left(\frac{d\Sigma_s}{d\Omega} \right)_i \frac{T_w^0}{1 - T_w^0} \frac{1}{f}. \tag{1.76d}$$

(vii) From Eq. (1.76d), the absolute macroscopic differential cross-section can then be obtained as:

$$\left(\frac{d\Sigma_s}{d\Omega}\right)_i = N_{s,i}\, f \left(\frac{1 - T_w^0}{T_w^0}\right) \frac{1}{4\pi\, t_w}, \qquad (1.76e)$$

where t_w is equal to 1 mm and the multiple scattering correction factor is $f \approx 1.37$ for the neutron wavelength of 4.5 Å used in the small angle neutron scattering experiment.

1.2.3 An illustrative example: The structure of protein–detergent complexes

Introduction

It is been well known (Tanford, 1968) that anionic surfactant sodium dodecyl-sulfate (SDS) is one of the most potent denaturants of water-soluble proteins. Since hydrophobic effects are the major driving force for the folding of water-soluble proteins into their native forms (Kautzmann, 1959) it is generally agreed in the literature (Reynolds and Tanford, 1970) that the unfolding of proteins by SDS is caused by binding of SDS molecules to the hydrophobic portions of the polypeptide chain. However, there is no unanimity in the literature about the precise structure of the resultant protein/SDS complexes thus formed in aqueous solutions (Makino, 1979). A similar problem of the interactions of SDS with water-soluble polymers and its resultant formation of polymer/SDS complexes has been important in industrial applications and the structure of such complexes has been discussed (Cabane and Duplessix, 1982). Understanding the structure of protein/detergent complexes is important even from a practical point of view. It is common practice in protein biochemistry to determine molecular weights of proteins by performing electrophoresis of protein/SDS complexes in polyarcry-lamide gels, following the classic work of Weber and Osborn (1969). The precise physical understanding of the basic principle of SDS gel electrophoresis at the molecular level depends critically on knowing the precise structure of protein/SDS complexes (Guo and Chen, 1990).

A well-known study of the structure of protein/SDS complexes using viscometry for a number of proteins by Reynolds and Tanford (1970) concluded with a proposal of a prolate ellipsoidal model for the complexes with a high content of α-helical structure. The size of the semi-minor axis of the ellipsoid is constant for several of the proteins studied and is equal approximately to 18 Å, while that of the major axis is variable, proportional to the length of the polypeptide chain. Lundahl *et al.* (1986) advanced another theoretical model for protein/SDS complexes. They suggested that in a complex, SDS molecules form a flexible, capped cylindrical micelle around which the hydrophilic segments of the polypeptide chain are helically wound. The bindings between polypeptide segments and the detergent molecules are mainly due to the hydrogen bonds. The hydrogen bond is between the nitrogen atom of the peptide bond and the sulfate oxygen of the detergent molecule.

In a less-known study by Shirahama *et al.* (1974), the authors noticed that the free electrophoresis mobilities of the saturated polypeptide/SDS complexes were virtually constant, regardless of the molecular weight of the polypeptide, and that the mobility was compatible with that of the SDS micelle. They therefore proposed a model of the complexes called the *necklace* model, which assumed that the polypeptide chains are flexible in solution and micelle-like clusters of SDS are distributed along the unfolded polypeptide chains. In addition, the NMR study by Tsujii and Takagi (1975) confirmed that SDS molecules in bovine serum albumin (BSA)/SDS complexes are in a micelle-like environment and clusters of surfactant molecules are formed on the protein, which supported the necklace model.

In 1986 Chen and Teixeira (1986) undertook a small angle neutron scattering (SANS) study of BSA/lithium dodecylsulfate (LDS) complexes in a buffer solution near the pI value of BSA. These authors adapted the necklace model of Shirahama (1974) and developed a *fractal model* to account for the spatial correlation between the strings of micelles linked together by the unfolded polypeptide chains. The resultant neutron cross-section calculated with the fractal mode was in excellent agreement with SANS data, confirming the validity of the necklace model.

In this example, we show some further SANS measurements of a globular protein – BSA (MW = 66 114), and their complexes with SDS (Guo *et al.*, 1990). By using the fractal model developed earlier, we are able to extract a set of parameters that characterise the structure and conformation of the protein/SDS complexes. We obtained the average micelle size and its aggregation number, the fractal dimension that characterises the topological nature of the conformation of the unfolded polypeptide chains, the average numbers of micellar binding sites on a polypeptide chain, and the correlation length, which gives a measure of the extent of the unfolded chains.

SANS cross-section

As mentioned above, it was found in two previous experiments on protein/detergent interaction that the data can be explained reasonably well by the so-called necklace model. This model assumed for the protein/SDS complex a structure consisting of a string of micelle-like clusters randomly distributed along the unfolded polypeptide chain and suspended in random orientations in solution. Suppose all the micelle-like clusters can be regarded as a uniform ellipsoid of revolution. One can then derive a theoretical cross-section formulae for the SANS data analysis, called the fractal model.

The scattering intensity distribution for a collection of uniform ellipsoidal micelles in the necklace model can be written generally as

$$I(Q) = N_1 \overline{N}(b_m - V_m \rho_s)^2 \tilde{P}(Q) S(Q), \tag{1.77a}$$

where N_1 is the number density of the total surfactant molecules in solution, $\overline{N}$ the average aggregation number of micelle-like clusters, b_m, the scattering length of the surfactant molecule, V_m its volume, and ρ_s the scattering length density of D_2O solvent. $\tilde{P}(Q)$ denotes the normalised particle structure factor of an ellipsoid, and can be written as:

$$\bar{P}(Q) = \int_0^1 d\mu \left| \frac{3 j_1(u)}{u} \right|^2, \tag{1.77b}$$

where $j_1(x) = x^{-2} \sin(x) - x^{-1} \cos(x)$, is the first-order spherical Bessel function. The variable u is a function of the semi-major and semi-minor axis of the prolate spheroid a and b, and μ is the direction cosine between the symmetry axis of the spheroid and the Q vector, namely,

$$u = Qb \left[\left(\frac{a}{b} \right)^2 \mu^2 + (1 - \mu^2) \right]^{\frac{1}{2}}. \tag{1.77c}$$

The integration over μ in Eq. (1.77b) essentially performs the orientational average of the ellipsoid with respect to the Q vector. $\bar{P}(Q)$ is a bell-shaped function with the peak centres around $Q = 0$. For small Q such that $Qb << 1$, it is essentially unity.

The $S(Q)$ function is the inter-particle structure factor that takes into account the inter-micellar positional correlations. $S(Q)$ can be defined, in general, in terms of the micellar pair correlation function $g(r)$ by

$$S(Q) = 1 + n \int_o^\infty dr 4\pi r^2 \frac{\sin Qr}{Qr} g(r), \tag{1.77d}$$

where n is the number density of micelles in solution. We can simply derive the asymptotic behaviour of $S(Q)$ at small Q (a range of Q covered by a small angle scattering experiment) by the following argument. In a D-dimensional space, the number $N(r)$ of individual scatterers within a sphere of radius r is given by $N(r) = (r/R)^D$. The physical meaning of D can be seen easily as a fractal dimension of the packing of the micellar clusters. For example, $D = 3$ means that the spheres are in a compact arrangement similar to three-dimensional packing of molecules in liquids. But in general D can be a fractional number less than three for packing with open structures. The derivative of the above equation can be related to the asymptotic form of the pair correlation function $g(r)$. From the definition of $g(r)$ one can derive a relation:

$$g(r) = \frac{D}{4\pi n} \frac{1}{R^D} r^{D-3}. \tag{1.77e}$$

In addition, for the necklace model of the protein/detergent complexes, one expects that the correlation among micelles has a limited range. Therefore we assume the micelle–micelle correlation to have a finite range ξ, which is a measure of the extent of the unfolded polypeptide chain. Thus, we introduce a cutoff factor $\exp(-r/\xi)$ into Eq. (1.77e), following an idea of Sinha *et al.* (1984), to get

$$g(r) = \frac{D}{4\pi n} \frac{1}{R^D} r^{D-3} \exp(-r/\xi). \tag{1.77f}$$

Upon substitution of Eq. (1.77f) into Eq. (1.77d) and taking the Fourier transform, we get

$$S(Q) = 1 + \frac{1}{(QR)^D} \frac{D\Gamma(D-1)\sin[(D-1)\tan^{-1}(Q\xi)]}{[1 + (Q\xi)^{-2}]^{[(D-1)/2]}}, \tag{1.78a}$$

where R is an equivalent micellar radius, $R = (ab^2)^{1/3}$, for the ellipsoidal micelle, D the fractal dimension of the micellar distribution in space, and ξ the correlation length. A special case of Eq. (1.78a) is when $D = 2$; it simplifies to

$$S(Q) = 1 + 2\left(\frac{\xi}{R}\right)^2 \frac{1}{1 + Q^2\xi^2}, \tag{1.78b}$$

which is just the Ornstein–Zernike form of the structure factor familiar in critical scattering from fluid. One notable feature of Eq. (1.78a) is that when $Q\xi \ll 1$, the structure factor approaches the limiting value

$$\lim_{Q\to 0} S(Q) = 1 + \Gamma(D+1)\left(\frac{\xi}{R}\right)^D. \tag{1.78c}$$

In Eqs. (1.78a) and (1.78c), $\Gamma(x)$ denotes the gamma function. Since the micelle size R is more or less fixed by the surfactant tail length, as the protein unfolds, ξ increases and hence the value of $S(Q \to 0)$, as well as the intensity, increases sharply. In the opposite limit where $Q\xi \gg 1$, $S(Q)$ approaches unity. In the intermediate Q range $(1/\xi < Q < 2/R)$, the structure factor has the form $S(Q) \approx (QR)^{-D}$, which is a general feature of the scattering from a fractal object.

Results of the data analyses

In the previous section, we derived a formula for the calculation of the neutron cross-section. We now apply it to analyse the experimental data in this section and to verify that the model is valid for characterisation of the protein/detergent complexes.

In Eq. (1.78a), the unknown parameters are D, ξ, N, a and b. Considering $\overline{N} = N_1/(N_pN)$, where N_p is the number density of protein molecules in solution and N is the number of micelles on a chain, we substitute $\overline{N}$ and rewrite Eq. (1.78a) as

$$I(Q)\left(\text{cm}^{-1}\right) = \frac{N_1^2}{N_pN}(b_m - V_m\rho_s)^2\,\tilde{P}(Q)\,S(Q).\qquad(1.79)$$

In this form, the unknown parameters are D, ξ, N, a and b. Therefore, the independent parameters used in data fittings are ξ, D, N, a and b. In addition, SANS data for pure proteins in buffer solutions were fitted to an ellipsoidal model without considering the protein/protein interaction. This is sufficient for dilute protein solutions. The fitted parameters are $a = 71\,\text{Å}$ and $b = 19\,\text{Å}$ for BSA (bovine serum albumin).

Since there are five unknown parameters in formula, it is impossible to extract them by a fitting program simultaneously. So the fitting procedure should be carried out in steps. The procedure we adopted is as follows:

(i) From the experiment data in the large Q range and the known surfactant tail length, we can determine a, b and an approximate N.

(ii) We then fix a, b and N to fit the data in the entire Q range using two other parameters, D and ξ. According to the result obtained, we modify a, b and N slightly, if necessary, to make the fit better.

(iii) By systematic fitting of all data, we finally fix values of a, b and N to extract the remaining D and ξ. Once these independent parameters are determined, other quantities, such as $\overline{N}$, the average SDS molecules per micelle-like cluster, can be calculated.

In order to illustrate the dramatic change in SANS intensity distribution from proteins to protein/detergent complexes, we compare, in Figure 1.8, SANS intensity distributions of 1/1 samples with 1% pure protein solution of BSA. It is clear from the figure that both the size and the number of the scattering unit in the 1/1 samples are markedly different from the 1% protein solutions since $I(Q \rightarrow 0)$ is proportional to the number and molecular weight of the scattering units. We also find no evidence for the inter-micellar correlations of the three-dimensional short-range type, characteristic of a liquid-like system, since the scattered intensity distributions do not show an interaction peak at around $Q = 0.05\,\text{Å}^{-1}$. Absence of the interaction peak confirms that micelles are not freely suspended in solutions. This is certainly the case for the 1/1 sample since the detergent concentration is below the saturation binding level. In fact, an ultracentrifugal photograph method

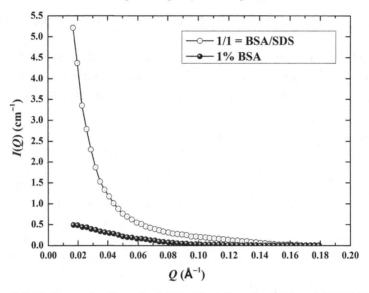

Figure 1.8 The scattering intensity $I(Q)$ of 1 wt% native BSA and 1/1 BSA-SDS
(1 g/dL BSA and 1 g/dL SDS) BSA/SDS complex solution in 0.1 M acetate buffer
solution with pD = 5.5 and 0.5 M NaCl. Open and solid circles denote SANS data
points and solid lines represent the results of the best theoretical fittings.

(Nelson, 1971) showed that the maximum bindings of SDS to BSA (disulfide bond
intact), at 0.4 M ionic strength, to be 1.9 g (g SDS/g protein) respectively.

BSA/SDS complex ($I = 0.6\,M$)

BSA is a transport protein that has the ability to bind hydrophobic ligands such
as fatty acids. It has 17 disulfide bonds and 68% helix content, which indicates
that α-helices are the major structural components. It has been suggested that BSA
has three domains consisting of six paired sub-domains. This implies that there
are six high-affinity binding sites for fatty acids and each sub-domain contains a
hydrophobic face and a polar face.

Figure 1.9 shows a set of four intensity distributions measured in Saclay
(France). These are samples with BSA/SDS wt ratios 1/1.0, 1/1.5, 1/2.0 and 1/3.0.
We present them in log $I(Q)$ vs. log Q plots. The figure shows that the theoretical
calculations (solid lines) are in good agreement with the experimental data (symbols) over the entire Q range for all samples. For a sufficiently large Q data (i.e.
$Q > 0.16\,\text{Å}^{-1}$), $S(Q) \rightarrow 1$, and the shape of $I(Q)$ depends only on $P(Q)$. The
parameters contained in the $P(Q)$ function, such as a and b, are thus more sensitive
to this portion of data.

In Figure 1.9, we show that data in this portion of Q are well fitted by assuming an ellipsoidal micelle. Thus, the assumption that SDS molecules aggregate into

Table 1.2 *Fitted parameters for BSA/SDS complexes (I = 0.6 M)*

BSA/SDS	D	x in Å	a in Å	b in Å	N	AGG
1/1	2.00	82	23	18	8	29
1/1.5	1.90	127	25	18	8	43
1/2	1.75	152	26	18	11	42
1/3	1.65	208	27	18	18	39

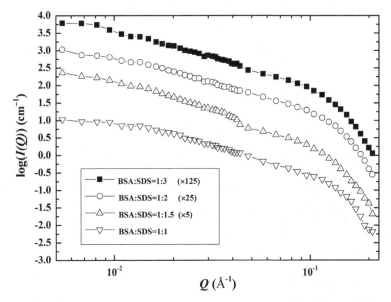

Figure 1.9 The log of scattering intensity $\log(I(Q))$ vs. $\log(Q)$ measured at the LLB in Saclay, France for four samples of BSA-SDS complexes having different BSA/SDS wt ratios. Shown are the SANS data point and their fittings using the fractal model described in the text. Notice that the top three curves are shifted by some scaling factors indicated in the figure.

micelle-like clusters is well established. In a Q range between 0.01 and 0.1 Å^{-1}, the fractal nature of micellar packing can be seen easily by plotting $\log I(Q)$ vs. $\log Q$. For fractal aggregates we have $\log I(Q) \approx D \log Q$ in the intermediate Q range $(1/\xi < Q < 2/R)$. The slope of the straight line is thus related to the value of the fractal dimension D and the magnitude of $I(Q)$ is sensitive to ξ. Table 1.2 shows that D decreases from about 2.1 to 1.65 and increases from 82 to 210 Å successively, with increasing SDS concentration. This is consistent with an intuitive picture that the conformation of SDS/BSA complexes transforms gradually from a compact globule to a more open random coil conformation as the denaturation

BSA:SDS = 1:2   BSA:SDS = 1:1

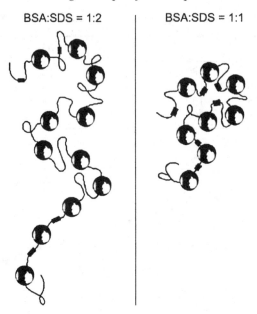

Figure 1.10 Pictorial representation of the *necklace* model of the protein-detergent complex in an aqueous buffer solution having different values of BSA/SDS weight ratio. The spheres with black stripes represent the SDS-polypeptide micelle-like cluster and the black wiggle-lines in-between them the unfolded polypeptide chain of the protein BSA. Short cylinders represent the collapsed structure (alpha helix).

proceeds. The ξ value of $1/3$ complex is consistent with the rms end-to-end distance of BSA in 6 M GuHCL, in which proteins are believed to have a random coil conformation.

This shows that ξ is correlated with the extent of the unfolding of the polypeptide chain. Moreover N, the numbers of micelle in a complex, increases during this process. This can be interpreted as a result of a further unfolding of the polypeptide chain. Interestingly, the N ($N = 8$) we extracted from the $1/1$ sample is compatible with the number of high affinity binding sites of a BSA molecule ($= 6$). The calculated micelle aggregation number generally varies from 30 to 45, smaller than ordinary SDS micelles in water. To our knowledge, this is the first time that this information has been reported. However, a similar result was reported from a study of synthetic polymer/SDS complexes. The reason smaller for micelle-like clusters in protein/SDS complexes is unknown. One may conjecture that the side chains of amino acid residues probably have a strong effect on micellisation. Then, one possible explanation would be that the hydrophobic portion of the polypeptide chain behaves like a nucleus of micellisation to effectively stabilise the micelle-like cluster having a smaller aggregation number.

Conclusion

In summary, our analyses of the absolute intensities of SANS data established the following picture of the protein/SDS complexes in high ionic strength solution ($I = 0.6$ M):

(i) SDS molecules in protein/SDS complexes form micelle-like clusters, each of which can be approximated as an ellipsoidal micelle.

(ii) The packing of the micelle-like clusters in a protein/SDS complex has a fractal-like feature having two characteristic parameters: a fractal dimension D and a correlation length ξ.

(iii) D and ξ can be used to characterise the conformation of denatured proteins.

(iv) The average number of micelle-like clusters per polypeptide chain N, the average aggregation number of the micellar cluster $\overline{N}$, and the dimensions of the micelle a and b can be determined.

(v) The saturation binding supposedly occurring at SDS/protein ratio $w \approx 1.4$ is not a clear-cut transition. We believe that even after the binding ratio of 1.4 is reached, SDS molecules continue to bind or be absorbed into the partially unfolded polypeptide chain.

Based on the above results, a schematic two-dimensional picture is drawn in Figure 1.10 to represent the structure of the protein/SDS complexes. The details of the local structure of the micelle-like clusters are not fully clear based only on these fitted parameters.

2

Statistical mechanics of the liquid state

In this chapter, we describe the elements of liquid theory. Our intention is not to present all aspects of liquid theory in its complete form, but instead only those that will be useful and sufficient for discussing its different applications for presentations in subsequent chapters. It will be addressed in the context of understanding problems that arise in complex fluids and colloidal science discussed in subsequent chapters.

In Section 2.1 we introduce the concept of the pair correlation function and the structure factor which are fundamental quantities when discussing applications of liquid theories to the analysis of scattering data, including light scattering, X-ray scattering and neutron scattering. We would like to remind the reader here that the pair correlation function and thus the structure factor are intimately connected with the statistical thermodynamics of the system. In this sense the study of the structure factor in the liquid state using scattering techniques is to investigate some aspect of the thermodynamics of the liquid system.

Then we discuss the solution of Ornstein–Zernike equation in its different approximations. In particular, in Section 2.2.3, we illustrate the use of the Baxter method, an elegant analytical method for solving hard-sphere and adhesive hard-sphere systems in the Percus–Yevick approximation. These two systems are the model systems for simple liquids as well as for the colloidal solutions. In Section 2.2.4 we present an analytical solution for the case of a narrow-square-well potential that avoids some aspects of the unphysical features of the Baxter solution of the adhesive hard-sphere system. We shall show in a later chapter the applications of this analytical solution for studying the kinetic glass transition in a micellar system.

We then skip the background introduction to the liquid theories of ionic solutions such as the classical Debye–Hückel theory and the Poisson–Boltzmann theory of ionic solutions, as well as the mean spherical approximation solutions of the so-called primitive model of ionic solution already available in Blum (1980).

Instead, we shall concentrate on presenting in Section 2.3.1 an analytical solution to the mean spherical approximation of one-Yukawa and two-Yukawa models for charged macroion solutions. In Section 2.3.2 we present the so-called generalised one-component macroion model of charged colloidal solutions, which will be useful for the discussion of protein solutions at different pH values.

2.1 Pair correlations and structure factors in liquids[1]

2.1.1 The canonical ensemble

A quantitative description of the static structure of a simple liquid can be made through a series of atomic distribution functions defined in terms of the phase space distribution function derivable from the Hamiltonian of an N-particle system. The Hamiltonian of the system can be written as:

$$H = \sum_{i=1}^{N} \frac{1}{2M} p_i^2 + U_N, \tag{2.1}$$

where the first term is the kinetic energy and the second the potential energy of the system. We then introduce the concept of an ensemble and the associated statistical averages, which define measurable quantities in statistical mechanics.

A canonical ensemble is a collection of systems characterized by the same values of N, V and T. The N-particle phase space distribution function $f_0^{(N)}$ in this ensemble is given by

$$f_0^{(N)}(\underline{r}^N, \underline{p}^N) = \frac{1}{N!} \frac{1}{h^{3N}} \frac{e^{-\beta H}}{Q_N(V, T)}, \tag{2.2a}$$

where $Q_N(V, T)$ is the canonical partition function defined as

$$Q_N(V, T) = \frac{1}{N!} \frac{1}{h^{3N}} \int \int d\underline{r}^N d\underline{p}^N \exp\left[-\beta H\left(\underline{r}^N, \underline{p}^N\right)\right]. \tag{2.2b}$$

The factor $N!$ takes account of the indistinguishability of the particles and the h^{3N} factor is inserted to make $Q_N(V, T)$ a non-dimensional quantity, even though the system we are treating is a classical one (i.e. described by classical mechanics).

The following relation can establish the link between statistical mechanics and thermodynamics:

$$A(V, T) = -k_B T \ln Q_N(V, T), \tag{2.2c}$$

where $A(V, T)$ is the Helmholtz free energy of the system. We can get other thermodynamic functions by differentiation of A. Since

$$dA = -p dV - S dT, \tag{2.3a}$$

[1] Chen (1971); Freidman (1985); Hansen and McDonald (2006).

pressure

$$p = -\left(\frac{\partial A}{\partial V}\right)_T, \tag{2.3b}$$

entropy

$$S = -\left(\frac{\partial A}{\partial T}\right)_V, \tag{2.3c}$$

and internal energy

$$U = \left[\frac{\partial\left(\frac{A}{T}\right)}{\partial\left(\frac{1}{T}\right)}\right] = \left[\frac{\partial\,(\beta A)}{\partial\beta}\right]_V = -\frac{\partial}{\partial\beta}\ln Q_N\,(V, T)\,|_V\,. \tag{2.3d}$$

Hence, denoting $\beta = 1/k_B T$, we can write the dimensionless free energy simply as

$$\beta A = -\ln Q_N. \tag{2.3e}$$

Let us take as an example the partition function of an ideal gas. The Hamiltonian in this case is

$$H_o = \sum_{i=1}^{N} \frac{1}{2M} p_i^2, \tag{2.4a}$$

so

$$\int\int d\underline{r}^N d\underline{p}^N e^{-\beta H_o} = V^N\left[\int d^3 pe^{-\frac{\beta p^2}{2M}}\right]^N = V^N\left[\left(\frac{2\pi M}{\beta}\right)^{\frac{3}{2}}\right]^N, \tag{2.4b}$$

$$Q_N^{id} = \frac{V^N}{N!h^{3N}}\left(\frac{2\pi M}{\beta}\right)^{\frac{3N}{2}} = \frac{V^N}{N!}\left[\frac{\sqrt{2\pi Mk_B T}}{h}\right]^{3N}. \tag{2.4c}$$

We define a thermal de Broglie wavelength Λ as $\Lambda = h(2\pi Mk_B T)^{-1/2}$, which has a value $\approx 0.3\,\text{Å}$ for argon at 84 K. This definition of the thermal de Broglie wavelength together with Eq. (2.4c), gives (using $\ln N! \doteq N \ln N - N$)

$$Q_N^{id} = \frac{V^N}{N!\Lambda^{3N}} = \frac{1}{N!}\left(\frac{V}{\Lambda^3}\right)^N, \tag{2.5a}$$

$$A^{id} = -\frac{1}{\beta}\left[N\ln V - \ln N! - 3N\ln\Lambda\right]$$

$$= -\frac{1}{\beta}\left[N\ln V + N - N\ln N - 3N\ln\Lambda\right], \tag{2.5b}$$

$$\frac{\beta A^{id}}{N} = \ln\left(\frac{N}{V}\right) + 3\ln\Lambda - 1, \tag{2.5c}$$

$$U^{id} = 3N \left. \frac{\partial \ln \Lambda}{\partial \beta} \right|_V = \frac{3}{2} N k_B T, \tag{2.5d}$$

$$p^{id} = -\left(\frac{\partial A^{id}}{\partial V} \right)_\beta = \frac{N}{\beta} \frac{N}{\frac{N}{V}} \frac{1}{V^2} = \frac{N k_B T}{V} = \rho k_B T, \tag{2.5e}$$

$$S^{id} = -\left(\frac{\partial A^{id}}{\partial T} \right)_V = N k_B \left[1 - \ln \left(\frac{N}{V} \right) - 3 \ln \Lambda \right]. \tag{2.5f}$$

The partition function of the system of interacting particles can then be written as

$$Q_N = \frac{1}{N!} \frac{1}{\Lambda^{3N}} Z_N(V, T) = Q_N^{id} \frac{Z_N(V, T)}{V^N}, \tag{2.6a}$$

where

$$Z_N(V, T) = \int d\underline{r}^N e^{-\beta U(\underline{r}^N)}. \tag{2.6b}$$

Then the free energy A can be decomposed into two parts

$$A = A^{id} + A^{ex}, \tag{2.7a}$$

with

$$A^{id} = -k_B T \ln Q_N^{id},$$

and

$$A^{ex} = -k_B T \ln \frac{Z_N(V, T)}{V^N} = -kT \ln \left[\frac{1}{V^N} \int d\underline{r}^N e^{-\beta U(\underline{r}^N)} \right]. \tag{2.7b}$$

Since

$$\mu^{id} = \left(\frac{\partial A^{id}}{\partial N} \right)_{V,T},$$

so

$$\beta \mu^{id} = \left. \frac{\partial \left(\beta A^{id} \right)}{\partial N} \right|_{V,T}, \tag{2.7c}$$

and

$$\beta A^{id} = N \left(\ln \rho + 3 \ln \Lambda - 1 \right),$$

$$\beta \mu^{id} = \left. \frac{\partial \left(\beta A^{id} \right)}{\partial N} \right|_{V,T} = \ln \frac{\rho}{\Lambda^3} - 1. \tag{2.7d}$$

2.1.2 Chemical potential and the grand canonical ensemble

When the system is open and can exchange particles with its surroundings, the chemical potential μ becomes relevant. Instead of using the Helmholtz free energy $A(V, T)$ to describe the thermodynamic potential of the system, one has to use the grand potential Ω, which is defined as

$$\Omega (V, T, \mu) = A - N\mu = -pV,$$

$$d\Omega = -pdV - SdT - Nd\mu, \, dA = -pdV - SdT,$$

$$p = -\left(\frac{\partial \Omega}{\partial V}\right)_{T,V}, \tag{2.8}$$

$$N = -\left(\frac{\partial \Omega}{\partial \mu}\right)_{V,T},$$

$$S = -\left(\frac{\partial \Omega}{\partial T}\right)_{V,\mu}.$$

The probability that the system has N particles distributed at $\underline{r}^N$, $\underline{p}^N$ is given by

$$f_0\left(\underline{r}^N, \underline{p}^N, N\right) = P(N) \, f_0^{(N)}\left(\underline{r}^N, \underline{p}^N\right) = \frac{1}{N!h^{3N}} \frac{e^{\beta N\mu} e^{-\beta H}}{\Xi(V, T, \mu)}. \tag{2.9}$$

The grand partition function $\Xi(V, T, \mu)$ is given as

$$\Xi = \sum_{N-0}^{\infty} e^{\beta N\mu} Q_N(V, T) = \sum e^{\beta N\mu} \frac{1}{N!\Lambda^{3N}} Z_N(V, T)$$

$$= \sum \frac{1}{N!}\left(\frac{e^{\beta\mu}}{\Lambda^3}\right)^N Z_N(V, T) = \sum \frac{1}{N!} z^N Z_N(V, T), \tag{2.10}$$

where $z = e^{\beta\mu}/\Lambda^3$ is the activity. The link between statistical mechanics and thermodynamics is made through a relation

$$\Omega = -k_B T \ln \Xi(V, T\mu) = -p\overline{V},$$

$$P(N) = \int d\underline{r}^N \int d\underline{p}^N f_0\left(\underline{r}^N, \underline{p}^N, N\right) = \frac{1}{\Xi} e^{\beta N\mu} Q_N(V, T)$$

$$= \frac{1}{\Xi}\left(\frac{z^N}{N!}\right) Z_N(V, T). \tag{2.11}$$

The average number of particles in the system is

$$<N> = \sum_{N=0}^{\infty} NP(N) = \sum N\frac{1}{\Xi}\left(\frac{z^N}{N!}\right) Z_N(V, T)$$

$$= \frac{1}{\Xi} z\frac{\partial}{\partial z} \underbrace{\sum \left(\frac{z^N}{N!}\right) Z_N(V, T)}_{\Xi} = \frac{\partial \ln \Xi}{\partial \ln z}. \tag{2.12}$$

It can be shown that

$$< N^2 > - < N >^2 = \frac{\partial < N >}{\partial (\beta \mu)} = \frac{\partial^2 \ln \Xi}{\partial^2 (\beta \mu)},$$ (2.13a)

so

$$\frac{< N^2 > - < N >^2}{< N >} = \frac{1}{< N >} \frac{\partial < N >}{\partial (\beta \mu)}.$$ (2.13b)

From Eq. (2.13b),

$$\frac{\sqrt{< N^2 > - < N >^2}}{< N >} = \frac{\sqrt{< \Delta N^2 >}}{< N >}$$ (2.13c)

$$= \frac{1}{\sqrt{< N >}} \left[\frac{1}{< N >} \frac{\partial < N >}{\partial (\beta \mu)} \right]^{1/2} \to 0, \ (N \to \infty).$$

Therefore, in the limit of a macroscopic system (the thermodynamic limit), the number fluctuation becomes negligible and calculations using the canonical ensemble and grand canonical ensemble become identical.

We now show that the expression in Eq. (2.13b) is proportional to the compressibility of the fluid. Since

$$\Omega = A - N\mu = -pV,$$
$$TdS = dU + pdV - \mu dN,$$

$$d\Omega = dA - Nd\mu - \mu dN = -pdV - SdT - Nd\mu,$$
$$= pdV - Vdp \to -SdT - Nd\mu = -Vdp.$$

So for an isothermal change, from the Gibbs equation $SdT = Vdp + Nd\mu = 0$,

$$Nd\mu = Vdp.$$ (2.14a)

If the change takes place also at constant volume and temperature, then dp and $d\mu$ are proportional to dN:

$$dp = \left(\frac{\partial p}{\partial N} \right)_{V,T} dN, \ d\mu = \left(\frac{\partial \mu}{\partial N} \right)_{V,T} dN, \ \chi_T = \frac{1}{\rho} \left(\frac{\partial \rho}{\partial p} \right)_{N,T}.$$ (2.14b)

Substituting into Eq. (2.14a) we get

$$\rho \left(\frac{\partial \mu}{\partial \rho} \right)_{V,T} = N \left(\frac{\partial \mu}{\partial N} \right)_{V,T} = V \left(\frac{\partial p}{\partial N} \right)_{V,T} = \left(\frac{\partial p}{\partial \rho} \right)_{N,T} = \frac{1}{\rho \chi_T}.$$ (2.14c)

Recalling Eq. (2.13a), we obtain the famous compressibility theorem:

$$\frac{< N^2 > - < N >^2}{< N >} = \frac{1}{< N >} \frac{\partial < N >}{\partial (\beta \mu)} = \frac{1}{\beta} \rho \chi_T.$$ (2.14d)

2.1.3 Statistical thermodynamic definition of the static pair correlation function

Within the canonical ensemble (fixed N), the n-particle distribution function is defined as

$$n_N^{(n)}(1, 2, \ldots, n) = \frac{N!}{Z_N(N-n)!} \int e^{-\beta U\{N\}} d\{N-n\}, \qquad (2.15a)$$

where

$$Z_N(V, T) = \int_v \cdots \int_v e^{-\beta U\{N\}} d\{N\}, \qquad (2.15b)$$

is called the configurational function. The factor

$$\frac{1}{Z_N} \int \cdots \int e^{-\beta U\{N\}} d\{N-n\} = \int f_0^{(N)} \left(\underline{r}^N, \underline{p}^N\right) d\underline{p}^N d\underline{r}^{(N-n)}, \qquad (2.15c)$$

expresses the probability, when we observe the configuration of the system of N particles with given $U(N)$, that the particle 1 will be at $d\underline{r}_1$, particle 2 at $d\underline{r}_2$, and particle n at $d\underline{r}_n$, irrespective of the configuration of the remaining $(N-n)$ particles. We also integrate out all the momentum variables, because we do not ask questions on how the particles are moving. The second factor takes into account that particles are indistinguishable and there are N choices for the particle in $d\underline{r}_1$, $N-1$ for $d\underline{r}_2$, $\ldots$, and $(N-n+1)$ for $d\underline{r}_n$ or a total of

$$N(N-1)(N-2)\cdots(N-n+1) = \frac{N!}{(N-n)!}, \qquad (2.15d)$$

possibilities of taking n particles from the total of N to put into the given configuration. Therefore the physical meaning of the function $n_N^{(n)}(1, 2, \ldots, n)$ $d1 d2 d3 \cdots dn$ is that it is the probability of finding a particle in $d\underline{r}_1$, a particle in $d\underline{r}_2$ a particle in $d\underline{r}_n$. This function depends, besides on $\underline{r}_1, \underline{r}_2, \ldots, \underline{r}_n$, on two thermodynamic variables V and T.

From the definition in Eqs. (2.15a) and (2.15c), we have the normalisation condition

$$\int \int \cdots \int n_N^{(n)}(1, 2, \ldots, n) \, d\{n\} = \frac{N!}{(N-n)!}. \qquad (2.16a)$$

In particular,

$$\int_v n_N^{(1)}(1) \, d1 = N \Rightarrow n_N^{(1)} = \rho, \qquad (2.16b)$$

$$\int_v \int_v n_N^{(2)}(1, 2) \, d1 d2 = N(N-1). \qquad (2.16c)$$

In a homogeneous and isotropic system, like a fluid

$$d1d2 \Rightarrow d(12) \, d2n_N^{(2)}(1, 2) = n_N^{(2)}(12),$$

so

$$\int_v n_N^{(2)}(1, 2) \, d(12) = \frac{1}{V} N (N - 1) = \rho (N - 1). \tag{2.17a}$$

If the distribution of particles were completely random, the probability of finding the particle 1 in $d1$, 2 in $d2$, ..., and n in dn is

$$\frac{d1}{V} \frac{d2}{V} \cdots \frac{dn}{V} = \frac{1}{V^n} d\{n\}. \tag{2.17b}$$

Then

$$n_N^{(2)}(1, 2, \ldots, n) \, d\{n\} = \frac{d\{n\}}{V^n} \frac{N!}{(N-n)!} = \left(\frac{N}{V}\right)^n \left[1 + O(1/N)\right] d\{n\}. \tag{2.17c}$$

We therefore define an n-particle correlation function $g_N^{(n)}(1, 2, \ldots, n)$ such that

$$n_N^{(n)}(1, 2, \ldots, n) = \left(\frac{N}{V}\right)^n g_N^{(n)}(1, 2, \ldots, n), \tag{2.18a}$$

so that in the random limit

$$g_N^{(n)}(1, 2, \ldots, n) \to 1. \tag{2.18b}$$

For a liquid of simple molecules

$$g_N^{(1)}(1) = 1. \tag{2.18c}$$

From Eq. (2.17a), so

$$\int n_N^{(2)}(12) \, d12 = \int \left(\frac{N}{V}\right)^2 g_N^{(2)}(12) \, d(12) = \frac{N}{V}(N - 1). \tag{2.19}$$

This suggests a physical interpretation

$$\rho g_N^{(2)}(12) \, d(12) = \text{number of particles at } (\vec{r}_{12}, \vec{r}_{12} + d\vec{r}_{12}), \tag{2.20}$$

if there is a particle at r_1, in agreement with the earlier definition in Chapter 1, Eq. (1.58c).

Generalisation to the grand canonical ensemble

If we now generalise our definition of the n-particle distribution function to an open system then we can write

$$n^{(n)}(1, 2, \ldots, n) = \sum_{N \geq n}^{\text{all} N} n_N^{(n)}(1, 2, \ldots, n) P(N). \tag{2.21}$$

From Eq. (2.16a) we have the normalisation condition

$$\int_V n^{(n)} (1, 2, \ldots, n) \, d\{n\} = \sum_{allN} \frac{N!}{(N - n)!} P(N) = \left\langle \frac{N!}{(N - n)!} \right\rangle, \qquad (2.22a)$$

$$n = 1, \quad \int_V n^{(1)} (1) \, d1 = < N > \Rightarrow n^{(1)} (1) = \frac{< N >}{V} = \rho, \qquad (2.22b)$$

$$n = 2, \quad \int_V \int_V n^{(2)} (12) \, d1 d2 = V \int_V n^{(2)} (12) \, d(12) = < N (N - 1) > . \qquad (2.22c)$$

From Eqs. (2.22b) and (2.22c) we can evaluate the integral

$$\int \int \left[n^{(2)} (1, 2) - n^{(1)} (1) \, n^{(2)} (2) \right] d1 d2 = < N^2 > - < N >^2 - < N > . \qquad (2.23)$$

Knowing $d1 d2 \Rightarrow d(12) \, d2$; $n_N (1, 2) = n_N (12)$ we get

$$\int \int \left[n^{(2)} (1, 2) - n^{(1)} (1) \, n^{(2)} (2) \right] d1 d2 = V \int \left[n^{(2)} (12) - \left(\frac{< N >}{V} \right)^2 \right] d(12). \qquad (2.24)$$

Introducing the pair correlation function $g^{(2)} (1, 2)$ by

$$n^{(2)} (12) = \left(\frac{< N >}{V} \right)^2 g^{(2)} (12), \qquad (2.25)$$

we get from Eqs. (2.23), (2.24) and (2.25) finally an equation:

$$V \int \left(\frac{< N >}{V} \right)^2 \left[g^{(2)} (12) - 1 \right] d(12) = < N^2 > - < N >^2 - < N > . \qquad (2.26)$$

On the other hand, the compressibility theorem of Eq. (2.14d) gives

$$< N^2 > - < N >^2 = < N > \frac{< N >}{V} k_B T \chi_T,$$

so we obtain a useful relation:

$$V \left(\frac{< N >}{V} \right)^2 \int_V \left[g^{(2)} (r) - 1 \right] d^3 r = < N > \left[\frac{< N >}{V} k_B T \chi_T - 1 \right]. \qquad (2.27)$$

In terms of the structure factor $S(k)$, we have arrived at an important boundary condition called the *fluctuation theorem*:

$$S(k \to 0) = 1 + \rho \int_0^\infty (g(r) - 1) \, 4\pi r^2 dr = \rho k_B T \chi_T. \qquad (2.28)$$

This equation is valid even when the potential is not pair-wise additive.

2.1.4 Energy equation of states

To relate the internal energy E to the pair correlation function, we proceed as follows.

Recall from Eq. (2.3d)

$$E = < H >= -\frac{\partial}{\partial \beta} \ln Q_N (V, T)|_V,$$

$$Q_N (T, V) = Q_N^{id} \frac{Z_N (T, V)}{V^N} = \frac{1}{N! h^{3N}} \left(\frac{2\pi M}{\beta}\right)^{3N/2} Z_N (T, V),$$

$$\ln Q_N (T, N) = const. - \frac{3N}{2} \ln \beta + \ln Z_N,$$

$$E = \frac{3N}{2} k_B T - \frac{\partial}{\partial \beta} \ln Z_N \qquad (2.29)$$

$$= \frac{3N}{2} k_B T - \frac{1}{Z_N} \frac{\partial}{\partial \beta} \int d\underline{r}^N e^{-\beta U(\underline{r}^N)}$$

$$= \frac{3N}{2} k_B T + \frac{1}{Z_N} \int d\underline{r}^N U \{N\} e^{-\beta U(\underline{r}^N)}$$

$$= \frac{3N}{2} k_B T + < U >.$$

With a pair approximation for the potential:

$$U (\underline{r}_1, \underline{r}_2, \ldots, \underline{r}_N) = 1/2 \sum_{i,j} u (r_{i,j}), \qquad (2.30)$$

we get a relation between the internal energy and the $g(r)$:

$$< U > = \frac{1}{Z_N} \int d\underline{r}_1 d\underline{r}_2 \cdots \underline{r}_N = 1/2 \sum_{i,j} u (r_{i,j}) e^{-\beta U}$$

$$= \frac{1}{2} \frac{N (N-1)}{Z_N} \int_v \cdots \int d\underline{r}_1 d\underline{r}_2 \cdots d\underline{r}_N = u (r_{i,j}) e^{-\beta U}$$

$$= \frac{1}{2} \int \cdots \int d\underline{r}_1 d\underline{r}_2 u (r_{12}) \left[\frac{N}{(N-2)!} \frac{1}{Z_N} \int d\underline{r}_3 d\underline{r}_4 \cdots d\underline{r}_N e^{-\beta U}\right]$$

$$= \frac{1}{2} \int \int d\underline{r}_1 d\underline{r}_2 u (r_{12}) n_N^{(2)} (1, 2) \qquad (2.30a)$$

$$= \frac{1}{2} V \left(\frac{N}{V}\right)^2 \int d^3 r u (\underline{r}) g^{(2)} (\underline{r})$$

$$= \frac{1}{2} V \rho^2 \int_0^\infty dr 4\pi r^2 u (r) g (r)$$

$$= \frac{1}{2} N \int_0^\infty dr 4\pi r^2 \rho g (r) u (r).$$

Thus internal energy per unit volume can be written as

$$\frac{E}{V} = \frac{3}{2}\rho k_B T + \frac{1}{2}\rho^2 \int_0^\infty dr 4\pi r^2 u(r) g(r). \tag{2.31}$$

This equation expresses E/V as a function of ρ and T under the pair approximation.

2.1.5 Virial equation of states

The *virial theorem* for a system in equilibrium states that the average value of the virial of the external and internal forces exerted on the particle system is equal to minus twice the average value of the kinetic energy of the system, namely

$$-2 < T > = < \sum_i^N \underline{r}_i \cdot \underline{F}_i > = -2 \cdot \frac{3}{2} N k_B T,$$

$$< \sum_i^N \underline{r}_i \cdot \underline{F}_i >_{ext} = -3pV, \tag{2.32}$$

$$< \sum_i^N \underline{r}_i \cdot \underline{F}_i >_{int} = -\frac{1}{2} < \sum_{i,j} r_{ij} \frac{du}{dr_{ij}} >, \tag{2.32a}$$

so

$$-3N k_B T = -3pV - 1/2 < \sum_{ij} r_{ij} \frac{du}{dr_{ij}} >,$$

$$pV = N k_B T - \frac{1}{6} < \sum_{ij} r_{ij} \frac{du}{dr_{ij}} >$$

$$= N k_B T - \frac{1}{6} < \sum_{J=1} r_{ij} \frac{du}{dr_{ij}} >$$

$$= N k_B T - \frac{1}{6} N\rho \int_0^\infty dr 4\pi r^2 g(r) r \frac{du}{dr},$$

using the pair approximation, or

$$p = \rho k_B T - \frac{1}{6}\rho^2 \int_0^\infty dr 4\pi r^2 g(r) r \frac{du}{dr}. \tag{2.32b}$$

This equation gives p as a function of ρ and T. To prove that

$$< \sum_{i=1}^N \underline{r}_i \cdot \underline{F}_i >_{ext} = -3pV. \tag{2.32c}$$

Figure 2.1 Geometry used to prove Eq. (2.32c).

we notice that the pressure can be defined in the following way (Figure 2.1). When $\underline{r}_i$ is at the wall of the vessel,

$$\vec{F}_i = -p\hat{n}dS,$$

$$< \sum_{i=1}^{N} \underline{r}_i \cdot \underline{F}_i > = \oint \oint \vec{r} \cdot (-p\hat{n}) \, dS = -p \oint \oint \vec{r} \cdot \hat{n} dS \qquad (2.33)$$

$$= -p \int_v di v \vec{r} dV = -3pV.$$

To calculate the free energy $A(V, T)$ from $u(r)$ and $g(r)$ we use the fact that

$$d\left(\frac{A}{N}\right) = -pdv = -pd\left(\frac{1}{\rho}\right), \qquad (2.34)$$

at constant T. Therefore we can integrate the virial equation of states with respect to ρ at constant T to get A/T, or we can also use a relation

$$E = U = -\left(\frac{\partial \beta A}{\partial \beta}\right)_V, \quad d\left(\frac{A}{T}\right) = Ed\left(\frac{1}{T}\right), \qquad (2.34a)$$

at constant density ρ. In this case we integrate the energy equation of states with respect to T at constant ρ. In both these methods, we need to know $g(r)$ as a function of ρ at constant T or as a function of T at constant ρ.

Example: Hard-sphere system
We can explicitly evaluate the integral in Eq. (2.32b) in the case of the hard-sphere system.

In this case, introduce $y(r) = e^{\beta u}g(r)$. This function is continuous across d

$$\frac{\beta p}{\rho} = 1 - \frac{2}{3}\pi\rho\beta \int_0^\infty dr r^3 u' y(r) e^{-\beta u}$$

$$= 1 + \frac{2}{3}\pi\rho \int_0^\infty dr r^3 y(r) \frac{d}{dr} e^{-\beta u} \qquad (2.35)$$

$$= 1 + \frac{2}{3}\pi\rho \int_0^\infty dr r^3 y(r) \delta(r-d),$$

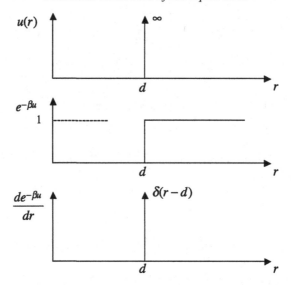

Figure 2.2 A hard sphere of potential $u(r)$; trends of the two other related quantities are indicated in the y-axis.

so

$$\frac{\beta p}{\rho} = 1 + \frac{2}{3}\pi\rho d^3 y\,(d) = 1 + \frac{2}{3}\pi\rho d^3 g\,(d)\,, \tag{2.35a}$$

namely, the pressure of a hard-sphere system is determined by its radial distribution function at contact $g(d)$ (see Figure 2.2).

2.1.6 Virial expansion

The virial expansion

$$\frac{\beta P}{\rho} = 1 + B_2\,(T)\,\rho + B_3\,(T)\,\rho^2 + B_4\,(T)\,\rho^3 + B_5\,(T)\,\rho^4 \tag{2.36}$$

for hard spheres is

$$B_2 = \frac{2}{3}\pi d^3,\; B_3 = \frac{5}{8}\left(\frac{2}{3}\pi d^3\right)^2,\; \ldots,\; B_7.$$

In the grand canonical ensemble the n-th order density correlation function is defined as

$$n^{(2)}\,(1, 2, \ldots, n) = \sum_{N \geq n}^{\infty} P\,(N)\,n_N^{(n)}\,(1, 2, \ldots, n)$$

$$= \frac{1}{\Xi}\sum_{N \geq n}^{\infty}\frac{z^N}{(N-n)!}\int e^{-\beta U_N\left(\underline{r}^N\right)d\{N-n\}}. \tag{2.37}$$

For an ideal gas, $U_N\left(\underline{r}^N\right) = 0$,

$$n^{(n)}\left(1, 2, \ldots, n\right) = \frac{\sum_{N \geq n} \frac{z^N}{(N-n)!} V^{N-n}}{\sum_{N=0} \frac{z^N}{N!} V^N} = z^n,$$

(2.38)

since

$$\beta \mu^{id} = \ln \rho + 3\ln \Lambda, \qquad z = 1 \frac{e^{\beta \mu^{id}}}{\Lambda^3}.$$

(2.39)

We can likewise define $g^{(n)}\left(1, 2, \ldots, n\right)$ as

$$g^{(n)}\left(1, 2, \ldots, n\right) = \frac{1}{\rho^n} n^{(n)}\left(1, 2, \ldots, n\right)$$

(2.40)

$$= \frac{1}{\Xi} \sum_{N \geq n}^{\infty} \frac{z^N}{(N-n)!} \int e^{-\beta U_N(\vec{r}^N)} d\{N - n\}.$$

Let $m = N - n$, or

$$\Xi \left(\frac{\rho}{z}\right)^n g^{(n)}\left(1, 2, \ldots, n\right) = e^{-\beta U_n} + \sum_{m=1}^{\infty} \frac{z^m}{m!} \int e^{-\beta U_{m+n}\{m\}}.$$

(2.40a)

In the limit $\rho \to 0$, $z \to 0$, $\rho/z \to 1$, $\Xi \to 1 = e^{zV}$

$$g^{(n)}\left(1, 2, \ldots, n\right) = e^{-\beta U_n}.$$

Note especially

$$g^{(n)}\left(1, 2\right) = e^{-\beta U_2(\underline{r}_1, \underline{r}_2)} = g\left(r\right) = e^{-\beta u(r)},$$

$$y\left(r\right) = e^{-\beta U(r)} g\left(r\right) \xrightarrow[\rho \to 0]{} 1 \quad \text{for all } r.$$

(2.40b)

The second virial coefficient $B_2(T)$

This is defined in Eq. (2.36), to the first order of ρ, as

$$\frac{\beta p}{\rho} = 1 + B_2\left(T\right)\rho$$

$$= 1 - \frac{2\pi}{3}\beta\rho \int_0^{\infty} dr r^3 u'\left(r\right) g_{id}\left(r\right)$$

$$= 1 - \frac{2\pi}{3}\beta\rho \int_0^{\infty} dr r^3 u'\left(r\right) e^{-\beta u},$$

so

$$B_2\left(T\right) = -\frac{2\pi}{3}\beta \int_0^{\infty} dr r^3 u'\left(r\right) e^{-\beta u}$$

$$= \frac{2\pi}{3} \int_0^{\infty} dr r^3 \frac{d}{dr}\left(e^{-\beta u}\right)$$

$$= \frac{2\pi}{3} \int_0^\infty dr r^3 \frac{d}{dr} \left(e^{-\beta u} - 1\right)$$

$$= \frac{2\pi}{3} \left[r^3 \left(e^{-\beta u} - 1\right) \Big|_0^\infty - \int_0^\infty 3r^2 \left(e^{-\beta u} - 1\right) dr \right] \qquad (2.41)$$

$$= -\frac{2\pi}{3} \int_0^\infty 3r^2 \left(e^{-\beta u} - 1\right) dr$$

$$= -\frac{1}{2} \int_0^\infty 4\pi r^2 \left(e^{-\beta u} - 1\right) dr$$

$$= -\frac{1}{2} \int_0^\infty 4\pi r^2 f(r)\, dr = \int_0^\infty 2\pi r^2 (1 - e^{-\beta u(r)})dr,$$

where $f(r) = e^{-\beta u} - 1$ is called the Mayer function. The Mayer function for hard spheres is

$$f(x) = e^{-\beta u} - 1 = \begin{cases} -1, & r < d, \\ 0, & r < d, \end{cases} \qquad (2.42)$$

so the second virial coefficient is

$$B_2^{HS}(T) = -2\pi \int_0^d r^2 (-1)\, dr = \frac{2}{3}\pi d^3 = 4V_{HS}. \qquad (2.42a)$$

The virial series is

$$\frac{\beta p}{\rho} = 1 + \frac{2}{3}\pi \rho d^3 + \cdots = 1 + 4\eta. \qquad (2.43)$$

The most accurate equation of state for hard spheres is

$$\frac{\beta p}{\rho} = \frac{1 + \eta + \eta^2 - \eta^3}{(1 - \eta)^3} = 1 + 4\eta + 10\eta^2 + \cdots. \qquad (2.43a)$$

This is called the Carnahan–Stirling equation of state (Carnahan and Starling, 1969). Compare Eq. (2.43a) with the rigorous hard-sphere equation of state, Eq. (2.35a):

$$\frac{\beta p}{\rho} = 1 + 4\eta g(d). \qquad (2.43b)$$

We get

$$g(d) = \frac{1 - 1/2\eta}{(1 - \eta)^3} \doteq 1 + \frac{5}{2}\eta = 5.69 \quad \text{at } \eta = 0.49. \qquad (2.44)$$

We also see from Eq. (2.43b) that the second virial coefficient corresponds only to a zeroth-order expansion of $g(d)$ in η.

The Carnahan–Stirling formula, Eq. (2.43a), corresponds to summation of the virial series to an infinite order. The exact virial coefficients are known (from analytical and Monte Carlo simulation) up to seventh order (see table below). Even a six-term expansion with the given coefficient will underestimate $(\beta P / \rho - 1)$ by 8% at $\eta = 0.5$.

Virial coefficient	Numerical value
B_2	$\frac{2}{3}\pi d^3 = b$
B_3	$\frac{5}{8}b^2$
B_4	$0.28695\,b^3$
B_5	$0.1103 \pm 0.0003 b^4$
B_6	$0.0386 \pm 0.0004 b^5$
B_7	$0.0138 \pm 0.0004 b^6$

The seven-term series reads

$$\frac{\beta p}{\rho} = 1 + 4\eta + 10\eta^2 + 18.365\eta^3 + 28.24\eta^4 + 39.5\eta^5 + 56.5\eta^6. \tag{2.45}$$

Starting from $p = -\left(\frac{\partial A}{\partial V}\right)_T$, by noting $\rho = N/V$, we can transform as

$$\frac{\beta p}{\rho} = \rho\left(\frac{\partial\,(\beta A/N)}{\partial\rho}\right)_T = \rho\left(\frac{\partial\,(\beta A^{id}/N)}{\partial\rho}\right)_T + \rho\left(\frac{\partial\,(\beta A^{ex}/N)}{\partial\rho}\right)_T.$$

Knowing $\frac{\beta A^{id}}{N} = \ln\rho + 3\ln\Lambda - 1$ we have

$$\rho\left(\frac{\partial\,(\beta A^{id}/N)}{\partial\rho}\right)_T = \eta = \frac{\pi}{6}\rho d^3.$$

So

$$\left(\frac{\beta p}{\rho} - 1\right) = \rho\left(\frac{\partial\,(\beta A^{ex}/N)}{\partial\rho}\right)_T,$$

from which we get

$$\int_0^\eta (\beta p/\rho - 1)\,\frac{d\eta}{\eta} = \frac{\beta A^{ex}}{N}. \tag{2.46}$$

Using the Carnahan–Stirling formula we get

$$\frac{\beta p}{\rho} - 1 = \frac{1 + \eta + \eta^2 - \eta^3}{(1-\eta)^3} - 1 = \frac{4\eta - 2\eta^2}{(1-\eta)^3},$$

$$\int_0^\eta \frac{4\eta - 2\eta^2}{(1-\eta)^3}\frac{d\eta}{\eta} = \frac{\beta A^{ex}}{N}. \tag{2.47}$$

Integration of Eq. (2.47) gives the excess free energy βA^{ex} per particle:

$$\frac{\beta A^{ex}}{N} = \frac{\eta \, (4 - 3\eta)}{(1 - \eta)^2}$$
$$\doteq \eta \, (4 - 3\eta) \left(1 + 2\eta + 3\eta^3\right) \doteq 4\eta + 5\eta^2. \qquad (2.48)$$

We see

$$\frac{\beta p}{\rho} - 1 = \eta \left(\frac{\partial \, (\beta A^{ex}/N)}{\partial \eta}\right)_T = 4\eta + 10\eta^2 + \cdots. \qquad (2.48a)$$

In agreement write the exact virial series given in Eq. (2.45). Starting from $G = A + pV$ one can derive

$$\frac{\beta G}{N} = \frac{\beta A}{N} + \frac{\beta p}{\rho},$$

or

$$\frac{\beta G^{ex}}{N} = \frac{\beta A^{ex}}{N} + \left(\frac{\beta p}{\rho} - 1\right),$$

so

$$\frac{\beta G^{ex}}{N} = \frac{\eta \, (4 - 3\eta)}{(1 - \eta)^2} + \frac{4\eta - 2\eta^2}{(1 - \eta)^3}$$
$$= \frac{8\eta - 9\eta^2 + 3\eta^3}{(1 - \eta)^3} = \beta \mu^{ex}, \qquad (2.49)$$

note

$$\frac{\beta G^{id}}{N} = \beta \mu^{id} = \ln \rho + 3 \ln \Lambda,$$
$$\frac{\beta A^{id}}{N} = \ln \rho + 3 \ln \Lambda - 1.$$

It is important to know the $Q \to 0$ limit of the scattered intensity in experiments with colloidal particles. We use the compressibility theorem of Eq. (2.28) to write

$$\frac{1}{S \, (Q \to 0)} = \frac{1}{\rho k_B T \chi_T}.$$

Now

$$\chi_T = -\frac{1}{V} \left(\frac{\partial V}{\partial p}\right)_T = \frac{1}{\rho} \left(\frac{\partial \rho}{\partial p}\right)_T, \quad \text{or} \quad \frac{\beta}{\rho \chi_T} = \frac{\partial \, (p\beta)}{\partial \rho}\bigg|_T,$$

so

$$\frac{1}{S \, (Q \to 0)} = \frac{\beta}{\rho \chi_T} = \frac{\partial \, (p\beta)}{\partial \rho}\bigg|_T, \qquad (2.50)$$

now

$$\frac{\beta p}{\rho} = \frac{1 + \eta + \eta^2 - \eta^3}{(1 - \eta)^3},$$

$$\beta p = \rho \frac{1 + \eta + \eta^2 - \eta^3}{(1 - \eta)^3},$$

$$\begin{aligned}
\left.\frac{\partial (\beta p)}{\partial \rho}\right|_T &= \frac{1 + \eta + \eta^2 - \eta^3}{(1 - \eta)^3} + \rho \frac{\partial}{\partial \rho} \frac{1 + \eta + \eta^2 - \eta^3}{(1 - \eta)^3} \\
&= \frac{1 + \eta + \eta^2 - \eta^3}{(1 - \eta)^3} + \eta \frac{\partial}{\partial \eta} \frac{1 + \eta + \eta^2 - \eta^3}{(1 - \eta)^3} \\
&= \frac{1 + \eta + \eta^2 - \eta^3}{(1 - \eta)^3} \eta \left[\frac{1 + 2\eta - 3\eta^2}{(1 - \eta)^3} + \frac{3 \left(1 + \eta + \eta^2 - \eta^3\right)}{(1 - \eta)^4} \right] \\
&= \frac{1 + 4\eta + 4\eta^2 - 4\eta^3 + \eta^4}{(1 - \eta)^4} \approx e^{8\eta} = 1 + 8\eta + \cdots.
\end{aligned}$$ (2.51)

2.1.7 Potential of mean force

We have just seen that in the dilute limit, Eq. (2.40b),

$$g(r) \xrightarrow[\rho \to 0]{} e^{-\beta u(r)}.$$ (2.52)

Let us therefore write, for an arbitrary density ρ:

$$g(r) = e^{-\beta W(r)},$$ (2.52a)

$$W(r) = kT_B \ln g(r).$$ (2.52b)

$W(r)$ is called the potential of the mean force. We can understand its name from the following theorem.

.

Reversible work theorem

$W(r)$ is the reversible work done in a process in which two tagged particles are brought into the system from an infinite separation to a relative separation r. The process is performed at constant V, T, N, so that $W(r, \beta, \rho)$ is also the free energy change of the system when we bring two particles from infinity to the separation r. To prove this system, particles 1 and 2 are tagged and held fixed. We then calculate the average force acting between 1 and 2, averaging over all configuration of particles $3, 4, \ldots, N$

$$\text{Average force} = -\left\langle \frac{d}{d\vec{r}_1} U_N\left(\underline{r}^N\right)\right\rangle_{1,2},$$

$$= \frac{-\int d\underline{r}_3 d\underline{r}_4 \cdots d\underline{r}_N dU_N/d\vec{r}_1 e^{-\beta U_N(\underline{r}^N)}}{\int d\underline{r}_3 d\underline{r}_4 \cdots d\underline{r}_N e^{-\beta U_N(\underline{r}^N)}}, \qquad (2.53)$$

$$= k_B T \frac{d/d\vec{r}_1 \int d\underline{r}_3 d\underline{r}_4 \cdots d\underline{r}_N e^{-\beta U_N}}{\int d\underline{r}_3 d\underline{r}_4 \cdots d\underline{r}_N e^{-\beta U_N}},$$

$$-\left\langle \nabla_{r1} U_N\left(\underline{r}^N\right)\right\rangle_{1,2} = k_B T \, d/d\vec{r}_1 \ln \int d\underline{r}_3 d\underline{r}_4 \cdots d\underline{r}_N e^{-\beta U_N}.$$

But we can multiply any constant to the integral without changing the $\frac{d}{dr}\ln$ of it. So

$$-\left\langle \nabla_{r1} U_N\left(\underline{r}^N\right)\right\rangle_{1,2} = k_B T \, d/d\vec{r}_1 \ln \left[\frac{N(N-1)\int d\underline{r}_3 d\underline{r}_4 \cdots d\underline{r}_N e^{-\beta U_N}}{\int d\underline{r}_1 d\underline{r}_2 \cdots d\underline{r}_N e^{-\beta U_N}}\right]$$

$$= k_B T \, d/d\vec{r}_1 \ln g^{(2)}\left(\underline{r}_1, \underline{r}_2\right)$$

$$= \frac{d}{d\vec{r}_1} W\left(\underline{r}_1 - \underline{r}_2; \rho, T\right) \qquad (2.54)$$

$$= \frac{d}{d\vec{r}_1} W\left(\underline{r}_1 - \underline{r}; \rho, T\right).$$

Therefore $W(r; \rho, T)$ is the work done in bringing two particles into the system with a separation r from infinity:

$$W\left(\underline{r}; \rho, T\right) = -k_B T \ln g\left(\underline{r}\right). \qquad (2.55)$$

We can easily understand that, in the limit $\rho \to 0$, $g(r) \to \exp[-\beta u]$ so

$$W\left(\underline{r}; \rho, T\right) \xrightarrow[\rho \to 0]{} u(r). \qquad (2.55a)$$

2.2 The Ornstein–Zernike equation and its approximate solutions[2]

We begin by introducing a total correlation function $h(r)$ by

$$h(r) = g(r) - 1. \qquad (2.56)$$

Recall that the structure factor $S(Q)$ is related to $h(r)$ by

$$S(Q) = 1 + \rho \int_0^\infty dr 4\pi r^2 \frac{\sin Qr}{Qr} h(r) = 1 + \rho \hat{h}(Q). \qquad (2.57)$$

---

[2] Egelstaff (1967); Freidman (1985); Chandler (1987).

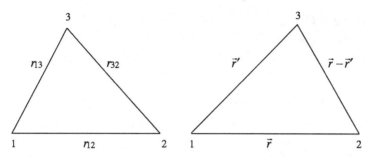

Figure 2.3 Schematic diagrams showing the relative positions of particles 1, 2 and 3.

Denote $h(r)$ by $h(12)$, we then introduce the Ornstein–Zernike (OZ) direct correlation function $c(12)$ by the relation (Ornstein and Zernike, 1914)

$$h(12) = c(12) + \int c(13)d3\rho h(32) = c(12) + \rho \int c(13)h(32)d^3r_3. \quad (2.58)$$

Figure 2.3 illustrates the relative position of particles 1, 2 and 3. Ornstein and Zernike argued that the total correlation $h(r)$ was the sum of a direct effect $c(r)$ of particle 1 on particle 2 plus an indirect effect of all other particles. The latter effect is expressed through the convolution of $\rho h(r)$ and $c(r)$. A virtue of the function $c(r)$ is that in many circumstances it is short range in comparison with $h(r)$. For example, in PY approximation of hard-sphere system, $c(r) = 0$ for $r > d$, while $h(r)$ oscillates and extends to many ds. The OZ equation essentially defines $c(r)$ in terms of $h(r)$ namely,

$$h(r) = c(r) + \rho \int h(|\vec{r} - \vec{r}'|)c(\vec{r}')d^3r'. \quad (2.59)$$

In order to solve for $h(r)$ one needs another so-called closure relation between $h(r)$, $c(r)$ and the dimensionless inter-particle potential $\beta u(r)$. In this regard we have two popular approximate closure relations, HNCA and PYA, which we introduce below.

HNCA (hypernetted-chain approximation) and PYA (Percus–Yevick approximation)

In the HNC approximation one writes

$$c(r) = -\beta u(r) + h(r) - \ln(1 + h(r)), \quad (2.60)$$

or on solving for $g(r) = h(r) + 1$, one gets

$$g(r) = e^{-\beta u(r)}e^{h(r)-c(r)}. \quad (2.60a)$$

Since according to OZ equation $h(r) - c(r) = \rho h * c$, which is first order in ρ, we can expand the exponential term in Eq. (2.60a) to get

$$g(r) = e^{-\beta u(r)}[1 + h(r) - c(r)],$$

or

$$c(r) = (1 - e^{\beta u})g(r). \tag{2.60b}$$

Equation (2.60b), considered as a closure, is called the Percus–Yevick (PY) approximation.

2.2.1 PY solution for a hard-sphere system

The PY equation is analytically soluble in the important case of a hard sphere fluid of arbitrary density. Define a Mayer function by $f(r) = e^{-\beta u(r)} - 1$. This function decays rapidly to zero with increasing r for all short-range potential. Recall the $y(r)$ function defined as

$$y(r) = e^{\beta u}g(r). \tag{2.61}$$

The PY approximation can be written as

$$c(r) = (1 - e^{\beta u})g(r) = y(r)f(r). \tag{2.61a}$$

For hard spheres of diameter d, Eq. (2.61a) means

$$\begin{aligned} c(r) &= -y(r) & r < d & \quad f(r) = -1, \\ c(r) &= 0 & r > d & \quad f(r) = 0. \end{aligned} \tag{2.61b}$$

The OZ equation reads

$$h(r) - c(r) = \rho \int h(|\vec{r} - \vec{r}'|)c(r')d^3r'. \tag{2.61c}$$

We express h and c in terms of y as follows:

$$\begin{aligned} h &= g - 1 = e^{-\beta u}y - 1, \\ c &= fy = (e^{-\beta u} - 1)y, \end{aligned} \tag{2.61d}$$

and $h - c = y - 1$.

Then we can perform the following manipulation:

$$\rho \int h(|\vec{r} - \vec{r}'|)c(\vec{r}')d^3r'$$

$$= \rho \int y(r')f(r') \left[e^{-\beta u(|\vec{r} - \vec{r}'|)}y(|\vec{r} - \vec{r}'|) - 1 \right] d^3r'$$

$$= -\rho \int y(r') f(r') d^3 r' + \rho \int f(r') e^{-\beta u(|\vec{r}-\vec{r}'|)} y(r') y(|\vec{r}-\vec{r}'|) d^3 r'$$

$$= \rho \int_{r' < d} y(r') d^3 r' - \rho \int_{|\vec{r}-\vec{r}'| > d} y(r') y(|\vec{r}-\vec{r}'|) d^3 r'. \tag{2.61e}$$

So the OZ equation becomes

$$y(r) = 1 + \rho \int_{r' < d} y(r') d^3 r' - \rho \int_{|\vec{r}-\vec{r}'| > d} y(r') y(|\vec{r}-\vec{r}'|) d^3 r'. \tag{2.62}$$

This equation for the function $y(r)$ can be solved by a Laplace transform method for $r < d$ (Thiele, 1963; Wertheim, 1963, 1964). The result is (noting that $c(r) = -y(r)$ for $r < d$):

$$c(x) = \begin{cases} 0, & r > d, \\ -\lambda_1 + 6\eta\lambda_2 \left(\frac{r}{d}\right) - \frac{1}{2}\eta\lambda_1 \left(\frac{r}{d}\right)^3, & r < d, \end{cases} \tag{2.62a}$$

where

$$\eta = \frac{\pi\rho d^3}{6}, \qquad \lambda_1 = \frac{(1+2\eta)^2}{(1-\eta)^4}, \qquad \lambda_2 = \frac{(1+\eta/2)^2}{(1-\eta)^4}. \tag{2.62b}$$

Since $c(r) = -y(r)$ for $r < d$, and at the low-density limit $g(r) \approx \exp(-\beta u)$, so $y(r) \approx 1$ and $c(r) \approx -1$. We see that as $\eta \to 0$ our solution $c(r)$ correctly tends to λ_1, that is -1. One can define a Fourier transform of $c(r)$ by

$$\hat{c}(Q) = \int_0^\infty dr 4\pi r^2 \frac{\sin Qr}{Qr} c(r). \tag{2.63}$$

We can compute this function explicitly: let $k = Qd$, then we have

$$\hat{c}(k) = \frac{4\pi d^3}{k^6} \left\{ \lambda_1 k^3 (\sin k - k \cos k) - 6\eta\lambda_2 k^2 (2k \sin k - (k^2 - 2) \cos k - 2) \right.$$

$$\left. + \frac{1}{2}\eta\lambda_1 [(4k^3 - 24k) \sin k - (k^4 - 12k^2 + 24) \cos k + 24] \right\}. \tag{2.64}$$

The structure factor $S(k)$ is computed as

$$S_{HS}(k) = (1 - \rho\hat{c}(k))^{-1}. \tag{2.65}$$

We note here an important relation between $\hat{h}(k)$ and $\hat{c}(k)$. Since

$$h(r) = c(r) + \rho \int h(|\vec{r}-\vec{r}'|) c(r') d^3 r,$$

$$\hat{h}(k) = \hat{c}(k) + \rho\hat{h}(k)\hat{c}(k),$$

$$\hat{h}(k) = \frac{\hat{c}(\hat{k})}{1 - \rho\hat{c}(k)}, \tag{2.66}$$

$$S(k) = 1 + \rho\hat{h}(k) = 1 + \frac{\rho\hat{c}(k)}{1 - \rho\hat{c}(k)} = \frac{1}{1 - \rho\hat{c}(k)}.$$

It is important to realise that we only need to calculate $S(k)$. Since from the compressibility theorem we know that

$$S(k = 0) = \rho k_B T \chi_T,$$

therefore

$$1 - \rho\hat{c}(k = 0) = \frac{1}{\rho k_B T \chi_T} = \frac{\beta}{\rho\chi_T} = \beta\left(\frac{\partial p}{\partial \rho}\right)_{T,N}. \tag{2.67}$$

From Eq. (2.64) we can calculate $\hat{c}(k = 0)$ and put it into Eq. (2.67) to obtain

$$\frac{\beta p^c}{\rho} = \frac{1 + \eta + \eta^2}{(1 - \eta)^3}. \tag{2.68}$$

Again from

$$g(d) = y(d^+) = -c(d^-) = \lambda_1 - 6\eta\lambda_2 + \eta\lambda_1/2 = \frac{1 - \frac{3\eta}{2} + \frac{\eta^3}{2}}{(1 - \eta)^4},$$

and the hard-sphere virial equation of state

$$\frac{\beta p^V}{\rho} = 1 + 4\eta g(d), \tag{2.69}$$

we get

$$\frac{\beta p^V}{\rho} = \frac{1 + 2\eta + 3\eta^2}{(1 - \eta)^2},$$
$$y(r) = e^{\beta u} g(r) y(d^+) = g(d^+), \tag{2.69a}$$
$$c(d^-) = -e^{\beta u} g(d^-) = -y(d).$$

Notice that $\frac{\beta p^c}{\rho}$ and $\frac{\beta p^V}{\rho}$ are not equal because the PY approximation is not exact and thus leads to a thermodynamic inconsistency. But a very good equation of state (the Carnahan–Stirling equation of state) can be obtained by averaging the two results of Eq. (2.68) and Eq. (2.69a), according to

$$\boxed{\frac{\beta P}{\rho} = \frac{2}{3}\frac{\beta P^c}{\rho} + \frac{1}{3}\frac{\beta P^V}{\rho} = \frac{1 + \eta + \eta^2 - \eta^3}{(1 - \eta)^3}.} \tag{2.70}$$

Note that Eq. (2.67) can be summarised as follows:

$$(k_BT)^{-1}\left(\frac{\partial p}{\partial \rho}\right)_T = \frac{1}{1 + \rho \int h(r)d^3r} = 1 - \rho \int c(r)d^3r. \qquad (2.71)$$

At $\eta = 0.49$, corresponding to the liquid state near the triple point, the PY (compressibility) pressure is roughly 30% greater than the PY (virial) pressure. But the PY pair correlation function $g(r)$ is almost indistinguishable from the accurate molecular dynamics (MD) result except at $r = d$. Note that in all cases except near the critical point $h(r)$ decays at least as fast as $u(r)$ as $r \to \infty$, and then we have

$$[c(r) = (1 - e^{\beta u})g(r) \to 1 - e^{\beta u} \to -\beta u]$$

$$c(r) \to -\beta u(r), \quad \text{as} \quad r \to \infty. \qquad (2.72)$$

2.2.2 HNC approximation for Coulomb liquids

For charged systems, one has an interaction potential (between i and j species)

$$u_{ij}(r) = \begin{cases} \infty, & r < d_{ij}, \\ \frac{Z_iZ_je^2}{\varepsilon r}, & r > d_{ij}, \end{cases} \qquad (2.73)$$

where $d_{ij} = (d_i + d_j)/2$, ε is the dielectric constant of the solvent. The approximate equation to solve is

$$h_{ij}(r_{ij}) = c_{ij}(r_{ij}) + \sum_k \rho_k \int h(r_{ik})c(r_{kj})d^3r_k, \qquad (2.73a)$$

with the HNC closure:

$$c_{ij}(r_{ij}) = h_{ij}(r_{ij}) - \ln(1 + h_{ij}(r_{ij})) - \beta u_{ij}. \qquad (2.73b)$$

This set of equation can be solved numerically by the so-called Rossky and Dale (1980) algorithm, and in general will give satisfactory $h_{ij}(r_{ij})$ (Freidman, 1985), but there are no analytical solutions which one can discuss meaningfully. Since asymptotically one can show

$$c_{ij}(r_{ij} \to \infty) = -\beta u_{ij}, \qquad (2.73c)$$

and $h_{ij}(r) < 1$ at large distance, we can linearise Eq. (2.73b) by putting $r_{ij} \to \infty$ and obtain an MSA closure

$$\boxed{c_{ij}(r_{ij}) = -\beta u_{ij}(r_{ij}).} \qquad (2.73d)$$

With MSA one can have analytical solution, which we will discuss in Section 2.2.3 and Section 2.2.4.

2.2.3 Module – The Baxter method for the hard-sphere and adhesive hard-sphere systems[3]

From computer simulations, it is known that the PY approximation is quite accurate for hard spheres. If $c(r)$ can be assumed to vanish beyond some range a, then the OZ relation can be transformed into a form suitable for computation. For a potential $u(r)$ of finite range a, the PY approximation is

$$c(r) = \left[1 - e^{\beta u}\right] g(r), \tag{2.74}$$

which predicts that for $r > a$

$$c(r) = 0. \tag{2.74a}$$

This condition allows us to make a *finite-range* transformation as follows. Starting from the definitions

$$\tilde{c}(k) = \int_0^\infty dr 4\pi r^2 \frac{\sin(kr)}{kr} c(r),$$

$$\tilde{h}(k) = \int_0^\infty dr 4\pi r^2 \frac{\sin(kr)}{kr} h(r), \tag{2.74b}$$

we can use integration by parts to obtain

$$\tilde{c}(k) = 4\pi \int_0^\infty dr C(r) \cos(kr),$$

$$\tilde{h}(k) = 4\pi \int_0^\infty dr H(r) \cos(kr), \tag{2.74c}$$

where

$$C(r) = \int_r^\infty dt\, t\, c(t),$$

$$H(r) = \int_r^\infty dt\, t\, h(t). \tag{2.74d}$$

Now using the condition of Eq. (2.74a) in Eqs. (2.74b) and (2.74d), we can write

$$\frac{1}{S(k)} \equiv \tilde{A}(k) = 1 - \rho \tilde{c}(k) = 1 - 4\pi\rho \int_0^a dr C(r) \cos(kr). \tag{2.75}$$

For a fluid with no long-range order, $h(r) \to 0$ for $r \to \infty$, $\tilde{h}(k)$ is finite for all real values of k. But

$$1 + \rho \tilde{h}(k) = \frac{1}{1 - \rho \tilde{c}(k)}. \tag{2.75a}$$

---

[3] Baxter (1967, 1968a,b).

So that $1 - \rho\tilde{c}(k)$ will never be zero and hence the right-hand side of Eq. (2.75) also can not be zero. When this is so, a Fourier transform theorem (Baxter, 1968a) guarantees that $\tilde{A}(k)$ can be factorised as

$$\tilde{A}(k) = \tilde{Q}(k)\tilde{Q}(-k) = \left|\tilde{Q}(k)\right|^2, \tag{2.76}$$

where $\tilde{Q}(-k)$ is the complex conjugate of $\tilde{Q}(k)$. This is called the Wiener–Hopf factorisation of $\tilde{A}(k)$, and the function $\tilde{Q}(k)$ is regular and has no zeros unless k has a negative imaginary part. The function $1 - \tilde{Q}(k)$ is therefore Fourier integrable along the real axis and a real function $Q(r)$ can be defined as

$$2\pi\rho Q(r) = \frac{1}{2\pi} \int_{-\infty}^{\infty} dk e^{-ikr}[1 - \tilde{Q}(k)]. \tag{2.76a}$$

From which it follows that $Q(r) = 0$ for $r < 0$ and

$$\tilde{Q}(k) = 1 - 2\pi\rho \int_0^{\infty} dr e^{ikr} Q(r), \tag{2.76b}$$

where $Q(r) = 0$ for $r > a$ also. Comparing the two different expressions for $\tilde{A}(k)$, one in terms of $C(r)$ in Eq. (2.75) and the other in terms of $Q(k)$ in Eq. (2.76), one arrives at a useful relation between $C(r)$ and $Q(r)$ in the following form:

$$C(r) = Q(r) - 2\pi\rho \int_0^a dt Q(t)Q(t - r), \quad 0 < r < a. \tag{2.77}$$

From this equation, combined with the fact that $C(a) = 0$ and $Q(r) = 0$ for $r < 0$, it can be deduced that $Q(a) = 0$. It is also possible to obtain a relation between $Q(r)$ and $H(r)$. To do this, note from Eqs. (2.75), (2.75a) and (2.76) that

$$\tilde{Q}(k)\left[1 + \rho\tilde{h}(k)\right] = \left[\tilde{Q}(-k)\right]^{-1}. \tag{2.78}$$

Now invert the Fourier transforms by multiplying Eq. (2.78) by $\exp(-ikr)$ and integrating with respect to k from 0 to ∞. When r is positive, the integration on the right-hand side can be closed around the lower half k-plane, which is regular, has no zeros, and can be seen from Eq. (2.76b) to tend to 1 at ∞. From Cauchy's theorem in complex-variable theory, it follows that this integral vanishes.

Using the expressions for $\tilde{Q}(k)$ in Eq. (2.76b) and $\tilde{h}(k)$ in Eq. (2.74c) on the left-hand side of Eq. (2.78), and performing the inverse Fourier transform, we get

$$- Q(r) + H(r) - 2\pi\rho \int_0^a dt Q(t)H(|t - r|) = 0. \tag{2.79}$$

To get the relation between $c(r)$ and $h(r)$ and $Q(r)$, we differentiate Eq. (2.77) and Eq. (2.79) respectively to obtain the central results of the finite-range transform theory of Baxter, expressed by the following two equations:

$$rc(r) = -Q'(r) + 2\pi\rho \int_r^a dt\, Q'(t)Q(t-r), \tag{2.80}$$

$$rh(r) = -Q'(r) + 2\pi\rho \int_0^a dt\,(r-t)h(|r-t|)Q(t), \tag{2.80a}$$

where $Q'(r)$ denotes $dQ(r)/dr$ in these two equations.

Q-function for simple hard spheres

For a hard-sphere potential we have the rigorous condition $h(r) = -1$ for $0 < r < a$. Substituting this condition into Eq. (2.80a) leads to

$$r = Q'(r) + 2\pi\rho \int_0^a dt\,(r-t)Q(t). \tag{2.81}$$

This equation can be seen to have a general form:

$$Q'(r) = Ar + B, \tag{2.81a}$$

with

$$A = 1 - 2\pi\rho \int_0^a dt\, Q(t), \tag{2.81b}$$

$$B = 2\pi\rho \int_0^a dt\, t Q(t). \tag{2.81c}$$

Carrying out integration of both sides of Eq. (2.81a) from r to a and remembering the boundary $Q(a) = 0$, we get

$$Q(r) = \frac{A}{2}(r^2 - a^2) + B(r-a). \tag{2.82}$$

Substituting $Q(r)$ from Eq. (2.82) into Eq. (2.81b) and Eq. (2.81c), we can solve for A and B in terms of the packing fraction $\eta = \pi\rho a^3/6$ as

$$A = \frac{1+2\eta}{(1-\eta)^2}, \quad B = -\frac{3a\eta}{2(1-\eta)^2}. \tag{2.82a}$$

From Eqs. (2.75) and (2.75a) we have

$$\tilde{A}(0) = \left|\tilde{Q}(0)\right|^2 = \frac{1}{S(0)} = \beta\left(\frac{\partial p}{\partial \rho}\right)_T = \frac{(1+2\eta)^2}{(1-\eta)^4}. \tag{2.83}$$

Integrating this equation with respect to ρ, using Eq. (2.76b), leads to the PY compressibility equation of state given by

$$\frac{\beta p^c}{\rho} = \frac{1+\eta+\eta^2}{(1-\eta)^3}. \tag{2.83a}$$

The PY direct correlation function can also be obtained by integrating Eq. (2.80) using the known $Q(r)$ from Eq. (2.82). We get, denoting $x = r/a$,

$$c(x) = \begin{cases} -A^2 + 6\eta(A + B/a)^2 x - \frac{1}{2}\eta A^2 x^3, & x < 1, \\ 0, & x > 1. \end{cases} \tag{2.84}$$

It should be noted that in this Q-method, the structure factor $S(k)$ is more directly calculable from the relations:

$$S(k) = \frac{1}{\left|\tilde{Q}(k)\right|^2}. \tag{2.85}$$

PY approximation for sticky hard spheres

Consider the square-well potential defined by

$$\beta u(r) = \begin{cases} \infty, & \text{if } r < \sigma, \\ -\ln\left(\frac{1}{12\tau}\frac{a}{a-\sigma}\right), & \text{if } \sigma < r < a, \\ 0, & \text{if } r > a. \end{cases} \tag{2.86}$$

The parameters in this potential are chosen so that τ is a dimensionless measure of the temperature, being zero at zero temperature and large at high temperatures (normal case). But it is possible to interpret τ as inversely proportional to the temperature for some applications. Note that $\tau \to \infty$ is the limiting case for the simple hard-sphere potential.

The PY approximation can be solved analytically for this potential provided one takes the limit $\sigma \to a$, τ and a being held fixed. In this limit the potential consists of a hard core together with an infinitely deep and narrow attractive well. But the divergence in attraction is only logarithmic.

The Mayer function $f(r) = e^{-\beta u} - 1$ for this potential is

$$f(r) = \begin{cases} -1 + \frac{a}{12\tau}\delta(r - a_-), & \text{if } 0 < r < a, \\ 0, & \text{if } r > a. \end{cases} \tag{2.87}$$

So the second virial coefficient B_2 is

$$B_2 = -\frac{1}{2}\int_0^a dr 4\pi r^2 f(r) = \frac{2}{3}\pi a^3 - \frac{\pi}{6}\frac{a^3}{\tau} = \frac{2}{3}\pi a^3\left(1 - \frac{1}{4\tau}\right). \tag{2.88}$$

When $r > a$, the PY approximation gives $c(r) = 0$ so the finite range transformation can be applied. Also when $r < \sigma$, the hard-sphere condition applies and we have $h(r) = -1$. The behaviour of $h(r)$ in the interval (σ, a) can be shown (Baxter, 1968b) to be

$$h(r) = -1 + \frac{\lambda a}{12}\delta(r - a_-), \quad 0 < r < a, \tag{2.89}$$

where λ is an undetermined parameter. Substitution of Eq. (2.89) into Eq. (2.80a) leads to

$$Q'(r) = \alpha r + \beta a - \frac{\lambda a^2}{12}\delta(r - a_-),\tag{2.90}$$

in the range $(0, a)$, where α and β are given by

$$\alpha = 1 - 2\pi\rho \int_0^a dt\, Q(t),\tag{2.90a}$$

$$\beta = \frac{2\pi\rho}{a} \int_0^a dt\, t\, Q(t).\tag{2.90b}$$

Integrating Eq. (2.90) with respect to r from r to a, and using the condition $Q(a) = 0$, gives

$$Q(r) = \frac{\alpha}{2}(r^2 - a^2) + \beta a(r - a) + \frac{\lambda}{12}a^2.\tag{2.91}$$

Substituting $Q(r)$ into Eq. (2.90a) and Eq. (2.90b) gives explicitly

$$\alpha = \frac{1 + 2\eta - \mu}{(1 - \eta)^2}, \beta = \frac{1/2(-3\eta + \mu)}{(1 - \eta)^2},\tag{2.92}$$

$$\mu = \lambda\eta(1 - \eta).\tag{2.92a}$$

It can be shown that λ satisfies the quadratic equation

$$\lambda\tau = \frac{\eta\lambda^2}{12} - \frac{\eta\lambda}{1 - \eta} + \frac{1 + \eta/2}{(1 - \eta)^2}.\tag{2.92b}$$

When τ is greater than a value

$$\tau_c = \frac{(2 - \sqrt{2})}{6} = 0.0976.\tag{2.92c}$$

there are two real solutions for λ throughout the permissible density range $0 < \eta < 1$. For, there exists an intermediate range of densities within which there are no real solutions for λ, so that the system has to undergo a discontinuous transition between states of different density. The PY approximation therefore predicts a first-order phase transition for this system, with a critical temperature τ_c and a critical density

$$\eta_c = \frac{3\sqrt{2} - 4}{2} = 0.1213.\tag{2.92d}$$

For the values of τ and η, for which two real solutions exist, a thermodynamic consideration indicates that the physical solution has to be the smaller root. It can be shown also that

$$\tilde{Q}(0) = \alpha = \frac{1 + 2\eta - \mu}{(1 - \eta)^2}.\tag{2.93}$$

So the relation

$$\beta\left(\frac{\partial p}{\partial \rho}\right)_T = \left|\tilde{Q}(0)\right|^2,\tag{2.94}$$

can be integrated to give the compressibility equation of states

$$\frac{\beta p}{\rho} = \frac{1 + \eta + \eta^2 - \mu(1 + 1/2\eta) + \mu^3/36\eta}{(1-\eta)^3}; \quad \mu = \lambda\eta(1-\eta).\tag{2.94a}$$

The pure hard-sphere limit is obtained when $\tau \to \infty$, $\mu = \lambda = 0$. The structure factor $S(k)$, $k = Qa$, can be calculated from $\rho\tilde{c}(k)$, which is given by

$$\rho\tilde{c}(k) = -\frac{24\eta}{k^6}\left\{\lambda_1 k^3[\sin k - k\cos k] - 6\eta\lambda_2 k^2[2k\sin k - (k^2 - 2)\cos k - 2]\right.$$

$$\left. + \frac{1}{2}\eta\lambda_1[(4k^3 - 24k)\sin k - (k^4 - 12k^2 + 24)\cos k + 24]\right\}\tag{2.95}$$

$$- 2\eta^2\lambda^2\frac{1 - \cos k}{k^2} - 2\eta\lambda\frac{\sin k}{k},$$

where λ is the physical root (smaller one) of Eq. (2.92b) and the coefficients are given by

$$\lambda_1 = \frac{(1 + 2\eta - \mu)^2}{(1 - \eta)^4}, \quad \mu = \lambda\eta(1 - \eta),$$

$$\lambda_2 = \frac{3\eta(2 + \eta)^2 - 2\mu(1 + 7\eta + \eta^2) + \mu^2(2 + \eta)}{12\eta(1 - \eta)^4}.\tag{2.96}$$

Appendix: Mapping of a square-well potential (SWP) fluid to the adhesive hard-sphere potential (AHSP)

$$\text{AHSP} \quad B_2 = -2\pi \int_0^\infty \left(e^{-\beta u} - 1\right) r^2 dr = \frac{2\pi}{3}R^3 - \frac{\pi}{6\tau}R^3,\tag{2.97}$$

$$\text{SWP} \quad B_2 = -2\pi \int_0^{R-\Delta} r^2 dr + 2\pi \int_{R-\Delta}^R \left(1 - e^{\beta\varepsilon}\right) r^2 dr,$$

$$= \frac{2\pi}{3}(R - \Delta)^3 + \frac{2\pi}{3}\left(1 - e^{\beta\varepsilon}\right)\left[R^3 - (R - \Delta)^3\right]$$

$$= \frac{2\pi}{3}R^3\left(1 - \frac{3\Delta}{R}\right) + \frac{2\pi}{3}\left(1 - e^{\beta\varepsilon}\right)R^3\left[1 - \left(1 - \frac{\Delta}{R}\right)^3\right]\tag{2.97a}$$

$$= \frac{2\pi}{3}R^3\left\{\left(1 - \frac{3\Delta}{R}\right) + \frac{3\Delta}{R}\left(1 - e^{\beta\varepsilon}\right)\right\}$$

$$= \frac{2\pi}{3} R^3 - 2\pi R^3 \frac{\Delta}{R} e^{\beta \varepsilon}$$

$$\text{AHSP} \quad = \frac{2\pi}{3} R^3 - \frac{\pi}{6\tau} R^3 = 4V_{HS}\left(1 - \frac{1}{\tau}\right) \tag{2.97b}$$

$B_2^{SWP} = B_2^{AHSP}$ leads to a relation

$$- 2\pi R^3 \frac{\Delta}{R} e^{\beta \varepsilon} = -\frac{\pi}{6\tau} R^3. \tag{2.97c}$$

Therefore,

$$\frac{1}{t} = \frac{12D}{R} e^{\beta}. \tag{2.97d}$$

2.2.4 Module – The analytical structure factor S(Q) for a narrow square-well potential[4]

A square-well potential with a repulsive core of diameter R' and real particle diameter R is used to model a hard sphere with an adhesive surface layer that is introduced to describe the inter-particle attractive interaction between colloidal particles. The pairwise potential is defined as

$$V(r) = \begin{cases} +\infty, & \text{if } 0 < r < R', \\ -u, & \text{if } R' < r < R, \\ 0, & \text{if } r > R. \end{cases} \tag{2.98}$$

We define here $\varepsilon = (R - R')/R$ as the fractional well width parameter and $T^* = k_B T/u$ as the effective temperature. Then the inter-particle structure factor is a function of four parameters: the real particle diameter R, the volume fraction ϕ, the fractional well width parameter ε, and the effective temperature T^*. It should be noted that the real particle diameter R is tied uniquely to the volume fraction ϕ through a relation $\phi = c\pi R^3/6$, where c is the number density of particles.

The Ornstein–Zernike (OZ) equation can be solved analytically for this square-well potential in the Percus–Yevick approximation to the first order in a series of small ε expansion (Liu *et al.*, 1996). The result of the structure factor is given as follows:

$$\frac{1}{S(Q)} - 1 = 24\phi \left[\alpha f_2(Q) + \beta f_3(Q) + \frac{1}{2}\phi\alpha f_5(Q)\right]$$

$$+ 4\phi^2\lambda^2\epsilon^2 \left[f_2(\epsilon Q) - \frac{1}{2}f_3(\epsilon Q)\right]$$

$$+ 2\phi^2\lambda^2 \left[f_1(Q) - \epsilon^2 f_1(\epsilon Q)\right]$$

---

[4] Liu *et al.* (1996); Foffi *et al.* (2000); Chen *et al.* (2002).

$$-\frac{2\phi\lambda}{\epsilon}\left\{f_1(Q) - (1-\epsilon)^2 f_1[(1-\epsilon)Q]\right\}$$
$$-24\phi\left\{f_2(Q) - (1-\epsilon)^3 f_2[(1-\epsilon)Q]\right\}, \qquad (2.99)$$

where $Q = kR$ and

$$\alpha = \frac{(1+2\phi-\mu)^2}{(1-\phi)^4},$$

$$\beta = -\frac{3\phi(2+\phi)^2 - 2\mu(1+7\phi+\phi^2) + \mu^2(2+\phi)}{2(1-\phi)^4},$$

$$\mu = \lambda\phi(1-\phi),$$

$$\lambda = \frac{6(\Delta - \sqrt{\Delta^2 - \Gamma})}{\phi},$$

$$\Delta = \tau + \frac{\phi}{(1-\phi)} = \frac{1}{12\varepsilon}\exp\left(-\frac{u}{k_B T}\right) + \frac{\phi}{(1-\phi)},$$

$$\Gamma = \frac{\phi(1+\phi/2)}{3(1-\phi)^2}, \qquad (2.100)$$

and

$$f_1(x) = \frac{1-\cos x}{x^2},$$

$$f_2(x) = \frac{\sin x - x\cos x}{x^3},$$

$$f_3(x) = \frac{2x\sin x - (x^2-2)\cos x - 2}{x^4}, \qquad (2.101)$$

$$f_5(x) = \frac{(4x^3 - 24x)\sin x - (x^4 - 12x^2 + 24)\cos x + 24}{x^6}.$$

This formula is accurate in the region of k around the first diffraction peak for $\epsilon < 0.05$ (Foffi *et al.*, 2000). The formula has been successfully applied to the analysis of SANS intensities of L64/water micellar solutions (Chen *et al.*, 2002).

Example of the structure factor calculated with Eq. (2.99)

Using Eq. (2.99), we can calculate the structure factor $S(Q)$ as shown by the solid line in Figure 2.4, which is compared with the MD simulation results (open circles) for the case ($\epsilon = 0.03$, $T^* = 1$ and $\phi = 0.475$). We can see that the agreement between the two is quantitative and excellent, all the way up to $qR = 120$.

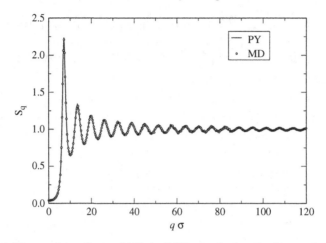

Figure 2.4 The structure factor $S(Q)$ (solid line) calculated using analytical solution of the narrow square well, Eq. (2.99), based on P approximation, compared to the MD simulation result (open circle) for the case $\epsilon = 0.03, T^* = 1$ and $\phi = 0.475$. (Note that $S(Q)$ is denoted S_q and qR is denoted as $q\sigma$.)

2.3 Theories of macroion solutions

As stated earlier, we will not repeat the general historical introduction to the liquid theories of ionic solutions available elsewhere. Instead, we shall present two modules to introduce an analytical solution in the mean spherical approximation of one-Yukawa and two-Yukawa potential models for charged macroion solutions. These are also called one-component macroion (OCM) theory. Then we will present the generalised one-component macroion (GOCM) theory of charged colloidal solutions. These two modules will be useful for the discussion of protein solutions of different concentrations at different pH values.

2.3.1 Module – The one-component macroion theory – one-Yukawa MSA[5]

In an ionic micellar solution or protein solution, the charged macroions are surrounded by a diffuse cloud of counterions, which maintains the charge neutrality of the solution. Counterions such as Na^+, K^+ or Cs^+ can be considered as point-like ions, in comparison to the large size of typical macroions. Under this condition, in sufficiently dilute systems of macroions, Verwey and Overbeek (1948) derived the potential of mean force of taking two macroions from an infinite separation to a separation r in the presence of other ions, which has a functional form:

$$u_{DLVO}(r) = \frac{Z^2 e^2}{e(1 + k\frac{d}{2})^2} \frac{e^{-k(r-d)}}{r}, \tag{2.102}$$

---

[5] Verwey and Overbeek (1948); Waisman (1973); Hayter and Penfold (1981).

where

$$k^2 = 8pe^2 I \frac{N_A}{ek_B T 10^3},$$ (2.102a)

and κ is the Debye screening constant which depends on the ionic strength I (in Mol/litre) $= \frac{1}{2} \sum_i \rho_i Z_i^2$ of the solution. Write the contact potential V_1 as

$$V_1 = \frac{Z^2 e^2}{\varepsilon d (1 + k/2)^2}.$$ (2.103)

We can rewrite $U_{DLVO}(r)$ in a simpler form, by putting $k = \kappa d$, $x = r/d$,

$$u_{DLVO}(x) = V_1 \frac{e^{-k(x-1)}}{x}, \qquad x > 1.$$ (2.104)

Since counterions cannot penetrate the macroions, we also demand $u(r) = \infty$, for $x < 1$.

It is usual in the last decade, to treat $u_{DLVO}(r)$ as a kind of effective potential function between macroions when one ignores the presence of the counterions. In this approximation, one can calculate the macroion–macroion structure factor $S_{OCM}(k)$ using the mean spherical approximation (MSA). This approach was introduced into colloid literature first by Hayter and Penfold (1981). They solved a one-component Ornstein–Zernike equation by an ansatz:

$$\begin{cases} c(x) = -\beta u_{DLVO}(x), & \text{if } x > 1, \\ g(x) = 0, & \text{if } x < 1. \end{cases}$$ (2.105)

The second equation in Eq. (2.105) is the hard core condition for the impenetrability of ions into macroions. Since $u_{DLVO}(x)$ is in Yukawa potential form, the analytical solution of OZ equation is possible, as was first shown by Waisman (1973). The direct correlation function inside the hard core is

$$c(x) = \begin{cases} A + Bx + \frac{1}{2}\eta A x^3 + \frac{C \sinh kx}{x} + \frac{F(\cosh kx - 1)}{x}, & x < 1 \\ -\beta V_1 \frac{e^{-k(x-1)}}{x}, & x > 1. \end{cases}$$ (2.106)

The coefficients A, B, C and F are given explicitly by Hayter and Penfold (1981). The structure factor is given by $\hat{c}(K)$, the Fourier transform of $c(x)$:

$$\hat{c}(K) = 4\pi d^3 a(K),$$ (2.107)

$$a(K) = \frac{A(\sin K - K \cos K)}{K^3} + \frac{B\left[\left(\frac{2}{K^2} - 1\right) K \cos K + 2 \sin K - \frac{2}{K}\right]}{K^3}$$

$$+ \eta A \frac{\left[\frac{24}{K^3} + 4\left(1 - \frac{6}{K^2} \sin K\right) - \left(1 - \frac{12}{K^2} + \frac{24}{K^4}\right) K \cos K\right]}{2K^3}$$

$$+ C \frac{K \cos hK \sin K - K \sin hK \cos K}{K(K^2 + k^2)} \tag{2.108}$$

$$+ F \frac{K \sin hK \sin K - K(\cos hK \cos K - 1)}{K(K^2 + k^2)}$$

$$+ F \frac{[\cos K - 1]}{K^2} - \beta V_1 \frac{[K \sin K + K \cos K]}{K(K^2 + k^2)}.$$

It is interesting to note the following limiting cases:

(i) $S(K \to 0) = -\frac{1}{A}$.

(ii) When $d \to 0$

$$S(K) \underset{d \to 0}{\to} \frac{1}{1 + \frac{\kappa_m^2}{K^2 + \kappa_m^2}}, \tag{2.109}$$

where

$$\kappa_m^2 = \frac{4\pi \beta Z^2 e^2 \rho}{\varepsilon}.$$

(iii) At low density of macroions

$$S(K) \underset{\rho \to 0}{=} 1 - \frac{\kappa_m^2}{(K^2 + \kappa_m^2)}, \tag{2.109a}$$

which is the Debye–Hückel result.

The two-Yukawa model and its applications[6]

A two-Yukawa (2Y) fluid is a system of spherical particles interacting with the spherically symmetric potential (expressed in units of thermal energy, $1/\beta$):

$$\beta u(r) = \begin{cases} \infty, & r \leq \sigma, \\ -\sigma K_1 \left[e^{-z_1(r-\sigma)}/r\right] + \sigma K_2 \left[e^{-z_2(r-\sigma)}/r\right], & r > \sigma. \end{cases} \tag{2.110}$$

By exploiting a theoretical method given by Blum and Høye (1978), a numerical method was developed by Liu *et al.* (2005a) to calculate an analytical structure factor with the MSA closure:

$$\begin{cases} h(r) = -1, & r \leq \sigma, \\ c(r) = -\beta u(r) & r > \sigma, \end{cases} \tag{2.111}$$

for the case of a two-Yukawa interaction potential. They were able to obtain an analytical expression for $c(r)$ inside the hard core. After some algebraic

[6] Chen *et al.* (2007); Liu *et al.* (2005a).

calculations, one is left with the following expression for the Fourier transform $c(Q)$ (Q being the magnitude of the wave-vector transfer):

$$\rho c(Q) = -24\varphi \frac{a(\sin Q - Q \cos Q)}{Q^3} + \frac{b\left[2Q \sin Q - (Q^2 - 2)\cos Q - 2\right]}{Q^4}$$

$$+ \frac{a\varphi\left[4(Q^2 - 6)Q \sin Q - (Q^4 - 12Q^2 + 24)\cos Q\right]}{2Q^6} \qquad (2.112)$$

$$+ \frac{12a\varphi}{Q^6} + \sum_{i=1}^{2} h_Q(K_i, Z_i, v_i),$$

where φ is the volume fraction of he macroions and

$$h_Q(K, Z, v) = \frac{v}{Z}\left(1 - \frac{v}{2KZe^Z}\right)\left(\frac{1 - \cos Q}{Q^2} - \frac{1}{Z^2 + Q^2}\right)$$

$$- \frac{v^2(Q \cos Q - Z \sin Q)}{4KZ^2(Z^2 + Q^2)} \qquad (2.113)$$

$$+ \frac{Q \cos Q + Z \sin Q}{Q(Z^2 + Q^2)}\left(\frac{v}{Ze^Z} - \frac{v^2}{4KZ^2 e^{2Z}} - K\right),$$

whose coefficients a, b, v_i ($i = 1, 2$) are calculated in terms of the known parameters $K_i, Z_i, \varphi\}$ ($i = 1, 2$), with automatic criteria of good root selection, as explained in detail in Liu et al. (2005a). The static structure factor $S(Q)$, which can be extracted from SANS intensity distribution, can be therefore calculated through the usual relation in Fourier space

$$S(Q) = \frac{1}{1 - \rho c(Q)}. \qquad (2.114)$$

The MATLAB code for the calculation of the structure factor is freely available from Drs Sow-Hsin Chen or Yun Liu.

Figure 2.2 gives some examples of two-Yukawa charged protein solutions.

2.3.2 Module – Generalised one-component macroion theory[7]

The idea of a generalised one-component macroion (GOCM) theory originated from a paper by Beresford-Smith and Chan (1982). These authors showed that it is possible to first write down a set of multicomponent-coupled OZ equations involving the correlation functions of the macroions and counterions, and then to contract these equations to obtain an effective one-component OZ equation for the macroion alone. In this process, one is naturally led to a definition of an effective direct correlation function, $c_{00}^{eff}(r)$, for the macroions. We shall outline the result of the GOCM in the following section (Chen and Sheu, 1990).

---

[7] Belloni (1986); Sheu (1987); Chen and Sheu (1990)

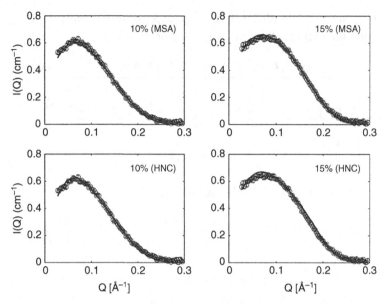

Figure 2.5 Comparison of the 2Y-MSA (top panels) and 2Y-HNC (bottom panels) fitting for two SANS spectra of cytochrome-C protein solutions at 10% and 15% volume fraction. Symbols represent experimental data, lines are fits with the 2Y model. Reprinted with permission of the International Union of Crystallography from Chen *et al.* (2007).

Generalisation of $u_{DLVO}(r)$ to finite concentration was made by Belloni (1986) and Sheu (1987). In GOCM one obtains the effective one-component direct correlation function, $c_{00}^{eff}(r)$, for the macroions by contracting a multicomponent OZ equations including macroions and counterions in the mean spherical approximation. The result, using the simplified notation $c_{00}^{eff}(r) \equiv c(x)$, is:

$$c(x) = -\frac{\beta Z^2 e^2}{\varepsilon} X^2 \frac{e^{-\kappa r}}{r}, \qquad r > d, \qquad (2.115)$$

$$(2.116)$$

$$g(x) = 0, \qquad r < d, \qquad (2.117)$$

where the factor X is given by $(k = \kappa d)$

$$X = \cosh \tfrac{k}{2} + V \left[\tfrac{k}{2} \cosh \tfrac{k}{2} - \sinh \tfrac{k}{2} \right],$$

$$V = \frac{\mu}{(k/2)^3} - \frac{\xi}{(k/2)},$$

$$(2.118)$$

$$\mu = \frac{3\eta}{(1-\eta)},$$

$$\xi = \frac{(G/2+\mu)}{(1+G/2+\mu)},$$

and G is the solution of an implicit equation,

$$G^2 = k^2 + \frac{t_0^2}{(1+G/2+\mu)^2},$$

(2.118a)

$$t_0^2 = \frac{24\beta Z e^2 \eta}{\varepsilon d}.$$

From the first equation of Eq. (2.118) it can be shown that in the limit $\rho \to 0$

$$X = \frac{e^{\kappa d/2}}{1 + \frac{\kappa d}{2}} = \frac{e^{k/2}}{1 + \frac{k}{2}}.$$

(2.118b)

So putting this result into Eq. (2.117) one gets

$$c(x) \underset{\rho \to 0}{\to} -\frac{\beta Z^2 e^2}{\varepsilon (1 + k/2)^2} \frac{e^{-k(x-1)}}{x} \cdot x > 1,$$

(2.119)

which is equivalent to putting $c(x) = -\beta u_{DLVO}(x)$. Thus GOCM is an extension of OCM (see Section 2.3.1) to a finite concentration of macroions and it is derived from an explicit liquid theory. In practice, one can say that GOCM is identical in form to OCM, so that the solution of OZ equation is the same except the charge parameter Z has a different value. In general Z_{GOCM} has lower value than Z_{OCM}. At 10% volume fraction of macroions Z_{OCM} is about 6% higher than Z_{GOCM}.

An example taken from analyses of SANS intensity from aqueous solutions of a globular protein cytochrome-C using GOCM theory[8]

In order to test the accuracy of the effective one-component direct correlation function given above we shall apply the GOCM theory to calculate the structure factors of concentrated protein solutions. Protein solutions are ideal testing cases for the theory because the protein charge Z can be estimated independently from a titration experiment. We shall illustrate this point with a series of SANS measurements made on the solutions of horse heart cytochrome-C.

The shape of cytochrome-C molecule is an oblate spheroid with a dimension. Since it is nearly spherical, we can take it to be a hard sphere of diameter $d = 32.6\,\text{Å}$ in the calculation of SANS intensity distribution $I(Q)$. The absolute intensity of a system of hard spheres can be written as $I(Q) = AP(Q)S(Q)$ where $P(Q)$ is the well-known particle structure factor of a hard sphere and $S(Q)$ is the inter-particle structure factor calculable by GOCM. The precise expression of the contrast factor A for a system of a pure protein molecules in a solvent of D_2O can be found in the paper of Wu and Chen (1987), which can be determined completely by a separate contrast variation experiment for a dilute protein solution.

---

[8] Wu and Chen (1987).

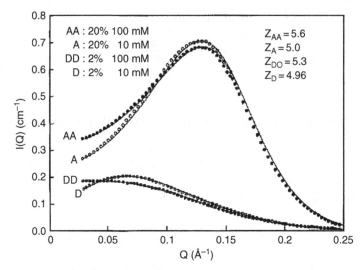

Figure 2.6 SANS intensity distributions (solid circles) and their model analysis curves made by GOCM (solid lines) for four protein solutions. AA and A are two high-protein concentration (20%) solutions containing 100 and 10 mM of sodium acetate respectively. DD and D are two low-protein concentration solutions containing the same respective salt. The protein charges so obtained agree reasonably well with those measured by independent titration experiments Chen and Sheu (1990). Graphic drawn with data from Wu and Chen (1987) with permission. Copyright 1987, AIP Publishing LLC.

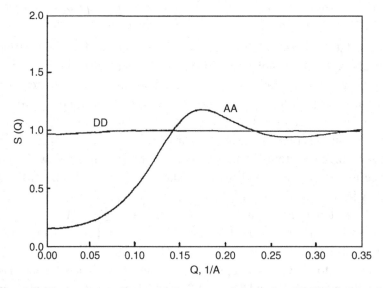

Figure 2.7 The structure factor $S(Q)$ extracted from the GOCM fitting of the SANS intensity distribution $I(Q)$ shown in Figure 2.6 above. It is striking to see that while the 2 % (DD) protein solution has a structure close to unity the 20%(AA) protein solution has a very structured $S(Q)$ function. Reprinted with permission from Wu and Chen (1987). Copyright 1987, AIP Publishing LLC.

Table 2.1 *GOCM fitted protein charge parameter Z_{GOCM} in comparison with the charge obtained from titration experiment Z_T.*

Protein solution	C (wt%)	η (vol. fraction)	[NaAc] (mM)	Z_T	Z_{GOCM}
AA	20	18.13	100	5.6 ± 0.6	5.1 ± 0.1
A	20	18.13	10	5.0 ± 0.5	5.0 ± 0.1
DD	2	1.89	100	5.3 ± 0.5	5.2 ± 0.1
D	2	1.89	18	5.0 ± 0.5	4.9 ± 0.1

Thus an absolute intensity measurement by SANS can give the inter-particle structure factor $S(Q)$ uniquely. Figure 2.6 shows an example of four absolute intensity distributions measurements – two high-concentration and two low-concentration cases respectively. Figure 2.7 shows the inter-particle structure factor for one of the high-concentration and one of the low-concentration cases extracted from the corresponding absolute intensity distributions as shown in Figure 2.6. We should notice that the results of GOCM analysis compare well with $S(Q)$ extracted from SANS measurements. In Table 2.1 we list the macroion charge Z_{GOCM} extracted from the GOCM analysis with the corresponding charges Z_T measured by the titration experiments. They agree to within the error bars of each method.

3

Aggregation and cluster formation

3.1 Introduction

Interactions between particles, both in molecular fluids and colloidal systems, are generally characterised by a strong short-range repulsion, which is responsible for excluded volume effects, followed by an attraction of variable strength. The latter is at the origin of cluster formation, a process that produces many different physical phenomena of great importance in the physics of simple and complex fluids. In atomic and molecular systems the most relevant effect of attraction is the appearance of critical points accompanying phase transitions, while in complex fluids, besides critical effects, peculiar phenomena develop such as aggregation, percolation, glass and sol–gel transitions. Recently the latter have been collectively named arrest phenomena, since their common feature is the pronounced slowing down of the dynamics. We first outline briefly the phenomenology and the approaches based on aggregation and percolation, which describe situations in which the attractive interaction is so strong that the colloidal particles adhere, leading to the formation of macroscopic clusters that eventually invade the physical sample. In the case of reversible aggregation the particles form the so-called physical gels. When the aggregation is irreversible, chemical gels are formed.

Thanks to the possibility of forming reversible or irreversible reactive links, during aggregation clusters of particles tend to coalesce and form larger aggregates. In the case of reversible bond formation a fragmentation process is also present. Aggregation is an ubiquitous process that can be observed in disparate situations at various length and time scales. Examples are polymer chemistry, aerosol systems, cloud physics, clusters of galaxies in astrophysics, etc. Although aggregation in colloidal suspensions has long been studied, it has become a subject of renewed interest in recent years because it is a non-equilibrium phenomenon, the final stage of which may lead, among other things, to the formation of a gel. We briefly summarise various aspects of clustering by introducing the Smoluchowski aggregation formalism, which is used in many different physical approaches to aggregation, and a brief summary of the salient aspects of percolation that are important for the physical phenomena we describe.

The relevant distinction in aggregation phenomena is between two limiting behaviours, aggregation limited by reaction, or bond formation, and aggregation limited by diffusion. In the sections of this chapter we treat some classical examples. In particular, the solution by Stockmayer and Flory of the reaction controlled reversible and irreversible Smoluchowski equations in Section 3.2. In Section 3.3 we consider in more detail a specific model of diffusion controlled irreversible aggregation and supramolecular ordering, while in Section 3.4 we treat the important topic of percolation in microemulsions. Critical phenomena in supramolecular system from the point of view of aggregation are finally summarised in Section 3.5.

The Smoluchowski approach

The coalescence of clusters is dominated essentially by two effects, the mobility phase related to the diffusive motion of the clusters and the bonding phase of the aggregation due to the mechanism of bonding reaction between clusters. Correspondingly it is possible to distinguish the following two limiting cases, which depend on the diffusion and aggregation time scales:

(i) aggregation limited by the reaction mechanism, when diffusion is very fast and bonding slow, so that the rate-controlling step is the chemical reaction leading to clustering;
(ii) aggregation limited by slow diffusion and high sticking probabilities of the aggregates.

In other words, the coalescence process is controlled by the slowest step. Indicating with C_k a cluster containing k particles, the reactions leading to aggregation start from smaller clusters C_j and C_{k-j} reacting with rates $K_{i,j}$ determined by the physical system under investigation. We consider two different types of aggregation, corresponding respectively to the specific reaction mechanism at the basis of clustering $K_{i,j}^c$, and the rate corresponding to the time spent by the clusters in diffusing between coalescence events, $K_{i,j}^d$. The total reaction rate can be written as

$$\frac{1}{K_{i,j}} = \frac{1}{K_{i,j}^c} + \frac{1}{K_{i,j}^d}. \tag{3.1}$$

In the case of aggregation based on diffusion, $1/K_{i,j}^d \gg 1/K_{i,j}^c$, and the reaction step is dominating so that the limiting step is diffusion, and $K_{i,j} \approx K_{i,j}^d$, while for the case in which the coagulation kinetics dominates, $1/K_{i,j}^c \gg 1/K_{i,j}^d$, since diffusion is efficient, then $K_{i,j} \approx K_{i,j}^c$.

The rate equation approach to clustering and gelation, due to Smoluchowski, is a mean-field theory which is capable of describing in a satisfactory way the time evolution of cluster distributions. The equations are in the form of a master equation in which the coalescence of clusters is determined by the probability of binary collisions between clusters multiplied by a reaction rate. The balance equations for aggregation are ($k = 1, 2, \ldots;\ j = 1, 2, \ldots, k - 1$)

$$C_j + C_{k-j} \xrightarrow{K_{j,k-j}} C_k \tag{3.2}$$

for the formation of k-clusters starting from smaller clusters. Clusters can also disappear due to the formation of larger aggregates, as in

$$C_j + C_k \xrightarrow{K_{j,k}} C_{j+k}. \tag{3.3}$$

Considering N particles and N_k the number of clusters containing k particles, the Smoluchowski rate equations for the cluster concentrations $c_k = N_k/N$ read

$$\frac{dc_k(t)}{dt} = \sum_{j=1}^{k-1} c_j K_{j,k-j} c_{k-j} - c_k \sum_{j=1}^{\infty} K_{k,j} c_j, \tag{3.4}$$

and are usually complemented with the initial conditions of only monomers present in the system

$$c_k(0) = \delta_{k,1}. \tag{3.5}$$

It is worth noting that the reaction steps included in the previous equations refer to cluster coalescence so that we do not expect the system to reach a steady state, but to keep growing until an infinite cluster forms, all the particles eventually belonging to it. This fact is usually accompanied by gel formation, turning irreversible aggregation into a case of gelation.

Diffusion dominated kernels $K_{j,k}$ of the Smoluchowski equations are the sum ($K_{j,k} \propto j+k$) and product ($K_{j,k} \propto j\,k$) kernels, together with the kernel describing the diffusive motion of the clusters. Reaction dominated kernels are represented by the well-known model for polycondensation or polymerisation denoted RA_f. This model was studied by Flory and Stockmayer in purely statistical terms using as a parameter the bond formation probability, without reference to the kinetic rate equations, and then shown to be a solution to the Smoluchowski equations at any time.

In order to allow for a steady state and consequently reversible aggregation, a step in which the aggregates break up in smaller units must be inserted in the rate equations. When disaggregation is allowed, then the inverse reactions of fragmentation read

$$C_j + C_{k-j} \xleftarrow[F_{j,k-j}]{} C_k, \qquad C_j + C_k \xleftarrow[F_{j,k}]{} C_{j+k}, \qquad (3.6)$$

with the rates $F_{j,k-j}$, $F_{j,k}$, and the Smoluchowski equations become

$$\frac{dc_k(t)}{dt} = \sum_{j=1}^{k-1} (c_j K_{j,k-j} c_{k-j} - F_{j,k-j} c_k) - \sum_{j=1}^{\infty} (c_k K_{k,j} c_j - F_{j,k} c_{j+k}). \qquad (3.7)$$

Percolation

We summarise the salient features of percolation theory (Stauffer and Aharony, 1992) with the aim to define some basic properties that will be used in the rest of the chapter. Percolation is a phenomenon important in many diverse physical situations, for example we can refer to mixtures of conducting and non-conducting spheres in a container and investigate the condition for conduction through the sample, or the passage of a liquid in a finely dispersed powder, or filtering through pores, and so on. The best way to define percolation is using a lattice in d dimensions, the sites of which can be occupied or not by a particle. The relation consisting in particles being connected can be easily defined, and the class of all the particles connected to a given one are aggregates or clusters of particles. In a similar fashion we can imagine the existence of bonds between neighbouring sites, leading to what is termed bond percolation, as opposed to the first type defined above, which is called site percolation. In other words, one considers the vertices or the edges of the lattice to be occupied or not. The procedure to introduce percolation on the lattice consists in associating an equal a-priori probability p to each site or bond, independently from the others. The main request in a percolation problem is to be able to evaluate the percolation threshold p_c, i.e. the limiting probability for the formation of an infinite cluster, the cluster spanning the macroscopic infinite sample. Other important quantities are the probability $P(p)$ that an occupied site belongs to the infinite cluster, the cluster size distribution, c_k, i.e. the number of clusters containing k particles, and the mean cluster size

$$S(p) = \frac{\sum_k k^2 c_k}{\sum_k k \, c_k}. \qquad (3.8)$$

In the one-dimensional case the percolation threshold is easy to evaluate to be $p_c = 1$, while the cluster size distribution is

$$c_k = k \, p^k \, (1 - p)^2. \qquad (3.9)$$

To proceed further let us consider a more complex situation. As stated earlier, the first model of percolation was introduced long ago by Flory and Stockmayer as a model of formation of large macromolecules when chemical bonds are allowed between the monomers, and are called percolation on the Bethe lattice (or Cayley tree). The Flory–Stockmayer model will be treated in detail in Section 3.2. It is very instructive to calculate directly the percolation threshold in this case, in fact it can be shown that it is given by the same value $p_c = 1/(f-1)$ of the Flory–Stockmayer model, and the same is true for the average cluster size $S(p)$ for $p < p_c$ and the corresponding cluster size distribution

$$c_k = k^{-5/2} e^{c(p)\, k}, \tag{3.10}$$

with

$$c(p) \approx |p - p_c|^2. \tag{3.11}$$

Then for the probability of the infinite cluster

$$P \approx |p - p_c|^\beta, \tag{3.12}$$

and

$$S \approx |p - p_c|^\gamma, \tag{3.13}$$

with $\beta = 1$ and $\gamma = 1$.

The important feature of percolation is the validity of scaling properties in the general case of lattices of any dimensionality d, with various values of the power-law indices and different forms of the scaling functions. The power laws are defined as follows:

$$c_k = k^{-\tau} f\left(\frac{k}{M_z}\right), \tag{3.14}$$

$$M_z = \frac{M_3}{M_2} = |p - p_c|^{1/\sigma}, \tag{3.15}$$

which introduce the two exponents τ and σ. The two indices introduced above

$$P \approx |p - p_c^*|^\beta, \tag{3.16}$$

$$S \approx |p - p_c|^\gamma, \tag{3.17}$$

are related to the previous one by

$$\beta = \frac{\tau - 2}{\sigma}, \qquad \gamma = \frac{3 - \tau}{\sigma}. \tag{3.18}$$

3.2 Module – Reaction controlled aggregation of colloidal particles

Reaction controlled irreversible aggregation

When the time spent during bond formation attempts is much larger than the diffusion time, the phenomenon changes drastically and the form of the Smoluchowski kernel is different from the one relevant for diffusive aggregation. One of the first attempts to describe aggregation in colloids was made by Stockmayer (1943) and Flory (1953) (FS) in 1943 using a statistical equilibrium approach. They studied the polycondensation or polymerisation, i.e. the bonding of monomers in the cases where they possess two, three or more adhesive points on their surface. The bonds are formed with a given probability between branching points of an infinite Bethe lattice (or Cayley tree). The lattice is defined starting from an origin (or root) site, surrounded by a number f of nearest neighbours, each of which has $f - 1$ neighbours, and so on. The importance of the Bethe lattice lies in the fact that it is effectively an infinitely dimensional system, since the volume and the surface of the tree are proportional (for finite dimensions the ratio of surface to volume is the volume to the power $-1/d$, while for the Bethe lattice it is finite). The model is called RA_f. The FS statistical approach on the Bethe lattice makes the assumptions following:

- since bonding takes place on a Bethe lattice, the clusters cannot form crosslinks, so there are no cycles, or loops, in the resulting structures;
- all the bonds have an equal possibility of linking, independently of the cluster to which they belong.

Since the monomers consist of f-functional reactive endgroups when two monomers react, the resulting dimer has $2f - 2$ reactive endgroups, a trimer has $3f - 4$ endgroups, and a general k-mer has $kf - 2(k - 1)$ endgroups, or

$$\sigma_k = (f - 2)k + 2, \tag{3.19}$$

free bonds, when loops are not allowed.

Considering N monomers, the total number of ways of partitioning them in clusters, where N_k is the number of clusters containing k units for $k = 1, 2, \ldots, N$, is given by

$$N! \prod_{k=1}^{k=N} \left(\frac{1}{k!}\right)^{N_k} \frac{1}{N_k!}. \tag{3.20}$$

If there are w_k ways to form a k-mer starting from k monomers, Ω gives the probability of the clusters' distribution as

$$\Omega = N! \prod_{k=1}^{k=N} \left(\frac{w_k}{k!}\right)^{N_k} \frac{1}{N_k!}, \tag{3.21}$$

and w_k was calculated by Stockmayer in 1943 as

$$w_k = \frac{f^k(fk - k)!}{\sigma_k!}. \tag{3.22}$$

The most probable distribution is obtained by maximising Ω with respect to N_k under the constraint of a constant number of monomers N and a number of clusters N_c:

$$\sum_{k=1}^{k=N} k N_k = N; \qquad \sum_{k=1}^{k=N} N_k = N_c. \tag{3.23}$$

The maximisation is performed by differentiating with respect to N_k the quantity $\ln \Omega + (\sum_{k=1}^{k=N} k N_k - N) \ln A + (\sum_{k=1}^{k=N} N_k - N_c) \ln \xi$, where A and ξ are Lagrange multipliers. The result for the cluster distribution is

$$c_k = A \frac{w_k}{k!} \xi^k, \tag{3.24}$$

where the Lagrange multipliers A and ξ are expressed in terms of N and N_c. The result is more transparently written in terms of the quantity

$$p = \frac{2(N - N_c)}{f N}, \tag{3.25}$$

which is the ratio of the number of bonded links to the total number of links, i.e. the bond probability. The relations are

$$A(p) = f N \frac{(1 - p)^2}{p}; \qquad \xi(p) = \frac{p(1 - p)^{f-2}}{f}. \tag{3.26}$$

Additional quantities can also be derived, for example the average number of monomers in a cluster

$$\frac{N}{N_c} = \frac{1}{1 - \frac{1}{2} f p}, \tag{3.27}$$

and the mean cluster size

$$S(p) = \frac{1}{N} \sum_{k=1}^{\infty} k^2 N_k = \frac{1 + p}{1 - (f - 1)p}. \tag{3.28}$$

The RA_f model has the striking feature that gelation occurs. Gelation means that a finite fraction of the total mass condenses into an infinite cluster. Thus beyond gelation, the system divides into two phases: the gel phase that consists of the infinite cluster, and the remaining sol phase of finite clusters. Note that the average cluster size becomes infinite as $p \to 1/(f - 1)$, a value that we define as the percolation threshold p_c, so that

$$S(p) = \frac{p_c(1 + p)}{p_c - p} \approx \frac{p_c(1 + p_c)}{p_c - p}, \tag{3.29}$$

the last equality being valid for $p \approx p_c$.

An analogous development can be performed on the cluster distribution N_k, by using the Stirling formula for large k

$$(ak + b)! \approx \sqrt{2\pi} \ e^{-ak} \ (ak)^{ak+b+\frac{1}{2}}, \tag{3.30}$$

in the expression for $w_k/k!$, so that

$$c_k(p) \approx \frac{p_c}{p} \left(\frac{1 - p}{1 - p_c} \right)^2 \left[\frac{p(1-p)^{f-2}}{p_c(1-p_c)^{f-2}} \right]^k k^{-\frac{5}{2}}. \tag{3.31}$$

A further development for p close to p_c produces

$$c_k(p) \approx k^{-\frac{5}{2}} e^{-c(p)k}, \tag{3.32}$$

with a power-law behaviour in the cluster size and a power law in the exponent of the scaling function in terms of the distance from p_c:

$$c(p) = \frac{(p_c - p)^2}{2 \, p_c^2 \, (1 - p_c)}. \tag{3.33}$$

What has the FS solution to do with the Smoluchowski rate equations? Clearly it is not obvious that the FS distribution, which is independent of time is a solution of the rate equation, unless we suppose a time-dependent bond probability $p(t)$ satisfying an appropriate first-order differential equation, a case that we will examine shortly. In the RA_f model, since all endgroups are equally reactive, the reaction rate between two clusters equals the product of the number of endgroups. Thus

$$K_{i,j} = k_1 \, \sigma_i \, \sigma_j = k_1 \left[4 + 2(f - 2)(i + j) + (f - 2)^2 \, i \, j \right], \tag{3.34}$$

where the proportionality coefficient is the rate constant for formation of a single bond k_1, which sets the unit of time.

It is then rather straightforward to show that the FS solution of the RA_f model satisfies the Smoluchowski equations provided that the bond probability follows the equation

$$\frac{dp(t)}{dt} = k_1 \, f(1 - p)^2, \tag{3.35}$$

with the solution

$$p(t) = \frac{k_1 \, f \, t}{1 + k_1 \, f \, t}. \tag{3.36}$$

The long time limit of the bond probability gives $p(\infty) \equiv p_{eq} = 1$ and as a consequence $c_k(\infty) = 0$ vanishing for any k, i.e. the absence of an equilibrium solution different from the trivial one.

The important point to note here is that the FS solution is valid at each instant, provided the bond probability is the appropriate function of time. In other words, at each time the distribution has always the same form provided by the FS solution.

Reaction controlled reversible aggregation

As stated earlier, in order to allow for a steady state and consequently reversible aggregation, a step in which the aggregates break up in smaller units must be inserted in the rate equations. When disaggregation is allowed the Smoluchowski equations have the aspect of Eq. (3.7).

We will consider the general case of aggregation-fragmentation, which describes both irreversible aggregation where the c_k vanish asymptotically, and reversible aggregation when c_k reaches a stationary solution. This case was treated by van Dongen and Ernst (1984). At this point, imposing the condition of detailed balance in order to get a steady state solution

$$c_j(\infty)K_{j,k-j}c_{k-j}(\infty) = F_{j,k-j}c_k(\infty), \tag{3.37}$$

the number of k-clusters created by aggregation must be equal to the number of clusters disappearing. When looking for a solution of the form FS,

$$N_k = A(p) \frac{w_k}{k!} \xi(p)^k, \tag{3.38}$$

one obtains

$$F_{j,k} = \lambda \frac{(j+k)!}{j!\,k!} \sigma_j \sigma_k \frac{w_j\,w_k}{w_{j+k}}, \tag{3.39}$$

with $\lambda = A(\infty)$, and the rate equations become of the same form as the irreversible aggregation, provided that the time is changed $t \to \tau$ according to the rule

$$d\tau = d t \left[1 - \frac{\lambda}{A(p)}\right]. \tag{3.40}$$

The form of the cluster distribution c_k is the same function of $p(t)$ as in the irreversible case, but with the evolution equation for $p(t)$ given by

$$\frac{1}{k_1}\frac{d p}{d t} = f\,(1-p)^2 - \lambda\,p, \tag{3.41}$$

with the equilibrium solution $p(\infty) \equiv p_{eq}$:

$$p_{eq} = 1 + \frac{\lambda}{2f} - \sqrt{\left(\frac{\lambda}{2f}\right)^2 + \frac{\lambda}{f}}. \tag{3.42}$$

The solution, with initial condition $p(0)$, is

$$p(t) = p_{eq} \frac{1 - \left[\frac{1-p(0)/p_{eq}}{1-p(0)\ p_{eq}}\right]e^{-\Gamma t}}{1 - p_{eq}^2\left[\frac{1-p(0)/p_{eq}}{1-p(0)\ p_{eq}}\right]e^{-\Gamma t}}, \tag{3.43}$$

where

$$\Gamma = 2\,k_1\,f\,\sqrt{\left(\frac{\lambda}{2f}\right)^2 + \frac{\lambda}{f}} = k_1\,f\,\frac{1 - p_{eq}^2}{p_{eq}}. \tag{3.44}$$

Note that in the limit $\lambda \to 0$ the irreversible aggregation solution is recovered.

From the physical point of view one can say that systems forming progressively larger and larger loopless branched aggregates tend to evolve in time via a sequence of states. These states are identical to the states explored in equilibrium at appropriate values of temperature. The equality in the fraction of formed bonds p (the extent of reaction in chemical language) provides the connection between time during reversible or irreversible aggregation and temperature in equilibrium.

Despite their relevance, the predictions of van Dongen and Ernst (1984) have seldom been tested experimentally or numerically to check their limit of validity. Recent numerical attempts (Corezzi *et al.*, 2008, 2009) have extensively investigated this point and are reported here. The model was originally inspired by polymerisation and represents two types of mutually reactive molecules, A and B, as hard homogeneous ellipsoids of revolution whose surface is decorated in a predefined geometry by f_A and f_B identical reactive sites. The study of the chemical version of this model showed that, because of the non-spherical particle shape and the location of the reactive sites, the formation of closed bonding loops in finite size clusters is highly disfavoured. Therefore, it offers the possibility to carefully check the theoretical predictions. One can verify how closely the chemical gelation process in a system of functionalised or patchy units, as well as the reversible evolution from a monomeric to an equilibrium bonded state, can be described as a progressive sequence of equilibrium states, closely connecting time and temperature. The system is a binary mixture composed of $N_A = 480$ pentafunctional ($f_A = 5$) ellipsoids of type A and $N_B = 1200$ bi-functional ($f_B = 2$) ellipsoids of type B, so that the number $N_A f_A$ of A-type reactive sites equals the number $N_B f_B$ of B-type reactive sites. A particles have mass m, revolution axis $a = 10\,\sigma$, and the other two axes $b = c = 2\,\sigma$; B particles have mass $3.4m$ and axes $a = 20\,\sigma$ and $b = c = 4\,\sigma$. The interaction potential is the hard ellipsoid potential V_{HE} supplemented by site–site square-well attractive interactions V_{SW} (of strength u_0 and width $\delta = 0.2\,\sigma$) between pairs of particles of different type. The unit of mass is m, and the unit of energy is u_0. Temperature is measured in units of the potential depth (i.e. $k_B = 1$), time t in units of $\sigma(m/u_0)^{1/2}$. The packing fraction is fixed at $\phi = 0.3$ and temperature is varied from $T = 0.3$ to 0.065. The percolation threshold was found to be at $p_c = 0.505 \pm 0.007$.

Figure 3.1 shows p_{eq} for all temperatures where equilibration was achieved, the system crossing from a monomeric state to a bonded state with more than

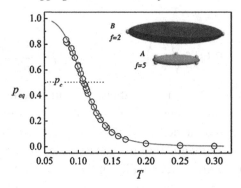

Figure 3.1 Temperature dependence of the fraction of bonds at equilibrium, $p(\infty)$. The solid line represents p_{eq} from Eq. (3.45). A one-parameter best-fit gives $\Delta S/k_B = 8.52$. The percolation threshold $p_c = 0.5$ is indicated by a horizontal line. The shape of the A and B particles are also reported, the centres of the small spheres locating the bonding sites on the surface of the hard-core ellipsoids. Reprinted with permission from Corezzi *et al.* (2009). Copyright 2009, American Chemical Society.

80% of the bonds formed. The sharp sigmoidal shape is well represented by the independent-bond mass-action law

$$\frac{p_{eq}}{\left(1 - p_{eq}\right)^2} = e^{\,\beta\,(u_0 - T\,\Delta S)}, \tag{3.45}$$

where u_0 and ΔS describe the energy and the entropy change associated to the formation of a single bond respectively.

The model allows to check, without fitting parameters, if the evolution of the system during equilibration and irreversible aggregation does follow a sequence of equilibrium steps, and in particular, how time in chemical gelation can be associated to a corresponding equilibrium temperature.

Figure 3.2 shows the cluster size distribution c_k at three different values of T, corresponding to p_{eq} values below, at, and above the percolation threshold p_c. Data are very well described by the FS predictions without fitting parameters and below and above percolation, confirming that bonding loops in finite size clusters can be neglected. Therefore the model allows to check if the evolution of the system during equilibration and irreversible aggregation follows a sequence of equilibrium steps, and in particular, how t in chemical gelation can be associated to a corresponding equilibrium temperature T. In particular, one compares the equilibrium quantities evaluated at selected elapsed times during the irreversible aggregation process. The specific time value is chosen in such a way that $p(t) = p_{eq}(T)$. The cluster size distributions, at each t are identical to those obtained in equilibrium at the same extent of reaction, demonstrating that the evolution of the system

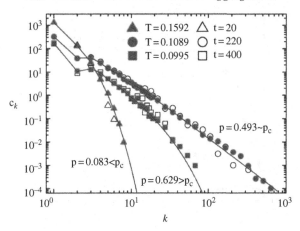

Figure 3.2 Cluster size distribution. Equilibrium results (closed symbols) are compared with results at different times t during irreversible aggregation (open symbols) such that $p(t) = p_{eq}(T)$. Solid lines indicate FS predictions. Reprinted with permission from Corezzi et al. (2009). Copyright 2009, American Chemical Society.

connectivity during irreversible aggregation follows a sequence of equilibrium states.

Figure 3.3 shows the time dependence of the bond probability $p(t)$, following a T-jump starting either from a high-T unbonded configuration $p(0) = 0$ or from a previously equilibrated configuration characterised by branched aggregates $p(0) \neq 0$. Lines are the theoretical predictions with the Flory post-gel assumption for values of $p(t)$ above percolation. The entire equilibration dynamics in this aggregating breaking branching system only depends on one parameter, k_1, which fixes the time scale of the aggregation process. The theoretical expressions represent rather well the numerical data, except for the two lowest studied temperatures ($T \leq 0.07$), where extensively bonded states are reached ($p > 0.8$). The disagreement can be traced to a failure of the Flory post-gel assumption or to a progressive role of the size dependence of the cluster mobility. The overall agreement between theory and simulation is further stressed by the T dependence of the single fit parameter k_1. In fact it shows the behaviour $k_1 \approx e^{-u_0/k_B T} T^{-1/2}$ which incorporates both an Arrhenius term and the thermal velocity component $\sim T^{-1/2}$, entering the attempt rate of bond breaking.

The results of this study clearly indicate that the irreversible evolution of a system of patchy or functionalised particles, in which bonding loops can be neglected, can be put in correspondence with a sequence of equilibrium states. For this class of aggregating systems, it is thus possible to convert irreversible aggregation time into an effective temperature and to envisage the evolution of a chemical gel as a

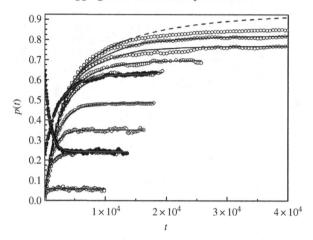

Figure 3.3 Time dependence of the fraction of bonds $p(t)$ during reversible aggregation at different temperatures ($T = 0.17$, 0.13, 0.12, 0.11, 0.10, 0.095, 0.09, 0.082 and 0.07) from the lower to the higher equilibrium value p_{eq}, starting from an unbounded configuration ($p(0) = 0$, open symbols) or from a generic equilibrated configuration ($p(0) \neq 0$, closed symbols). Lines represent solutions of the Smoluchowski equation for reversible aggregation. Reprinted with permission from Corezzi *et al.* (2009). Copyright 2009, American Chemical Society.

progressive cooling of the corresponding physical model. The elapsed time t can be uniquely associated to an equilibrium T, recalling the concept of fictive T in ageing glasses. Equally important is the possibility of interpreting cases in which during the formation of a chemical gel the corresponding thermodynamic path crosses a thermodynamic instability line, e.g., the gas–liquid coexistence line, producing an inhomogeneous arrested structure. Such thermodynamic lines have been recently calculated using statistical mechanics methods and computer simulations. Thus, it will become possible to interpret the stability and structural properties of chemical gels by connecting them to the thermodynamic properties and to the phase diagram of the corresponding physical models.

This study also demonstrates, for a realistic model of patchy particles, that the self-assembly kinetics of particles aggregating in loopless structures can be fully described with no fitting parameters by merging a thermodynamic approach providing p_{eq} with the Smoluchowski equations studied by van Dongen and Ernst, thus bringing reversible and irreversible aggregation of loopless branched systems to the same level of understanding as the equilibrium polymerisation case. This is made possible by the limited valence of the particles and by the specificity of the site–site interaction. The reduced valence, together with the non-spherical particle shape, contributes to depress the formation of closed bond-loops, while the specificity in the bonding interaction contributes to reduce the rate-controlling role of diffusion compared to the bonding process.

3.3 Module – Diffusion controlled irreversible aggregation and supramolecular ordering

The phenomenon of irreversible aggregation of colloidal particles from an initially stable suspension of monomers is related to the possibility of decreasing the potential barrier separating the minima of the DLVO potential. This is usually done by changing the electrostatic interactions by means of an electrolyte in solution, for example added salt. The colloidal particles are then capable of overcoming the barrier and stick irreversibly when colliding. If the barrier is still high enough the particles experience many collisions before being able to form a cluster. The two limiting regimes of aggregation have rather different features and give rise to diffusion limited cluster aggregation (DLCA) and reaction limited cluster aggregation (RLCA) respectively. Both cases are characterised by the predominance of the diffusive part of the motion with respect to the rapid bond formation when there is an encounter between clusters. The names refer to the fact that in the first case the reaction stage is much shorter than the time spent in the diffusive motion, which is therefore the limiting step in the aggregation. In RLCA the situation is reversed, the limiting step being due to the time it takes to the particles to coalesce.

There are experiments performed on various colloidal systems, such as silica, gold, polystyrene and others to determine the exponents characteristic of aggregation. It has been possible in particular to distinguish the two classes of DLCA and RLCA by measuring the fractal dimension d_f. The two regimes give rise to clusters of different fractal dimension D, to different cluster size distributions $c_k(t)$, and to different scaling laws for the average cluster size as a function of time t. In RLCA clusters of dimension $d_f \sim 2.1$ are formed, while $d_f \sim 1.8$ in the DLCA case.

DLCA describes a process of irreversible aggregation among non-interacting freely diffusing clusters which aggregate when they come in contact, the newly formed cluster continuing to diffuse. Differently from the process of supramolecular ordering of an unstable molecular fluid system, DLCA clusters are fractals. The original extensive studies on DLCA focused mainly on this aspect of the aggregation, but not on the spatial arrangement of the growing clusters or the existence of a characteristic length scale in the cluster spatial arrangement.

A step toward a deeper understanding of the phenomenon of aggregation was made by studying the correlation properties of the resulting objects and, as a consequence, their spatial ordering. A similar phenomenon of formation of large groups of connected units happens in systems quenched from the homogeneous one-phase region to the region of coexisting phases in the vicinity of a phase transition, a situation usually referred to as spinodal decomposition. The latter is characterised by scaling in time and space of the inter-cluster structure factor obtained using radiation scattering.

The process of supramolecular ordering, like the growth of the minority phase in unstable binary mixtures or the growth of crystalline regions in a supercooled system, is characterised in Fourier space by a wave vector q and time-dependent scattered intensity $I(q, t)$ which conveys information on the leading aggregation mechanism. Often $I(q, t)$ is characterised by a well-defined peak at the wave vector q_m. During the aggregation process, the growth of the aggregates in mass and radius manifests itself in the growth of the peak amplitude and in the shift of the peak position toward smaller and smaller wave vectors. The scattered intensity at different times can be scaled on a common master curve by plotting $I(q/q_m)$ versus q/q_m, suggesting an underlying scaling in space and time of the ordering process.

In the case of DLCA it was possible to show the existence of a characteristic length scale in the spatial arrangement of the clusters through light-scattering experiments on polystyrene colloids. In fact the experimentally measured scattered intensity displays a peak, which grows and shifts in q space during the aggregation kinetics. It was found that, after an initial regime, the measured scattered intensity seems to scale exactly as predicted for late-stage decomposition theories, if one substitutes the fractal dimension of the clusters d_f to the space dimension d in the scaling plot. The growth of the scattered intensity stops when the less and less dense fractal clusters completely fill the available space, leaving in the frozen scattered intensity state a memory of the growth process. The experimental observations were confirmed subsequently by experimental studies on cluster–cluster correlation in two-dimensional systems and by computer simulations. Even in the two-dimensional case, a growing and moving peak in the scattered intensity is observed. From the available real space images and from mass conservation, it was conjectured that the peak was a manifestation in Fourier space of the depletion region, which develops around the growing clusters.

We report here a theoretical interpretation of structure factor scaling during irreversible cluster–cluster aggregation (Sciortino and Tartaglia, 1995; Sciortino *et al.*, 1995). Let us consider a system consisting of an average number of clusters per unit volume $n_c = N_c/V$, each of mass M and containing the average number of monomers, evolving through diffusion and coalescing in time due to collisions. Each cluster is characterised by the average radius of gyration linearly related to its radius $R(t)$ so that its fractal structure is dictated by the relation

$$M(t) = M(0) \left[\frac{R(t)}{R(0)} \right]^{d_f}. \tag{3.46}$$

Since two clusters cannot approach closer than a distance of twice their diameter, we can fix a particle of radius $2R$ in the origin and consider the flux of point-like clusters impinging on it while diffusing. As aggregation proceeds, $n_c(r, t)$ varies as a function of space and time, and so do the mass of a cluster $M(t)$ and its radius $R(t)$ as a function of time. The total number density is therefore given by $n(r, t) = n_c(r, t) \, M(t)$ and obeys the usual diffusion equation

$$\frac{\partial}{\partial t} n(r, t) = D \nabla^2 n(r, t), \tag{3.47}$$

with the diffusion constant D supposed to depend on M as $D(M) = 2D_0 M^{-\gamma}$. The boundary and initial conditions, with n_0 constant, are

$$n(2R(t), t) = 0, \quad n(\infty, t) = n_0, \quad n(r, 0) = n_0. \tag{3.48}$$

Then for the diffusing clusters

$$\frac{\partial n_c(r, t)}{\partial t} = D \, \nabla^2 n_c(r, t) - \frac{n_c(r, t)}{M(t)} \frac{dM(t)}{dt}, \tag{3.49}$$

together with the boundary and initial conditions

$$n_c(2 \, R(t), t) = 0, \quad n_c(\infty, t) = n_0 \frac{M(0)}{M(t)}, \tag{3.50}$$

$$n_c(r, 0) = n_{c0} \quad for \quad r \geq 2 \, R(0). \tag{3.51}$$

Note that one of the boundary conditions is written for the moving boundary of the seed particle. At the same time the extended cluster increases in size because of the flux on the clusters colliding on its surface

$$\frac{dM(t)}{dt} = 4 \, \pi \, r^2 \, DM(t) \left(\frac{\partial n_c(r, t)}{\partial r} \right)_{r=2R(t)}. \tag{3.52}$$

The problem is similar to the well-known Stefan problem of moving boundaries. Its solution can be written explicitly in the case of compact clusters, when $d_f = d$, as a function of the variables r and τ derived from the relation $ds = D(M)dt$:

$$n_c(r, \tau) = \begin{cases} 0, & r \leq 2R(s), \\ n_{c0} \frac{M(0)}{M(s)} \left[1 - \frac{F(\frac{r}{2\sqrt{s}})}{F(\lambda)} \right], & r \geq 2R(s), \end{cases}$$

where

$$F(x) = \begin{cases} erfc(x), & d = 1, \\ Ei(-x^2), & d = 2, \\ \frac{e^{-x^2}}{x} - \sqrt{\pi} \, erfc(x), & d = 3, \end{cases}$$

in various dimensions. The constant λ appears in the solution for the radius

$$R(s) = \lambda s^{\frac{1}{2}}, \tag{3.53}$$

and is obtained by solving the equation obtained substituting the solution for $n_c(r, s)$ in the differential equation for $M(t)$

$$
\begin{cases}
\sqrt{\pi}\, erfc(\lambda)\, \lambda\, e^{\lambda^2} = 2R(0)\, n_{c0}, & d = 1, \\[2ex]
-Ei(-\lambda^2)\, \lambda^2\, e^{\lambda^2} = \pi\, [2R(0)]^2\, n_{c0}, & d = 2, \\[2ex]
2\lambda^2 e^{\lambda^2}\left[e^{-\lambda^2} - \lambda\sqrt{\pi}\, erfc(\lambda)\right] = \tfrac{4\pi}{3}\, [2R(0)]^3\, n_{c0}, & d = 3,
\end{cases}
$$

where the right-hand side depends only on the initial situation.

The scattered intensity can be calculated from the definition

$$S(q, s) = 1 + \int d\mathbf{r}\, e^{i\mathbf{q}\cdot\mathbf{r}}\, [n_c(r, s) - n_c(\infty, s)]. \tag{3.54}$$

At small q, one can expand the exponential term in the right-hand side of this equation in a series, the leading term in the small q expansion being q^2, as imposed by mass conservation. Moreover, since $n(r, s)$ is a function of the scaled variable $r/s^{1/2}$, also $S(q, s)$ scales in time as $q\, s^{1/2}$ or, by Eq. (3.53), as $q R(s)$. The total scattered intensity $I(q, t)$ measured experimentally can be approximated as the product of the previously calculated $S(q, t)$ and of the so-called cluster form factor $P(q, t)$, a well-known function for any d. $P(q, t)$ takes into account the intracluster contribution to the scattered intensity. Only in the absence of correlation among clusters, $I(q, t) \approx P(q, t)$, being $S(q, t) = 1$. $P(q, t)$ can be expanded at low q as $P(q, t)\, M(t)(1 - R_g q^2/3)$, and at high q values as $P(q, t) \approx q^{-(d+1)}$, where R_g is the gyration radius of the cluster. $P(q, t)$ is proportional to the mass of the scatterer, i.e. in our case to the mass of the average cluster and it is a function only of $q R$, i.e. of the same scaled variable of $S(q, t)$. This implies that the total scattered intensity will also be a scaled function of $q R$ and that a plot of $I(q R(t))/M(t)$ versus $q R(t)$ will show a remarkable data collapse, of the same kind as observed in late-stage spinodal decomposition. From the limiting behaviour of $S(q, t)$ and $P(q, t)$, at low q, the total scattered intensity goes as q^2, being controlled by $S(q)$, while at high q it goes as $q^{-(d+1)}$, being fixed by the decay of the form factor for objects with sharp interfaces. It is important to note that $P(q, t)$ does not depend on the number of clusters in solution at any time. Thus, differently from $S(q, t)$, $P(q, t)$ is independent from the initial number density n_0. As a consequence, the scaling function for $I(q, t)$ will depend on n_0. The smaller the initial number density, the larger the depletion region and the smaller the q vector at which a maximum in the scattered intensity will occur. On reducing the initial monomer concentration, the $I(q, t)$ maximum will eventually move out from the finite experimental window. From the solution of the differential equations the value of the exponent z, which controls the power-law growth of the mass, can be calculated. On going from τ to t and using $M \approx R^d$, one finds $M(t) \approx t^{\frac{d}{2+\gamma d}}$. In three dimensions and with the $\gamma = 1/d$ value for Stokes–Einstein diffusion, one recovers the Smoluchowski result $z = 1$.

Among the consequences of the solution for compact clusters the following properties are to be stressed:

(i) $n(r, t)$ monotonically increases from zero at the sticky boundary to $n(\infty, t)$, showing that a depletion region forms around each growing cluster.
(ii) Scaling in the compact aggregation case can be traced to the fact that both the radius of the growing clusters and the size of the depletion region grow at the same rate, determined by the time evolution of the scaled variable $r/s^{1/2}$.
(iii) The mass per cluster evolves in time as $M(t) \propto t^{\frac{d}{2+\gamma d}}$ where the exponent tends to one using $\gamma = 1/d$, the value for Stokes–Einstein diffusion, and the same result predicted by the Smoluchowski rate equation.
(iv) The structure factor shows no scaling behaviour for the initial stage of the aggregation, but scaling in $q\, s^{1/2}$ for late stages.

When the simultaneous growth of the cluster size and the extent of the depletion region proceed according to different time scales, scaling of the solution and the associated physical quantities is lost. This happens when the aggregates are fractal characterised by the dimension d_f which is smaller then the space dimension d. Fractal clusters tend to fill the space, being characterised by an average density, which decreases with the cluster size, i.e.

$$\rho(r) = \rho_0 \left[\frac{R(0)}{r} \right]^{d-d_f},$$

where ρ_0 and $R(0)$ are, respectively, the density scale constant and the monomer radius. When clusters reach a space-filling configuration, gelation occurs. One single cluster fills up all the available space. When the growing cluster is a fractal, a new length scale related to the average cluster size at the gelation point is expected to arise. The change in time of the average cluster density complicates the structure of Eq. (3.52). The relation between mass and radius gives now $dM \approx R^{d_f-1}dR$, an R dependence that no longer cancels the surface term in the right-hand side of Eq. (3.52). This extra R dependence reflects the fact that the radius grows faster than it would if d_f were equal to d. On the other end, the diffusion of the fractal clusters is still happening on an Euclidean substrate, so that no change in the characteristic space-time relations are expected in the evolution of the depletion region. The changes in the time dependence of the cluster growth, compared to the time dependence of the growth of the depletion region, brings as a consequence that the $n_c(r, t)$ profile no longer scales with $R(t)$. On increasing the time, the inter-cluster distance becomes comparable with the cluster size. In this late-stage regime the cluster radius appear to be the only typical length scale. Under such circumstances, which appear close to gelation, an apparent scaling in $qR(t)$ is again expected. The numerical solution of the equations shows also that during the initial stage of the growth process, (when the inter-cluster distance is much bigger than the cluster size) the flux of matter at R is proportional to R^{d-2}, as in the Euclidean case. Under such conditions, from Eq. (3.52) one has $R^{d_f-1} dR/ds \approx R^{d-2}$, which immediately gives $R \approx s^{\frac{1}{d_f - d + 2}}$ or $M \approx t^{\frac{d_f}{d_f(1+\gamma) - (d-2)}}$, the classical mean field exponent. When the inter-cluster distance becomes comparable to the cluster size, the numerical solution of the equations shows a faster and faster increase of the average mass, which diverges when the average density of the fractal cluster becomes equal to the initial monomer density. Before comparing the prediction of the theory with numerical and experimental results, we note that the model we propose can be expressed in the case of fractal growing clusters in terms of a scaled variable $x = r/R_f$, where R_f is defined as the radius of the cluster at gelation. R_f depends only on the monomer concentration at $t = 0$. We define R_f via

$$\rho_\infty (2R_f)^d = \int_{R(0)}^{R_f} dr \, \rho_0 \left(\frac{R(0)}{r} \right)^{d-d_f} r^{d-1},$$

where ρ_∞ is the density of the sample at $t = 0$. The important point is that it is possible to recast Eqs. (3.49) and (3.52) in the scaled variable $x = r/R_f$

$$\frac{\partial}{\partial t} n^*(r, t) = D^* \nabla_x^2 n^*(r, t) - \frac{n^*}{M} \frac{dM}{dt}, \tag{3.55}$$

$$\frac{dM}{dt} = D^* M \nabla_x n^*|_{2X(t)}, \tag{3.56}$$

where $D^* = D/R_f^2$, $X(t) = R(t)/R_f$, and $n^*(x, t)$ is a dimensionless number density. The corresponding initial and boundary conditions become $n^*(x, 0) = 1$ and $n^*(\infty, 0) = 1$. The initial monomer density now appears in the equations only through R_f. Thus, the model predicts that at constant ratio $R(t)/R_f$, one should observe the same scattering pattern, independently of n_0. This brings as a consequence the result that the scaling form of the total scattered intensity in the gel state should be the same independently from the initial monomer density, a prediction that has been recently confirmed experimentally.

The predictions of the model can be tested with the experimental results of Carpineti and Giglio (1992) on aggregation of polystyrene spheres in water. They observed the formation of clusters with $d_f = 1.9$ for $d = 3$. The scattered intensity shows a well-defined peak that moves in time. The kinetic process is separated in

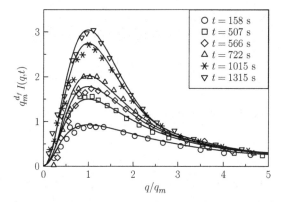

Figure 3.4 Comparison between experiments (symbols) and theory (full lines) in the $d = 3$ case. Scaled $I(q, t)$ at different times during the aggregation process in the non-scaling regime (Sciortino and Tartaglia, 1995). Copyright 1995, by The American Physical Society.

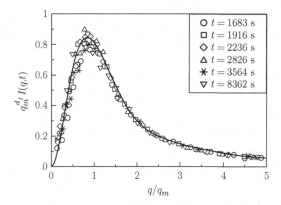

Figure 3.5 Comparison between experiments (symbols) and theory (full lines) in the $d = 3$ case. Scaled $I(q, t)$ at different times during the aggregation process in the scaling regime. The theory is obtained when the function $n(r)$ is approximated with a step function (Sciortino and Tartaglia, 1995). Copyright 1995, by The American Physical Society.

three regions: an initial region where no scaling in $q_m^{d_f} S(q/qm)$ is observed (symbols in Figure 3.4), an intermediate region where scaling is observed (symbols in Figure 3.5), and a saturation region where no further change in the dynamical structure factor is observed. On the basis of experimental work, it is not possible, of course, to know the radius growth law to scale data in qR. For this reason, experiments are presented in q/q_m, q_m being the experimentally determined position of the maximum in the total scattered intensity. To make contact with the experimental work, we present data also in q/q_m, even if such a choice may obscure the absence of true scaling for growing fractal clusters.

We have evaluated $S(q, t)$ from the numerical solution of the model equations with $d_f = 1.9$. For $P(q, t)$ the standard Fisher–Burford equation is used:

$$P(q) = \frac{P(0)}{\left[1 + (q\, R_g)^2\right]^{d_f/2}},$$

relating the gyration radius R_g in this expression to $R(t)$ according to

$$R_g^2 = \frac{3d_f + d}{5 - d_f + d}\, R^2(t).$$

This equation was shown to give a satisfactory representation of the experimental cluster form factor, its validity is limited to q vectors smaller than the inverse of the monomer size, above which one should observe the scattering arising from the monomer form factor. The experimental data are well within the region of validity of the equation. The volume fractions studied in the experiments range between 3×10^{-5} and 3×10^{-3}, supporting the possibility of writing the total scattered intensity as $I(q, t) = S(q, t)P(q, t)$. $I(t)$ at selected times in the non-scaling regime is shown in Figure 3.4 and compared with the experimental data. In the scaling region the scaled $q/q_m\, I(q/q_m)$ is within a few per cent coincident with the quantity obtained approximating $n(r, t)$ with a step function.

The growth of the cluster in this time region produces significant changes of $n(r, t)$ in real space, but only very minor changes in the normalised $n(r/R, t)/n(\infty, t)$ function. In this region, an apparent scaling of $q_m^{d_f}\, I(q/q_m)$ has again been observed, being the size the depletion region very close to the cluster radius. The final $q_m^{d_f}\, I(q/q_m))$, calculated assuming a step function shape for $n(r, t)$, is compared with the experimental data in the scaling regime in Figure 3.5.

The agreement between the experimental data and the prediction of the model is surprisingly good, both in the non-scaling and in the scaling regions. The apparent scaling in $I(q)$ observed in experiments can be related to the approximate scaling observed theoretically when one optimises the value of q_m instead of using its correct theoretical value, the differences between the two values of q_m being well within the experimental errors. This might explain why close to gelation the experimental uncertainties in the position of the maximum have suggested the presence of apparent scaling. In the late-stage regime, the apparent scaling allows the calculation of the z exponent from the experimental data, being $I(q, t) \approx M(t)$.

The calculated value is $z = 1$, consistent with the prediction of the model, when for γ the hydrodynamic regime value $1/d_f$ is chosen. The presence of a peak in the scattered intensity is independent from the initial monomer density. The initial density only modulates the position in q space at which the peak will appear. For very diluted concentrations, the depletion region is so large that the peak appears well below the experimental resolution of light-scattering experiments. On increasing concentration, the initial position of the $I(q, t)$ peak will move to higher q values. This explains why the intensity peak was only detected when fairly concentrated samples were investigated.

3.4 Module – Percolation in microemulsions

The structure of ternary microemulsions

Three-component systems made of water, oil and ionic or non-ionic surfactants form microemulsion systems that exhibit a rich and interesting phase behaviour around room temperature.

At a characteristic temperature where the surfactant has balanced affinities toward water and oil, called the hydrophile-lipophile balance (HLB) temperature, a typical symmetric ternary microemulsion system with equal volume fractions of water and oil, as the surfactant concentration is increased from zero to more than 8%, shows the following 2-3-1 phase progression:

(i) When the surfactant concentration is very low, the molecules are dispersed in water and oil as monomers. The system is naturally separated into two phases, with a lighter oil-rich phase on the top and a heavier water-rich phase on the bottom because of the high interfacial tension between water and oil. The two phases do not show organised microstructure.

(ii) At relatively higher surfactant concentrations, a three-phase coexistence, with a middle-phase microemulsion in coexistence with an oil-rich phase on the top and a water-rich phase on the bottom, is obtained simply because of the equal solubilisation power of the surfactant for water and oil. The middle-phase microemulsion shows an organised microstructure often described as being bicontinuous in both water and oil.

(iii) With further increase in the surfactant concentration, a minimum concentration will be reached whereby all the excess water and oil are solubilised into a single-phase microemulsion. This minimum concentration is usually between 5% and 8% for a good microemulsion system. The value of the minimum concentration is a measure of the amphiphilicity of the surfactant molecules at that temperature, being lower for higher amphiphilicity. In the vicinity of this minimum surfactant concentration, the microstructure of the one-phase microemulsion is disordered bicontinuous.

(iv) As the surfactant concentration further increases, the one-phase microemulsion transforms into a lamellar structure which may be called an ordered bicontinuous structure, and then to some other three-dimensional ordered structures. This disorder-to-order transition usually occurs at around 15% of the surfactant concentration.

Away from the HLB temperature, the phase progression as a function of surfactant concentration is different because at these temperatures the surfactant film, located in between water and oil, acquires a spontaneous curvature. Depending on the sign of the curvature, a two-phase coexistence, with either a water-in-oil (w/o) microemulsion on the top and a water-rich phase on the bottom, or with an oil-in-water (o/w) microemulsion on the bottom and an oil-rich phase on the top, becomes possible for moderate concentrations of the surfactant. Micro-structures of w/o or o/w microemulsions are likely to be a droplet type.

We now turn to a specific microemulsion system, the water/decane mixture with the surfactant sodium bis(2-ethylhexyl) sulfosuccinate (AOT), which is characterised by a bulky tail and a small head. It does not follow, however, the usual pattern of phase behaviour. Around room temperature, or more specifically from 10 to 50 °C, the surfactant film possesses a spontaneous curvature toward water due to the hydrophilicity–lipophilicity imbalance of the AOT molecules. Thus, at room temperature one finds a large one-phase region called the L2 phase, where even with equal volume fractions of water and oil, the microemulsion, instead of being bicontinuous like other three-component microemulsion systems made of non-ionic surfactants, consists of water droplets, coated by a monolayer of AOT and dispersed in decane. With this microstructure, the microemulsion is basically an insulator, with a conductivity of the order of $10^{-6}\,\Omega^{-1}\mathrm{m}^{-1}$ because the water droplets are separated from each other. The average radius of the water droplets is determined essentially by the molar ratio of water to AOT in the system, called W. An approximate empirical relationship between the radius (in Å) and W is $\langle R \rangle = \frac{3}{2}W$. Thus, for $W = 40$ the average water-droplet radius is about 60 Å. This water-in-oil droplet structure is maintained even if the volume fractions of water and oil are equal, provided the temperature is below 25 °C. This case is in sharp contrast to the common situation that, for equal water and oil volume fractions, the microstructure of one-phase microemulsions were generally found to be bicontinuous. Even for the AOT/water/decane system, when a small amount of salt (NaCl) is added, the common 2-3-1 phase progression is obtained at around the hydrophile–lipophile balance temperature of 40 °C and a SANS experiment in the one-phase channel around this temperature conclusively showed that the microstructure is bicontinuous. The persistent droplet structure in the ternary AOT/water/decane system can be used to produce an interesting coexistence of a critical phenomenon at low volume fraction and high temperature, and a percolation phenomenon at lower temperatures but at all volume fractions.

Figure 3.6 shows the $(T–\phi)$ phase diagram of the AOT/H$_2$O/decane system when the water-to-AOT molar ratio is $W = 40.8$. In the $(T–\phi)$ diagram, one sees

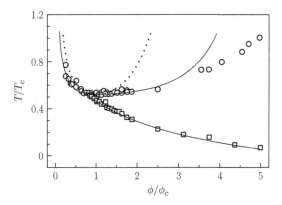

Figure 3.6 Phase diagram of the AOT/water/decane system, showing phase coexistence, spinodal lines, critical point and percolation loci. Symbols are experimental data (Chen *et al.*, 1994). ©IOP Publishing. Reproduced with permission. All rights reserved.

a one-phase (L2) region below 40°C. In the interval of ϕ between 0 and 0.4, there is a cloud-point curve separating the one-phase droplet microemulsions from two-phase droplet microemulsions. The critical volume fraction is approximately 0.1 and the critical temperature is 40°C in H_2O-based microemulsions.

Electric conductivy in w/o microemulsions

The characteristic feature of the ternary microemulsion phase diagram is the existence of a percolation line which extends from the region of the critical point, all the way to higher volume fractions, gradually decreasing in temperature to about 23°C at $\phi = 0.7$. Below the percolation line the microemulsion is non-conducting but above the percolation line it becomes conducting. In crossing the line, the conductivity increases by over five orders of magnitude as shown in the inset of Figure 3.7 showing a typical result of a direct current conductivity measurement σ as functions of T.

The steeply rising sigmoidal curve can be used to define a set of loci (T_p, ϕ_p) in terms of their inflection points as shown in Figure 3.7 which reports the behaviour of σ near the threshold, for fixed ϕ ranging from $\phi = 0.098$ to $\phi = 0.65$, as

$$\sigma = A \left(\frac{T_p - T}{T_p}\right)^{-s'}, \qquad \sigma = B \left(\frac{T - T_p}{T_p}\right)^{t}, \qquad (3.57)$$

on approaching percolation respectively from below and above. The values of the two exponents derived from the experiments are $s' = 1.2 \pm 0.1$ and $t = 1.9 \pm 0.1$, the corresponding values for the regular percolation transition being $s = 0.73$ and $t = 1.9$. The percolation index s' was in fact introduced by the dynamic percolation

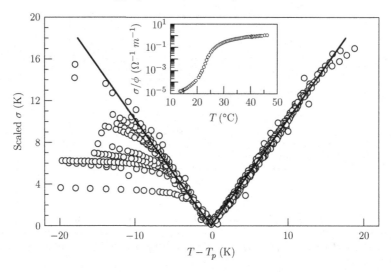

Figure 3.7 Scaled conductivity $T_p(\sigma/A)^{1/s'}$ and $T_p(\sigma/B)^{1/t}$ as a function of the distance from the percolation line $T - T_p$ at various volume fractions. The bottom curve is for $\phi = 0.098$, the top one for $\phi = 0.65$. The inset shows a typical measured conductivity sigmoidal curve as a function of temperature. The inflection point determines the percolation point (Cametti *et al.*, 1990). Copyright 1990, by The American Physical Society.

model that was formulated to describe the anomaly of conductivity in microemulsions. The physical mechanism underlying the conductivity in w/o systems is the formation of surfactant coated water clusters which are sufficiently close one to the other to permit hopping of charge carriers from one cluster to a neighbouring one and diffusive motion on a single cluster. As shown in Figure 3.7, above the percolation threshold the data at different volume fractions follow the asymptotic power law very closely, while below threshold the power law is well obeyed at high volume fractions, but poorly followed for low values of ϕ.

Conductivity in the Baxter adhesive spheres model

A reasonable model for a microemulsion in the L2 phase is to regard it as spherical colloidal particles of average diameter σ interacting via a short-range temperature-dependent attractive pair potential. This pair potential can, for example, be a square-well potential with a hard-core diameter σ, plus an attractive well of depth ϵ and width Δ. Correlations in an arbitrary square-well potential cannot be derived analytically, except for the limiting case of the Baxter model, in which ϵ tends to infinity and Δ to zero in such a way that a contribution of the attractive tail to the second virial coefficient exists. Analytically, when the particle distance r is between σ and $\sigma + \Delta$ the interaction is $\beta u = -ln(\sigma + \Delta)/(12\tau\Delta)$ with $\Delta \rightarrow$

and the dimensionless parameter τ is called the stickiness parameter. To the lowest order in $\Delta/(\sigma + \Delta) << 1$, one gets the relation

$$\frac{1}{\tau} = \frac{12\Delta}{\sigma + \Delta} \exp(\beta\epsilon).$$

For AOT molecules in decane, the parameter Δ corresponds roughly to the length of the hydrocarbon tail. The tail could take an increasingly stretched conformation as temperature increases, thus increasing the interaction strength among droplets. As shown previously, Baxter showed that the Ornstein–Zernike equation using this sticky pair potential can be solved analytically in the Percus–Yevick approximation. One can then derive the compressibility equation of state,

$$\frac{\beta P}{\rho} = \frac{1 + \eta + \eta^2}{(1-\eta)^3} - \frac{\lambda\eta(1-\eta)(1+\eta/2) - \lambda^3\eta^3(1-\eta)^3/36}{(1-\eta)^3},$$

with density ρ, volume fraction η and parameter λ, the smaller real solution of the equation

$$\frac{\eta}{12}\lambda^2 - (\tau + \frac{\eta}{1+\eta})\lambda + \frac{1+\eta/2}{(1-\eta)^2} = 0.$$

From the compressibility equation of state. one finds the existence of a gas–liquid phase transition with a critical point occurring at $\eta_c = 0.1213$ and $\tau_c = 0.0976$, and the corresponding phase coexistence in the variables η and τ. The coexistence curve which can be deduced from the use of Baxter's potential is highly skewed toward the low-volume-fraction side, a feature which is often experimentally observed in micellar solutions and microemulsions. This is due to the interaction, which is short range and strong, and is in sharp contrast to the well-known van der Waals case, which is derived from an interaction that is long range and weak. To assess the degree of asymmetry on the gas and liquid sides, we have worked out the respective asymptotic behaviours

$$1 - \frac{\tau}{\tau_c} = 0.1584 \left(1 - \frac{\eta}{\eta_c}\right)^2 \qquad \text{for} \quad \eta < \eta_c,$$

and

$$1 - \frac{\tau}{\tau_c} = 0.0264 \left(\frac{\eta}{\eta_c} - 1\right)^2 \qquad \text{for} \quad \eta > \eta_c.$$

One of the nicest feature of Baxter model is, however, that one can also derive analytically the percolation loci in the (τ, η) plane. In order to derive an expression for the percolation line, it is necessary to use the pair connectedness function $P(r)$ introduced by Coniglio *et al.* (1977) to describe clustering in continuum percolation theory. Accordingly, given a particle located at the origin, $P(r)\,dr$ gives the probability of finding another particle in the spherical shell $(r, r + dr)$ and belonging to the same cluster. The pair connectedness was shown to satisfies an equation of the type of Ornstein–Zernike with a modified direct correlation function $c^+(r)$. By invoking the short-range nature of $c^+(r)$ and the sticky spheres condition, $P(r) = \lambda a \delta(r - a)/12$, Chiew and Glandt (1983) were able to show that the average cluster size is given by $s = 1/(1 - \lambda\phi)^2$. When s diverges it signals the onset of percolation and allows to determine the percolation locus

$$\frac{1}{\tau} = \frac{12(1 - \phi)^2}{19\phi^2 - 2\phi + 1}, \tag{3.58}$$

the temperature dependence of which can be obtained using the following relation between the parameter τ and the T:

$$\frac{\tau_c}{\tau} = 1 - \alpha \left(1 - \frac{T}{T_c}\right)^\gamma.$$

The result is shown in the phase diagram of Figure 3.6, and is in very good agreement with the experiments. The values $\alpha = 11$ and $\gamma = 0.94$ are derived heuristically by studying the low-Q behaviour of the form factor averaged structure factor $S(Q)$ for a system of sticky hard spheres of average diameter 100 Å and a polydispersity index $Z = 10$ at the critical volume fraction $\eta_c = 0.1213$. At sufficiently small Q, the Ornstein–Zernike functional form is obtained and one can thus extract the long-range correlation length ξ as a function of τ when approaching the critical point. By comparing ξ as a function of T and τ one can derive the functional dependence of the two variables. The fits to the experimental coexistence curve using the sticky-sphere model together with the percolation loci can then be performed and are shown in Figure 3.6, which shows a good agreement with the experimental data.

3.5 Module – Critical supramolecular systems

Critical phenomena in supramolecular systems are of primary interest in soft condensed matter, relevant examples being criticality in binary mixtures of water and non-ionic surfactant and multicomponent microemulsion systems. In particular, critical dynamics experiments of intensity correlation functions, both at short and long times, on the pseudo-binary AOT/water/decane water-in-oil microemulsion system have been analysed in great detail and are the starting point of our discussion. In the experiment we consider (Rouch et al., 1993), the critical temperature for the microemulsion, having a water to AOT molar ratio of 40.8, is $T_c = 39.85\,^\circ\text{C}$, and the critical volume fraction of the dispersed phase is $\phi_c = 0.10$. The important novelty was the use of a multiple sampling time digital corrector with 256 channels which allowed to measure, with a very good accuracy, both the short and long time behaviour of the intensity correlation functions. While the time delay between the first sixteen channels was 0.2 μs, as the delay time increases there is a progressive increase of the channel widths, and as a result, the signal to noise ratio was kept constant at both short and long times and a precise determination of the uncorrelated background to an accuracy of the order of 10^{-4} was possible. It is essential that the background be determined to this degree of accuracy before one can quantitatively assess the rather small deviation of the intensity correlation function from an exponential, one of the aims of this experiment. The temperature of the sample was kept constant with an accuracy of 1 mK for period of hours. It must be stressed that scattering by this microemulsion is very weak, even very close to the critical point, because of the good refractive index matching of the microemulsion droplets and the solvent. Therefore multiple scattering was not a limitation of the experiment.

Two scattered intensity squared correlation functions $C^2(t)$ obtained at $\Delta T = 2.53\,°C$ and $\Delta T = 0.05\,°C$ are shown in Figure 3.8 from which one can see the logarithmic spacing of the correlator channels. The background, which is not shown in the graphs, spans approximately 100 channels and was measured on a very large time interval from 100 ms up to 100 s. It was constant in this time interval and equal to 1.0000 ± 0.0004.

One of the most striking features, as far as critical phenomena in supramolecular systems are concerned, is the presence of large background effects in the relaxation rate of the fluctuations of the order parameter, when analysed in terms of standard mode-coupling theories of the transport coefficients introduced by Kawasaki (1986).

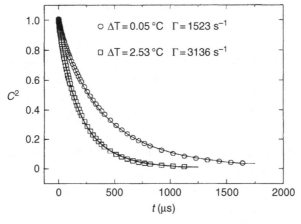

Figure 3.8 The measured squared correlation function at two temperatures, both at a scattering angle of 90°, with a logarithmic spacing of the correlator channels (Rouch *et al.*, 1993). Copyright 1993, by The American Physical Society.

A model proposed by Oxtoby and Gelbart (1974) introduced a simple modification of the mode-coupling equations, relating the order parameter fluctuations and the shear viscosity in binary systems, and obtained equations including the background line width with no adjustable parameters. It is in fact possible to account phenomenologically for these effects by introducing an upper Debye cutoff wave number $q_D^{-1} = 2R_H$, where R_H is the hydrodynamic radius of the microemulsion droplet. This approach is appropriate when the correlation function shows a single exponential decay, which is rather accurate in the case of molecular liquids. The situation is different in the case of supramolecular liquids, where the quasi-microscopic size of the aggregates provides a natural small length in the system, and this is the problem we will address next.

In Module 3.4 the same AOT microemulsion system was described, and it was shown that the behaviour of the low-frequency electrical conductivity and the droplet density correlation function, in the region of the phase diagram where the droplet volume fraction is large, are governed by a percolation phenomenon. Static and dynamic light-scattering experiments, performed in the vicinity of the well-defined percolation locus, were accurately accounted for by assuming that the microemulsion droplets are forming transient polydisperse fractal aggregates.The fractal dimension is $D = 2.5$ and the polydispersity index $\tau = 2.2$, values typical of percolation. In the high-volume-fraction and low-temperature regime, the droplet density time correlation function shows marked deviations from a single exponential decay.

It evolves continuously from a single exponential decay at short times to a stretched exponential at long times with the exponent given by $\beta = D/(D + 1)$. This behaviour is characteristic of strongly interacting systems with many degrees of freedom, involving different relaxation channels. A very important feature of the percolation loci in this system is that it extends from low temperatures, but very large volume fractions, to the vicinity of the critical point at low volume fractions. The percolation point at the critical volume fraction occurs only $0.2\,^\circ\mathrm{C}$ below the critical consolute point. Essentially for this system percolation occurs in the critical regime. This suggests that the behaviour of the microemulsion may be governed, even in the vicinity of the critical point, by clustering phenomena characterised by the indices D and τ.

The crucial point is to describe the system of aggregating units as a set of independent clusters, so that it is possible to write the structure factor of the system as the superposition of contributions from independent units. The equality between the expressions of the structure factor, in terms of units or in terms of clusters, underlines the requirement that clusters behave as independent units, i.e. with no correlation among them.

Describing the system as a collection of N_c independent clusters permits to write static and dynamic scattering properties in a straightforward way. It has been recognised for a long time that the simple definition of clusters in terms of groups of closely interacting monomers does not produce independent clusters. The issue was solved by Coniglio and Klein (1980) who showed, for the Ising model, that independent clusters can be generated by breaking clusters of nearby parallel spins into smaller ones.

Accordingly, nearby parallel spins are said to belong to the same cluster only with a probability $1 - \exp[-2\beta J]$ where J is the interaction energy. Clusters defined in this way have the remarkable property of percolating at the critical point and being characterised by thermal critical exponents. Therefore one is interested in the possibility of interpreting scattering data in terms of independent clusters, close to the critical point. In more precise words, we require a procedure to partition the N monomers in a system in N_c clusters containing N_n particles each, in such a way that

$$S(q) = \frac{1}{N} < |\sum_{i=1}^{N} e^{i\vec{q}\vec{r}_i}|^2 > = \frac{1}{N} < |\sum_{n=1}^{N_c}\sum_{i=1}^{N_u}|^2 > e^{i\vec{q}R_n + u_{nli}}, \qquad (3.59)$$

becomes equal to

$$S(q) = \frac{1}{N} < \sum_{n=1}^{N_c} |\sum_{i=1}^{N_n}|^2 >, \qquad (3.60)$$

where R_n and u_{nl} are respectively the position of the centre of mass and the relative displacement of monomer i in the n-th cluster. The equality between the two expressions underlines the requirement that clusters behave as independent units, i.e. with no correlation among them. Describing the system as a collection of N_c independent clusters allows writing static and dynamic scattering properties in a straightforward way. This effect is shown in Figure 3.9 where the structure factors, evaluated from a two-dimensional 128×128 Monte Carlo simulation of an Ising model at three selected temperatures, are interpreted using Coniglio and Klein's procedure. The intensity scattered by the collection of clusters is coincident with the intensity scattered by the whole system, considered as made up of individual interacting spins. Moreover the Lorentzian line shape of scattered light can be written as a sum of the scattering of polydisperse fractal clusters characterised by a polydispersity index τ and a fractal dimension D. The exponents τ and D are related to the thermal critical exponents of the coexistence curve β and correlation length ν by $D = d - \beta/\nu$ and $D(\tau - 1) = d$. The independent-cluster picture can

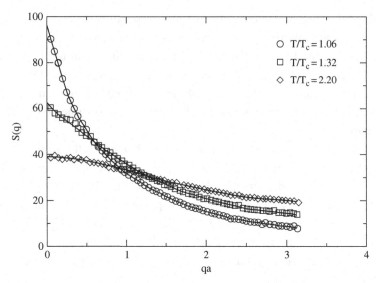

Figure 3.9 Structure factor evaluated from a two-dimensional 128×128 Monte Carlo simulation of an Ising model at three selected temperatures. The structure factor has been evaluated according to Eq. (3.59) (symbols) and Eq. (3.60) (lines), assuming that monomers are sitting in all up-spin states, defining cluster according to the Coniglio–Klein prescription. This result shows that it is possible to find a procedure to divide the spins into clusters such that the scattered intensity can be written either summing over the individual spins or summing over the independent clusters.

be generalised to the off-critical case if the asymmetry in the concentration of up and down spins is properly taken into account.

The starting point of the theoretical approach is the assumption that the slow dynamics of the droplet density fluctuation is dominated by diffusive motions of the independent percolation clusters. This assumption is expected to be good in the vicinity of the percolation threshold, where large fractal clusters are formed, and was used to explain the conductivity exponent below the percolation threshold. In the AOT/water/decane system, as one can see from the phase diagram of Figure 3.6, the critical point is very close to the percolation point. One therefore expects that the cluster structure and the cluster size distribution in the critical region are similar to those at the percolation point. When light scattering is used, the wavelength of visible light is much larger than the droplet sizes, hence for this q range, the particle form factor is nearly unity and can be ignored. In order to deduce the dynamic structure factor $S(q, t)$ of a collection of independent clusters, one first calculates the inter-particle structure factor $S_k(q)$ for a cluster containing k particles and having a radius of gyration R_k. By Fourier transforming the cluster pair correlation function

$$\rho[g_k(r) - 1] = \frac{D}{4\pi R_1^D} \frac{1}{r^{3-D}} \exp\left[-\frac{r}{R_k}\right],$$ (3.61)

one obtains

$$S_k(q) = k \frac{\sin[(D-1)\tan^{-1}(q R_k)]}{(D-1)q R_k (1 + q^2 R_k^2)^{(D-1/2)}},$$ (3.62)

where R_k, the radius of gyration of the k-cluster, is connected to the monomer radius R_1 by the relation

$$R_k = R_1 k^{1/D}.$$ (3.63)

At low q values the cluster structure factor can be approximated as

$$S_k(q) \approx k \exp\left[-\frac{1}{3}(q R_k)^2\right].$$ (3.64)

This approximation is essential in the theory since it leads to a simple analytical form for the average relaxation rate, which reduces to a well-known formula in the limit of small droplet size as will be discussed next. The normalised cluster size distribution function c_k used is borrowed from aggregation and percolation as (Stauffer and Aharony, 1992; Wang, 1989)

$$c_k = \frac{s^{\tau-2}}{\Gamma(2 - \tau, 1/s)} k^{-\tau} \exp\left[-\frac{k}{s}\right],$$ (3.65)

where $\Gamma(x, y)$ is the incomplete Euler gamma function and s the cutoff of the cluster size distribution, proportional to $k^{1/(3-\tau)}$. By assuming that the physical clusters are independent scatterers, the q-dependent scattered intensity $I(x)$ is a function of the scaling variable $x = q\xi$, with $\xi = (R_1/\sqrt{3}) s^{1/D}$

$$I(x) = \frac{\Gamma(3 - \tau, u)}{\Gamma(2 - \tau, (R_1/\xi)^D)} \left(\frac{\xi}{R_1}\right)^D (1 + x^2)^{-D(3-\tau)/2},$$ (3.66)

where $u = (R_1/\xi)^D (1 + x^2)^{D/2}$ is related to the reduced radius of the monomer. The scattered intensity turns out to be not only a function of the scaling variable x, but also of the size R_1 of the droplet. When R_1 becomes small compared to ξ this form reduces to the Ornstein–Zernike relation, since $D(3 - \tau)/2 = 1 - \eta/2$.

Turning to dynamical quantities, the averaged dynamical structure factor is then given by

$$S(q, t) = \frac{\int_1^\infty dk k c_k S_k(q) \exp\left[-D_k q^2 t\right]}{\int_1^\infty dk k c_k S_k(q)},$$ (3.67)

where D_k is the cluster diffusion coefficient

$$D_k = D_1 k^{-1/D}.$$ (3.68)

Explicitly $S(q, t)$ is a function of two reduced variables u and w

$$S(q, t) = \frac{1}{\Gamma(3 - \tau, u)} \int_u^\infty dz \, z^{2-\tau} \, e^{-z - wz^{-1/D}}, \tag{3.69}$$

where $w = D_1 R_1 q^3 t (1 + x^2)^{1/2}$ is the reduced time. This quantity also explicitly depends on the nature of the sample via both R_1 and the diffusion coefficient of the microemulsion droplet $D_1 = Rk_B T/(6\pi \eta R_1)$, where R is a fitting parameter. The normalised scattered-intensity autocorrelation function $C(q, t)$, the quantity which is experimentally measured in a dynamic light-scattering experiment, is given by

$$C(q, t) = 1 + |S(q, t)|^2. \tag{3.70}$$

Another quantity of experimental interest is the average relaxation rate $\Gamma(q)$ for the cluster diffusion is given by the first cumulant of $S(q, t)$

$$\Gamma(q) = -\lim_{t \to 0} \left[\frac{d}{dt} \ln S(q, t) \right]. \tag{3.71}$$

In order to compare in more details the predictions of this model with the classical theory of critical phenomena, one has to evaluate both the fractal dimension and the polydispersity index for the critical clusters. As far as percolation is concerned, numerical simulations gave universal values for the indices $D = 2.5$ and $\tau = 2.2$. These indices are connected by a hyperscaling relation Eq. (3.72) involving the space dimension d

$$\tau = \frac{d + D}{D}, \tag{3.72}$$

which implies $\tau = 2.2$ when taking $D = 2.5$, in agreement with the simulation result. As we mentioned before, the concentration fluctuation of the particle system can be represented by a cluster diffusion of independent clusters with a fractal dimension D given by

$$D = d - \frac{\beta}{\nu}, \tag{3.73}$$

where $\beta = 0.33$ and $\nu = 0.63$ are respectively the Ising universal exponents of the coexistence curve and the long-range correlation length, and $d = 3$ is the dimensionality of space. From Eq. (3.73) one gets $D = 2.5$ which, when introduced into the hyperscaling relation Eq. (3.72), leads to $\tau = 2.2$. These two values are equal to the fractal dimension and the polydispersity index of percolating clusters. To get more accurate estimates of D and τ, one can use a set of scaling laws known in the theory of critical phenomena to eliminate the factor β/ν on the right-hand side of Eq. (3.73). In fact in three dimensions $\beta/\nu = (1 + \eta)/2$, so that

$$D = \frac{5 - \eta}{2} \tag{3.74}$$

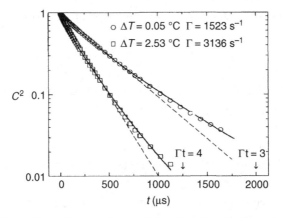

Figure 3.10 The measured squared correlation function showing deviations from exponential behaviour, represented by the dashed lines. The full lines refer to the theoretical expression of Eq. (3.69) (Rouch *et al.*, 1993). Copyright 1993 by The American Physical Society.

and the polydispersity index τ is then given by the hyperscaling relation as

$$\tau = \frac{11 - \eta}{5 - \eta}. \tag{3.75}$$

Taking the accepted value of the Fisher exponent $\eta = 0.03$, one obtains $D = 2.49$ and $\tau = 2.21$. These values of the indices will be used as input parameters when fitting experiments to theory. Figure 3.10 shows two samples of the measured $C^2(t)$ and the fits to the experimental data, spanning three decades from a delay time of $1\,\mu s$ up to more than 1 ms. The dashed lines in Figure 3.10 refer to an exponential fit. The values of the parameters are $\xi_0 = 10 \pm 2$ Å, $R_1 = 40 \pm 10$ Å and $R = 1.2 \pm 0.1$ in the temperature range studied. The renormalised critical exponent $\nu = 0.71$ was used since in reality the AOT/water/decane microemulsion is a ternary system. These values are in agreement, within the error bars with previous measurements. At sufficiently short time $\Gamma t \ll 1$, Eq. (3.69) can be analytically integrated and leads to an exponential decay for the photon correlation function

$$S(q, t \to 0) = 1 + \exp\left[-2\Gamma(q)t\right], \tag{3.76}$$

where Γ is the first cumulant defined above. At long times $\Gamma t \gg 1$, the integral can also be evaluated analytically by using the steepest-descent method. In this regime, the photon correlation function asymptotically approaches a stretched exponential form

$$S(q, t \ll 1) = 1 + \exp\left[-2(\bar{\Gamma}t)^\beta\right], \tag{3.77}$$

where the exponent $\beta = D/(D+1)$ is the universal number 0.713 and $\bar{\Gamma}$ is given by

$$\bar{\Gamma} = \beta^{1/\beta} D^{1/D} D_1 R_1 q^3 \left(1 + \frac{1}{x^2}\right)^{1/2}. \tag{3.78}$$

The theoretical model used for describing $C^2(t)$ accounts well for the small but sizeable deviation from a single exponential decay observed experimentally in the long time limit $\Gamma t > 2$.

From the preceding equations one can calculate specific quantities experimentally accessible through light-scattering experiments. The relaxation rate $\Gamma(q)$ can be cast in a universal form, depending only on a scaling parameter x and the non-dimensional parameter $x_1 = q R_1$ representing the lower limit of the cluster sizes. From the above definitions, one can express the reduced first cumulant $\Gamma^*(x, x_1)$ in terms of the two non-dimensional parameters by

$$\Gamma^*(x, x_1) = \frac{\Gamma(q)}{D_1 R_1 q^3}. \tag{3.79}$$

and is explicitly given by

$$\Gamma^* = \frac{3\pi}{8} \frac{\Gamma(3 - \tau, x_1^D)\Gamma(3 - \tau - 1/D, u)}{\Gamma(3 - \tau - 1/D, x_1^D)\Gamma(3 - \tau, u)} \left(1 + \frac{1}{x^2}\right)^{1/2}, \tag{3.80}$$

which depends on the droplet size. Very close to the critical point where ξ is much larger than R_1, the reduced first cumulant becomes a universal function of $x = q\xi$. It reduces exactly to the Perl and Ferrel (1972) mode-decoupling result when $R_1 = 0$, and in this case it is also numerically very close to the Kawasaki (1986) formula

$$\lim_{x_1 \to 0} \Gamma^*(x, x_1) = \frac{K(x)}{x^3}, \tag{3.81}$$

where $K(x)$ is the dynamic scaling function

$$K(x) = \frac{3}{4} \left[1 + x^2 + \left(x^3 - \frac{1}{x}\right) \tan^{-1}(x)\right], \tag{3.82}$$

which is known to account very well for light-scattering data near critical points of one-component fluids or binary mixtures of molecular liquids. $\Gamma^*(x, x_1 = 0)$ has the simple asymptotic behaviour

$$\Gamma^*(x, x_1 = 0) = \begin{cases} \frac{a}{x} & x \ll 1, \\ b & x \gg 1, \end{cases} \tag{3.83}$$

where a and b are known constants. The reduced line width, Eq. (3.80), is plotted as a function of the scaling variable x in Figure 3.11. Since it explicitly depends on the size of the droplets, it is plotted for different values of the parameter x_1. The lines, which are the theoretical predictions, correspond to different values of x_1, the one

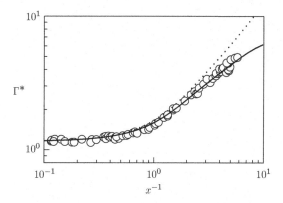

Figure 3.11 Reduced first cumulant of the measured correlation function of the AOT/water/decane system (Rouch *et al.*, 1993), compared with the theoretical expression given by Eq. (3.80). The dashed and solid lines correspond to $x_1 = 0$ and $x_1 = 0.07$ respectively. Symbols are measurements taken at 90° scattering angle and varying temperatures, ranging from 0.01 to 10 K from the critical point. Copyright 1993 by The American Physical Society.

for $x_1 = 0$ being very close to Kawasaki's result. A very good agreement between the theoretical model we propose and the experiment is obtained in the critical regime, the hydrodynamic regime, and even in the crossover regime corresponding to temperatures 10 °C away from the critical point. This can be achieved with the same values of ξ_0 and R_1 reported above. We remark that in previous analyses of critical phenomena in supramolecular aggregates the relaxation rate Γ is taken to be the sum of a critical part, which can be calculated using the mode-coupling theory, and a system-dependent background part which depends explicitly on the Debye cutoff parameter q_D. The experiment showed that the dynamical background was roughly proportional to the size of the aggregates, leading to a correction of the order of 10% of the total line width when close to the critical point. In the theoretical model the dynamical background term enters in a natural way into the theory and is physically provided by the lower bound of the integral defining the correlation function. However, the price one has to pay is the introduction of the multiplicative constant $R = 1.2 \pm 0.1$ in the definition of D_1, in a way similar to the renormalisation group approach to critical phenomena.

Part II
Structural arrest

4

The theory of slow dynamics in supercooled colloidal solutions

4.1 Introduction

In 1995, while discussing the future of science, P. W. Anderson made the following statement on the unsolved problems of physics:

The deepest and most interesting unsolved problem in solid state theory is probably the theory of the nature of glass and the glass transition. This could be the next breakthrough in the coming decade. (Anderson, 1995)

In recent years substantial progress has been made in the understanding of the glass transition, both experimentally and theoretically, but from the point of view of supercooled liquids, and in particular liquids of supramolecular aggregates, more than from the point of view of solid state physics. The general name given to these phenomena, commonly found in soft condensed matter, is dynamically disordered arrested phenomena. Examples belonging to this category are the liquid–glass, the sol–gel and the percolation transitions. The reason for this variety of behaviours is related to the large space and time scales of the aggregates, the various shapes the molecular aggregates can take and the corresponding specific interaction potentials, and the possibility of changing a large number of control parameters both at the level of the inter-particle potential and the state of the system. Moreover the length and time scales involved allow the use of powerful experimental techniques, such as neutron and light scattering and confocal microscopy, the latter allowing tracking of the trajectories of individual particles. Examples of systems where dynamical arrest was observed include thermoreversible physical gels, associating polymers at low concentration, micellar systems, star-polymer mixtures, colloidal gels and block-copolymers. All these colloidal systems show the typical slowing down of the dynamics leading to the arrest, with a mechanism which is in general very sensitive to changes in the external control parameters, and which persists over time scales that might be even longer than the typical observation time of the experiment. The general features of the slowed dynamics in different systems show many similarities suggesting the possibility of an underlying common mechanism with some degree of universality, even if the interactions can be rather different in the various systems and the final states show different physical properties. In other words, the dynamics becomes so slow that during the experimental observation time the system evolution is frozen in a structure which depends on the previous history of the sample (Sciortino and Tartaglia, 2005).

A pioneering set of experiments on the possibility of a glass transition in colloidal systems was performed by Pusey and van Megen (1986), who studied suspensions of spherical colloidal particles in a liquid and were able to determine various regions of the phase diagram (Figure 4.1). The system studied was realised through colloidal polymethylmethacrilate (PMMA), the particles being spherical with a core of (305 ± 10) nm coated with a steric stabiliser layer of poly-12-hydroxystearic acid of $(10-20)$ nm, and interacting through a steep repulsive potential due to the interpenetration of the polymer coatings. The particles were suspended in a medium which was a mixture of decalin and carbon disulphide, in the volume ratio 2.66:1, matching the index of refraction of the particles, which is 1.51. They behaved to a very good approximation as hard spheres and actually PMMA became one of the standard colloids to use in studies of the glass transition. The hard sphere is a model system for studying the structure and dynamics of liquids, crystals and glasses and the thermodynamic transitions between them, and it is often used as reference model for evaluating properties of more sophisticated and realistic potentials. After position randomisation the samples were allowed to

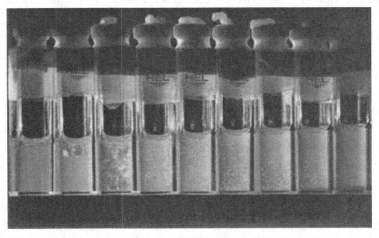

Figure 4.1 The PMMA samples of colloidal fluids, crystal and glass, moving from left to right (Pusey and van Megen, 1986). Reprinted by permission from Macmillan Publishers Ltd: Nature, copyright 1986.

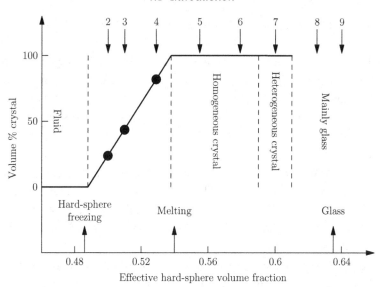

Figure 4.2 The empirical phase diagram for PMMA, the numbers refer to the numbering in Figure 4.1 (Pusey and van Megen, 1986). Reprinted by permission from Macmillan Publishers Ltd: Nature, copyright 1986.

evolve over a long period of time. With increasing particle concentration a progression was observed from colloidal fluid, to fluid and crystal phases in coexistence, to fully crystallised samples. At high concentrations very viscous samples were obtained in which full crystallisation did not occur for long times, and these were identified as amorphous colloidal glass. The empirical PMMA phase diagram is shown in Figure 4.2.

The phase diagram of hard spheres of uniform size depends only on the packing fraction ϕ, and computer simulation studies provided evidence that the hard-sphere fluid, despite the absence of any attractive part in the potential, starts crystallising when $\phi > 0.49$, while crystal and fluid coexist up to $\phi = 0.545$, and above this value the crystal phase is stable.

The experimental study of the crystallisation process in hard-sphere colloidal dispersions is an active field of research, revitalised by experiments in microgravity in order to eliminate sedimentation, convection and stress due to gravity. Interestingly enough, experiments have shown that crystallites grow faster and larger in microgravity and the coarsening between crystallites is suppressed by gravity. Particularly important appears the experimental evidence that crystallisation proceeds even for $\phi > 0.58$, a value for which, in normal gravity, no signs of crystal formations are observed even for extremely long observation times. Computer simulation studies confirm the role of gravity in affecting crystallisation, suggesting that gravity stabilises the glass state by reducing the mobility of the particles.

When hard-sphere configurations are generated at a high packing fraction, $\phi \geq 0.58$, both experimentally by shear melting of the crystal or numerically by compression algorithms, the relaxation time of the system becomes much longer than the experimentally available time and the system behaves as a non-ergodic system. In particular, in experiments on Earth, no signs of crystallisation have been observed in the range $0.58 < \phi < 0.64$ (except for an extremely slow ageing) leading to the hypothesis that, even for a one-component monodisperse hard-sphere system, a glass state can be defined. The comparison with the zero gravity experiment confirms that these glassy states are metastable, as compared to the crystal state, and are kinetically stabilised by a growth of the crystallisation time scale induced by gravitational effects.

From the point of view of the theoretical interpretation of the phenomena associated with the glass transition, different pathways have been followed. The first starts from the traditional theory of liquids with non-linear interactions among hydrodynamical modes added to the standard dynamical equations. The non-linear interactions were introduced in analogy to the case of critical phenomena in liquid systems, where dynamical anomalies are related to the non-linear coupling of the order parameter and hydrodynamic modes. This approach was developed over many years under the name of mode-coupling theory (MCT), and has determined the interpretation of the glass transition as related to an ideal kinetic transition, predicted by the theory but not observed in experiments (Götze, 2009). The salient mathematical features of the theory will be reviewed in the following sections.

Another line of research, again based on the theory of liquids, is the inherent structures theory of Stillinger and Weber (1984), which was widely studied, particularly using computer simulations. In this case the metastable states typical of the glass transition are studied directly in the potential energy space, the system being typically trapped in a metastable state. The latter approach allowed a sound physical understanding of glassy behaviour from the point of view of the classical theory of liquids.

A different approach originates from the successful replica approach to the physics of spin glasses, which originated from the mean field solution of the Sherrington–Kirkpatrick model proposed by Parisi (1980). This approach was extended more recently to structural glasses (Mézard and Parisi, 1998), where disorder is self-generated, as opposed to spin glasses where the system is dominated by quenched disorder. The theory developed is somewhat closer to the liquid approach and allows a qualitative unified physical picture of glass transition phenomena.

More recently theoretical research on the physics of glasses provided new insights into the problem; however, in the following sections, we limit ourselves to various types of experiments and numerical simulations in supramolecular liquid systems where the results were interpreted in terms of the MCT, which is capable of predicting in a detailed fashion many aspects of the behaviour of supercooled liquids in the vicinity of a glass transition.

In Section 4.2 we will summarise the relevant mathematics and salient features of MCT, which is to date the most successful theoretical effort to study the behaviour of supercooled liquids, and interpret the experimental and the simulation results. From a more general point of view there have been many efforts to go beyond MCT and identify its strength and weakness, but the theory maintains its validity and produces quantitative predictions that can be accurately compared to experimental results.

Section 4.3 treats first and briefly the prototypical cases of hard-sphere systems, where many experiments and simulations have been performed. A very interesting new phenomena was predicted when short-range attraction is added to repulsion, theoretically interpreted in terms of the higher-order singularities predicted by MCT. The predictions were later confirmed both experimentally and numerically.

Section 4.4 reports the effects of long-range repulsion on a typical colloid, with short-range repulsive and attractive interactions. The influence on the dynamical arrest introduces interesting new concepts, such as microphase separation, also referred to as the formation of patterned or modulated phases, or cluster phases.

4.2 Module – The mode-coupling theory of supercooled liquids

The approach to glass transition in colloidal systems that is widely considered the most successful is mode-coupling theory (MCT), dating back to the mid-1980s, which has been developed since then mainly by Götze (1991) and coworkers. MCT was shown to be capable to interpret in a quantitative way, to a 20% level of accuracy, experimental data close to a supercooled liquid–glass transition. The name 'mode coupling' was borrowed by the successful theories of dynamical critical phenomena, since in analogy with the latter the relevant variable is identified with the local density, and a non-linear (quadratic) coupling among density modes is explicitly considered in the kinetic evolution equations. In MCT the input static quantity is the wave vector k-dependent static structure factor S_k, but contrary to critical phenomena, close to the glass transition there is no static singularity leading to a diverging correlation length and to the Ornstein–Zernike anomaly of the structure factor. The added difficulty of the theory, compared to critical phenomena, is related to the fact that all length scales are equally important and have to be taken into account simultaneously. MCT predicts only a kinetic singularity in the evolution equations, which leads to an ergodic to non-ergodic transition, characterised by the non-vanishing of the long time limit of the density correlation functions.

The original derivation of MCT makes use of the projection operators technique of statistical mechanics, but other procedures have been used by many authors in order to derive the equations. Consider N particles of mass m, in the cubic volume V, with coordinates $\{\mathbf{r}^N\} = \{\mathbf{r}_1, \mathbf{r}_2, \ldots, \mathbf{r}_N\}$. The local density is defined as

$$\rho(\mathbf{r}) = \sum_{j=1}^{N} \delta(\mathbf{r} - \mathbf{r}_j),$$

and gives the number density $n = \langle \rho(\mathbf{r}) \rangle$. Its spatial Fourier transform with wave vector k is

$$\rho_k = \int_V d\mathbf{r} e^{i\mathbf{k}\cdot\mathbf{r}} \rho(\mathbf{r}) = \sum_{j=1}^{N} e^{i\mathbf{k}\cdot\mathbf{r}_j}.$$

MCT derives equations for the normalised time-dependent density correlators of the Fourier components of the particle density deviations $\delta\rho_k(t)$,

$$\Phi_k(t) \equiv \langle \delta\rho_k(t)\delta\rho_{-k}(0)\rangle / \langle |\delta\rho_k|^2\rangle,$$

starting only from the number density n and the structure factor $S_k = \langle |\delta\rho_k|^2\rangle / N$.

For Newtonian dynamics, the evolution equations read

$$\partial^2/\partial t^2 \Phi_k(t) + \omega_k^2 \, \Phi_k(t) + \int_0^t dt' \, M_k(t-t')\partial/\partial t' \Phi_k(t') = 0,$$

with initial conditions $\Phi_k(0) = 1$ and $\partial\Phi_k(0)/\partial t = 0$. Here $\omega_k^2 \equiv k^2/(m\beta S_k)$ are characteristic frequencies, with $\beta = 1/(k_B T)$ where T is the temperature and k_B the Boltzmann constant. The kernels $M_k(t)$ are expressed in terms of correlators of the fluctuating forces, which in colloids include the effect of the interactions with the solvent particles. The characteristic time scale of these interactions is much shorter than the time scale of the colloidal particles dynamics, therefore they can be approximated as a friction force of the form $\nu_k \partial \Phi_k(t)/\partial t$, with ν_k microscopic damping coefficients. In the case of colloidal systems the propagation of the density fluctuations is controlled by the Smoluchovski equation. In this case the friction term is large compared to the inertia term $\partial^2 \Phi_k(t)/\partial t^2$, which can be neglected and implies Brownian rather than Newtonian dynamics. The MCT equations for colloids are then of first rather than second order,

$$\nu_k \partial/\partial t \Phi_k(t) + \omega_k^2 \, \Phi_k(t) + \int_0^t dt' \, M_k(t-t')\partial/\partial t' \Phi_k(t') = 0, \qquad (4.1)$$

with the initial conditions $\Phi_k(t) = 1$. The ν_k term goes to a constant in the limit of k going to zero, differently from the k^2 behaviour characteristic of Newtonian dynamics. Within MCT, all features of the short-time dynamics enter the glassy dynamics only via an overall time scale t_0. The complete dependence of the glassy dynamics on k and from the distance from the glass line for Newtonian and stochastic dynamics is identical, except for the choice of t_0. Therefore, according to MCT it does not matter whether glassy dynamics is analysed in simulations for Newtonian or for stochastic dynamics.

The kernels $M_k(t)$ are decomposed as the sum of a regular term $M_k^{reg}(t)$, which contains normal effects in colloids such as hydrodynamic interactions, and mode-coupling contributions of the form $\omega_k^2 m_k(t)$. In the MCT calculations we will report, hydrodynamic interactions are not explicitly taken into account. In the limit of quadratic non-linearities, the memory kernel is given by

$$m_k(t) = n/V \sum_{k' \neq k} S_k S_{|k-k'|} S_{k'} \left| \mathbf{k} \cdot \mathbf{k}'/k^2 \, c_{k'} + \mathbf{k} \cdot (\mathbf{k} - \mathbf{k}')/k^2 \, c_{k-k'} \right|^2$$

$$\times \, \Phi_{|k-k'|}(t) \Phi_{k'}(t) \tag{4.2}$$

in terms of the structure factor S_k or the short-range correlation function $nc_k = 1 - S_k^{-1}$.

The glass transition predicted by MCT is obtained solving the $t \to \infty$ limit of the equations for the normalised correlators $\phi_k(t)$, the so-called non-ergodicity factors f_k

$$f_k = \lim_{t \to \infty} \Phi_k(t),$$

which have the form

$$\lim_{t \to \infty} m_k(t) = f_k/1 - f_k.$$

The solution to these equations admits not only the usual trivial solution $f_k = 0$, but also solutions with $f_k \neq 0$. The value of f_k at the transition point is denoted f_k^c. In MCT language, the transition is called a type B transition when f_k grows discontinuously on entering in the non-ergodic phase, and type A transition when f_k grows continuously from zero. The non-ergodic solutions $f_k \neq 0$ depend on a number of control parameters, usually volume fraction of the colloidal particles and attraction strength. The theory accounts also for so-called higher-order singularities related to bifurcation theory (named A_3, A_4), whose realisation requires a fine tuning of the inter-particle potential parameters. The A_3 bifurcation is the end point of a type B transition.

The existence of a singularity of purely kinetic origin – not related to any thermodynamic singularity, such as in the case of non-linear critical dynamics – is the most important prediction of the theory. The physical interpretation of the non-ergodicity transition is related to the well-known cage effect, the difficulty of a particle to move due to the crowd of the surrounding ones. Motion of the particle can only take place if a collective rearrangement of the particles forming the cage opens up a passage for the arrested particles.

MCT density correlators

One of the merits of MCT is to identify the universal features of the temporal decay of density correlators in terms of asymptotic power laws of approach to the ideal glass transition. In order to properly define the asymptotic laws it is first necessary to define a parameter ϵ which quantifies, in terms of a control parameter x, the distance from the kinetic transition at x_c,

$$\epsilon = x - x_c/x_c,$$

and it is usually related to the fractional distance from the transition expressed in terms of volume fraction ϕ ($x = \phi$) or temperature ($x = 1/T$). A convenient way of describing the universal characteristics of the decay is to introduce its relevant time scales and the behaviour of the time correlators in the various time ranges, namely:

(i) The region of the microscopic decay related to a short time scale t_0 of the order of the Brownian time scale in colloidal systems and of the order of the inverse phonon relaxation frequency in Newtonian systems.

(ii) The β-relaxation region corresponding to the decay towards a plateau f_k^c and the further decay below the plateau when $\epsilon < 0$, while there is ergodicity breaking for $\epsilon > 0$. In the vicinity of the plateau MCT proposes a general expression for the density correlators of the form

$$\Phi_k(t) = f_k^c + h_k\sqrt{\epsilon}g_\pm(t/t_\epsilon),$$

where the subscript in $g_\pm$ corresponds to the sign of ϵ, called the factorisation property, since the space and time dependencies separate, and scales with the characteristic time t_ϵ,

$$t_\epsilon = t_0|\epsilon|^{-1/(2a)},$$

which in turn scales with an exponent related to the quantity a. The non-ergodicity factor f_k^c, the critical amplitude h_k and the β-correlator $g_\pm$, are independent from ϵ. Given a particular system, $g_\pm$ is a function which can be determined knowing n and the inter-particle potential, which determines the structure factor S_k. The leading behaviour of the function $g_\pm(t/t_\epsilon)$ above or below the non-ergodicity plateau is given by the power law in time valid for $t/t_\epsilon \ll 1$,

$$g_\pm(t/t_\epsilon) \approx (t/t_\epsilon)^{-a},$$

with $0 < a \le 1/2$, while for $t/t_\epsilon \gg 1$ the well-known von Schweidler law is valid, with

$$g_-(t/t_\epsilon) \approx -(t/t_\epsilon)^b,$$

and $0 < b \le 1$. It is possible to relate both exponents a (for t_ϵ and $g_\pm$) and b through the following relation involving the Euler Γ function:

$$[\Gamma(1-a)]^2/\Gamma(1-2a) = [\Gamma(1+b)]^2/\Gamma(1+2b).$$

As for the case of critical phenomena, exponents and critical laws are expected to be robust with respect to changes of system, while actual value of the kinetic transition is significantly affected by the approximations employed in MCT.

(iii) The α-decay regime is the last stage of the decay, the cage break-up, and is characterised by a time scale τ which diverges on approaching the transition as a power law

$$\tau \approx |\epsilon|^{-\gamma}, \tag{4.3}$$

where the exponent γ is given by

$$\gamma = 1/2a + 1/2b.$$

The time scale τ enters the so-called time–temperature superposition relation

$$\Phi_k(t) = F_k(t/\tau),$$

where $F_k(t/\tau)$ is a master function of the scaled time t/τ which allows us to draw a master plot of the density correlators. A good approximation of this function is very often given by the stretched exponential function

$$\phi_k(t) = A_k \exp\left[-(t/\tau_k)^{\beta_k}\right].$$

A sketch of the typical shape of the density correlators, highlighting the different time regions described by MCT, is shown in Figure 4.3.

Many aspects of the predictions of the ideal MCT have been tested in detail in various systems, both experimentally and using computer simulation, with good results. More recently MCT showed also a relevant predictive power in the important field of glassy colloidal systems dominated by attractive inter-particle interactions, where a number of interesting new phenomena have been discovered and will be discussed at length in the next sections.

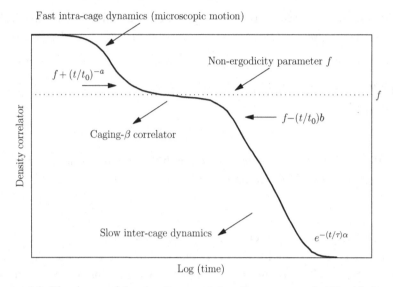

Figure 4.3 The decay of the density correlators in a supercooled liquid close to the glass transition. The initial microscopic non-universal intra-cage dynamics is followed by a time region where caging dynamics are observed (power laws). This region is followed by the slow inter-cages dynamics (stretched exponential function).

The study of the slow dynamics in colloidal systems has benefitted a lot from the possibility of a close comparison between theoretical predictions and experimental or numerical findings. Nevertheless the limit of validity of MCT remains controversial. Depending on the material, the location of the MCT glass line (which retains only the meaning of crossover from a power-law growth of the α-relaxation time to a super-Arrhenius dependence) can be very different from the location of the line at which arrest is observed on an experimental time scale (the calorimetric glass transition temperature).

The neglect of activated processes, especially in the most used formulation of MCT, which does not include them, is another strong objection to the theory. In fact, ignoring activated processes is probably a safe approximation only in the hard-sphere case and when excluded volume is the main driving force for caging. Activated processes may not be neglected when the attractive part of the interparticle potential plays a significant role in the caging process. Indeed, in the case of molecular (and apparently also network forming) liquids the ideal MCT predictions for the α relaxation properly describe only the first three to four orders of magnitude in the slowing down of the dynamics. Many attempts have been made to improve the ideal MCT, including the early extensions of the theory due to Götze and Sjögren (1987) to account for activated hopping processes. This is obtained through a coupling to current modes, besides the density modes, which destroy the ergodic–non-ergodic transition below the mode-coupling temperature. Unfortunately the full form of the extended MCT, which formally includes activated processes, cannot be compared to experimental or numerical data, except in a schematic version which neglects the wave vector dependence.

Extended MCT

More recently an extended version of MCT for glass transition was developed by Chong (2008). The activated hopping processes are incorporated via a dynamical theory, originally formulated to describe diffusion-jump processes in crystals and adapted in this work to glass-forming liquids. The dynamical-theory approach treats hopping as arising from vibrational fluctuations in the quasi-arrested state where particles are trapped inside their cages, and the hopping rate is formulated in terms of the Debye–Waller factors characterising the structure of the quasi-arrested state. The resulting expression for the hopping rate takes an activated form, and the barrier height for the hopping is *self-generated* in the sense that it is present only in those states where the dynamics exhibits a well-defined plateau and can be incorporated to develop an extended MCT (Chong, 2008; Chong *et al.*, 2009).

The latter is described by the same Eq. (4.1) using the Laplace transform of the original memory kernel $m_k(t) = m_k^{MCT}(t)$:

$$m_k(z) = i \int_0^\infty dt \ e^{izt} \ m_k(t), \quad (\text{Im } z > 0), \tag{4.4}$$

which is substituted by

$$m_k(z) = \frac{m_k^{MCT}(z)}{1 - \delta_k(z) \ m_k^{MCT}(z)}, \tag{4.5}$$

with an additional kernel $\delta_k(z)$, called the hopping kernel, which takes into account the effect from hopping processes. The Laplace transform $K_k(z)$ of the longitudinal current correlator can be used to write the density correlator as

$$\Phi_k(z) = -\frac{1}{z + K_k(z)}. \tag{4.6}$$

After some approximation one gets

$$K_k(z) = \delta_k(z) + K_k^{MCT}(z), \tag{4.7}$$

$$K_k^{MCT}(z) = -\frac{\omega_k^2/v_k}{z + m_k^{MCT}(z) \ \omega_k^2/v_k}. \tag{4.8}$$

which allows the following transparent physical interpretation:

(i) Cage-dominated regime. When $\delta_k(z) \sim 0$ then $K_k(z) \sim K_k^{MCT}(z)$ and the equations reduce to the usual MCT ones. On approaching the critical MCT temperature from above, $m_k^{MCT}(z)$ becomes larger due to the cage effect and the current correlator $K_k(z)$ vanishes at the dynamical temperature, leading to the sharp non-ergodic transition.

(ii) Hopping-dominated regime. In the presence of $\delta_k(z)$, on the other hand, the term $K_k^{MCT}(z)$ becomes negligible when $m_k^{MCT}(z)$ becomes large and there $K_k(z) \sim \delta_k(z)$. The hopping kernel takes over and hinders the currents from vanishing, thereby preventing the density fluctuations from becoming arrested completely.

The hopping processes are incorporated through $\delta_k(z)$ via a dynamical theory originally formulated to describe diffusion jump processes in crystals. The latter treats hopping as arising from vibrational fluctuations in the quasi-arrested state where particles are trapped inside their cages. The collective hopping kernel $\delta_k(z)$ is approximate using only of the self part $\delta_k^s(z)$, neglecting the distinct part describing possible correlated jumps, i.e. $\delta_k(z) = \delta_k^s(z)/S_k$. The self part of the hopping kernel is derived as

$$\delta_k^s(z) = i \, w_{hop} N_c \, \frac{1 - \sin(ka)/ka}{f_k^s}, \tag{4.9}$$

with w_{hop} a hopping rate depending on the state parameters of the system:

$$w_{hop} = \frac{1}{2\pi} \left(\frac{3}{5} \right)^{1/2} \omega_D \exp \left[-\frac{3mv_l^2 \Delta^2}{2k_B T} \right], \tag{4.10}$$

where v_L, the sound velocity in the quasi-arrested state and ω_D, the Debye frequency, are given by

$$v_L = \sqrt{\frac{k_B T}{m S_0 (1 - f_0)}}, \tag{4.11}$$

$$\omega_D = (6\pi^2 \rho)^{1/3} v_L. \tag{4.12}$$

In Eq. (4.12), Δ is a dimensionless measure of the critical size of the phonon-assisted fluctuation needed to cause a hopping. The details are given in Chong (2008) and Chong *et al.* (2009).

In the case of a Lennard-Jones system, with the structure factor evaluated within the Percus–Yevick approximation, the density correlators $\Phi_k(t)$ of the extended theory are reported in Figure 4.4 together with their MCT counterparts. The figure clearly shows in the case of MCT the existence of a plateau due to the ergodicity breaking and the appearance of the dynamical critical point. In the extended theory instead the plateau disappears in favour of a long time decay related to hopping phenomena.

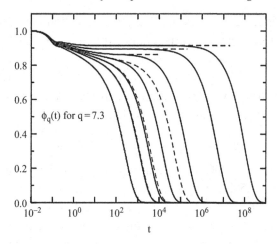

Figure 4.4 Density correlators $\Phi_k(t)$ at the peak position of the static structure factor for temperatures $T = T_c(1 - \epsilon)$ with $T_c = 1.637$ and $\epsilon = -0.10, -0.05, -0.03, -0.01, +0.01, +0.05$ and $+0.10$ (from left to right). The solid curves denote the results from the extended MCT. The dashed curves refer to the results from the idealised MCT, which exhibit the bifurcation of the long time limit above and below the critical dynamical temperature (Chong *et al.*, 2009). Copyright IOP Publishing. Reproduced with permission. All rights reserved.

4.3 Module – MCT for repulsive and attractive glasses

Hard-sphere systems

The pure hard-sphere (HS) system is one of the first being studied by MCT, the inter-particle potential depending exclusively on the particle size and the corresponding macroscopic control parameter being the volume fraction of the dispersed phase. Experimentally the transition from a supercooled liquid to a glass has been detected at $\phi_c = 0.58$, while MCT predicts a transition for $\phi_c = 0.516$. This is one of the drawback of MCT, i.e. a systematic underestimate of the location of the transition point. Obviously all the universal features of the transition described above are present in this system. Using as input the static structure factor S_k calculated in the Percus–Yevick (PY) approximation, MCT predicts close to the ideal transition at ϕ_c a relaxation time scaling as $|\phi - \phi_c|^{-\gamma}$ with $\gamma = 2.52$. A first test of exponents and scaling laws for HS was done by Götze and Sjögren (1987) based on the pioneering work of Pusey and van Megen described in Section 4.1. Exponents and critical laws are expected to have almost universal values, while the actual value of the transition can be significantly affected by the approximations used to derive MCT. Improvement on the S_k used as input to the theory (going from PY to Verlet–Weiss S_k or to the S_k calculated in computer simulations of the HS model) does shift the estimate of ϕ_c up to 0.546. In the spirit of MCT, the ideal glass transition is a kinetic transition, without an underlying thermodynamic transition. Extending the theory to include hopping effects, could explain the residual mobility beyond $\phi = 0.58$, in agreement with the observed extremely slow decay of correlation functions above ϕ_c and the crystallisation observed in the zero gravity experiments.

Binary HS systems are widely used, both experimentally and in simulations, to prevent the formation of crystalline solids. They have been analysed theoretically using MCT by Götze and Voigtmann (2003). Extending MCT to binary HS, the theory predicts two different scenarios. The speed-up of the dynamics reported by Williams and van Megen (2001) was found only for sufficiently large size disparity, say a size ratio $\delta < 0.65$. For $\delta > 0.8$, the opposite effect was predicted, i.e. mixing slows down the dynamics and the ideal-glass critical packing fractions decreases. One possible mechanism to avoid crystallisation is provided by disorder in the particle sizes, inducing concomitant disorder in the inter-particle potential. Above a certain polydispersity of about 7% a crystalline phase does not usually exist, while slowing down of the dynamics is certainly observed. In this respect, the glass dynamics does not require metastability with respect to a crystal phase.

There is also an increase of the height of the plateau of the density autocorrelation functions for small and intermediate wave vectors, reflecting a stiffening of the nearly arrested glass structure. These findings, which pose a challenge to theories of the glass transition, show, in particular, that the description of a glass-forming mixture by an effective one-component liquid cannot be possible for all properties of interest.

Hard spheres with attraction

In most cases, colloid–colloid interactions are characterised not only by the hard-core excluded volume interaction, but also by an additional one, arising from the polarisability of the particle and from the properties of the particle surface. The strength and range of these interactions is modulated by the solvent properties (temperature, salt concentration, pH, solvent composition). The effective colloid–colloid interaction can also be controlled by the addition of a third component (beside colloid and solvent), which acts as depletant agent. Often, the range of the interaction is significantly smaller than the particle size, giving rise to inter-particle effective potentials that have no counterpart in atomic or molecular systems.

An interesting case is generated when the hard-core potential is complemented by an attractive potential, the range of interaction of which is significantly smaller than the colloid size. The addition of an attractive component introduces the possibility of a separation into coexisting colloid-rich and colloid-poor phases, when the interaction strength overcomes a critical value. The separation of the homogeneous solution into two coexisting phases is the analogue of the liquid–gas coexistence in a one-component atomic or molecular system. Contrary to the atomic or molecular

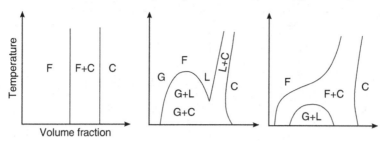

Figure 4.5 Phase diagram of colloidal particles. The first panel refers to hard spheres, with freezing at $\phi_f = 0.494$ and melting at $\phi_m = 0.545$. The middle panel shows the appearance of the liquid–gas coexistence and the associated critical point, when an attractive potential is added to repulsion. The last panel shows the disappearance of the liquid phase, which becomes metastable when the range of the interaction becomes sufficiently short range. Phases are labelled as fluid (F), crystal (C), gas (G) and liquid (L) (Anderson and Lekkerkerker, 2002). Reprinted by permission from Macmillan Publishers Ltd: Nature, copyright 2002.

liquid cases, for which the interaction range is always comparable to the particle size, in the case of very short-range potentials the two-phase coexistence is located on the metastable extension of the fluid free energy and technically no triple point (crystal, liquid, gas coexistence) exists. Coexistence of fluid and crystal phases is the equilibrium state. Indeed, it is found that the critical interaction strength (i.e. the strength at which a second-order critical point of coexisting colloid-rich and colloid-poor phases is found) decreases on increasing the range of interaction. When the interaction range becomes approximatively 20% of the colloid diameter, the critical point emerges into the stable fluid phase and a proper equivalent of the equilibrium liquid phase can be defined. A pictorial representation of the evolution of the phase diagram on increasing the range of interaction is shown in Figure 4.5 (Anderson and Lekkerkerker, 2002).

The addition of an attractive part to the hard-sphere potential brings in the possibility that slowing down of the dynamics can be driven not only by the increase in packing (the ϕ-route), but also by an increased role of the attraction between different particles (the T-route).

The first theoretical analysis of the role of a short-range attractive potential in the dynamics of colloidal suspension was presented in two MCT studies by Fabbian *et al.* (1999) and Bergenholtz and Fuchs (1999b). These studies showed that the arrest of density fluctuations can either be dominated by repulsion of the particle by its cage-forming neighbours (as in HS systems), but also by the formation of bonds (energetic cages) between particles. The original calculations, based upon Baxter's adhesive hard-sphere model (with a suitable cut-off at large wave vectors to account for the unphysical divergence due to excitations with large wave vectors), were

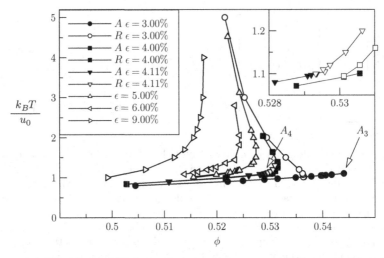

Figure 4.6 The ideal MCT phase diagram of the square-well system. The phase diagram shows cuts through the control parameter space, volume fraction ϕ and ratio of temperature to well depth $k_B T/u_0$, for fixed relative attraction-well width $\epsilon = \Delta/(\sigma + \Delta)$ ranging from 3% to 9%. The re-entrance of the glass lines appears only for $\epsilon < 0.09$ and extends the range of existence of the liquid phase. Below $\epsilon \approx 0.04$ the glass–glass line appears and increases its length as ϵ decreases. For these small values of ϵ two different solutions are found, commonly named the repulsive and the attractive one. The inset shows the location of the A_4 higher-order singularities. A_3 is the end point of the attractive line, while A_4 marks the location where the repulsive and the attractive lines join continuously, and it is the only higher-order singularity accessible from the equilibrium liquid (Dawson *et al.*, 2000). Copyright 2000 by The American Physical Society.

confirmed by more refined calculations based on the square-well model (Dawson *et al.*, 2000). The interactive square-well (SW) potential $V(r)$ for particles with separation distance r consists of a hard-core repulsion for $r < \sigma$, and it has the negative value $-u_0$ within the attraction shell $\sigma < r < \sigma + \Delta$. The SW kinetic MCT phase diagram is reproduced in Figure 4.6. The new theoretical predictions resulting from these analyses contained the following features:

(i) The possibility of melting a hard-sphere glass by progressively switching on an attractive interaction with a range smaller than about one tenth of the particle diameter. In the case where attraction is induced by depletion mechanisms, theory predicts a melting of a glass by further addition of a third component.

(ii) The vitrification of the melted hard-sphere glass upon further increase in the attraction strength.

(iii) The possibility of generating liquid states at packing fractions significantly higher than the packing fraction at which the HS system arrests.

(iv) The non-monotonic behaviour of the characteristic structural times as a function of the attraction strength.
(v) The possibility of a sharp transition between the HS and the bonding localisation mechanisms, for very small ranges of attractions. The transition is accompanied by a discontinuous changes of the elastic properties of the glass.
(vi) A decay of the density correlation function completely different from the one characteristic of the HS system, in the region where the two different localisation mechanisms compete with comparable strength.
(vii) The possibility of observing, for a specific choice of the interaction range, a fully developed logarithmic decay of density fluctuations (a higher-order singularity in the MCT classification scheme).

The basic physics behind these phenomena results from a competition between two different mechanisms constraining the particle motion. To grasp the origin of this competition, consider the HS system. When the volume occupied by the hard spheres becomes larger than 58% of the available volume, structural arrest is observed. In the resulting glass, particles are hindered from moving too much by the presence of neighbouring ones, and are *caged* by their neighbours. Only an extremely rare collective rearrangement, for example opening a channel out of the cage, makes particle diffusion possible. In such packed conditions, particles can only rattle within their own cage, getting no further than an average distance of around 0.1σ.

In the presence of an additional short-range inter-particle attractive interaction something different takes place. At high temperature, the attraction does not play any role and, if the volume fraction occupied by the particles is large enough, the material will behave like a HS glass. But if the range of the attraction Δ is smaller than 0.1σ, then on cooling particles will begin to stick together, effectively shrinking the confining cage size and producing a more inhomogeneous distribution of the empty space. The structure factor of the system reflects these changes. The height of the first peak decreases while the small wave vector value, a measure of isothermal compressibility and hence of the density fluctuations, increases. These empty regions form channels through which particles can diffuse and so the material starts to melt.

As a result, the glass has turned into a liquid on cooling. If the temperature is further lowered, inter-particle bonding will become stronger and longer lasting, thereby restoring the usual progressive slowing down of dynamics, which results in another structural arrest. The liquid turns back into a glass. In the temperature–volume fraction phase diagram, this glass–liquid–glass sequence results in a re-entrant (non-monotonic) glass transition line. It also means that some liquid state can be stabilised by the short-range attraction, and so – compared to the long-range attraction case – the range of stability of the liquid phase is increased. Within the MCT, the glass–fluid–glass transition arises from the progressive changes of the structure factor, the only input of the theory beside number density. The HS glass, stabilised by the large value of the first peak of the structure factor, melts due to the weakening of the structure factor in the first peak region. The new glass transition is instead driven by the increased large wave vector oscillation (arising from the well-defined bonding length and to the bonding localisation). The phenomena we just described are summarised in the schematic sketch of Figure 4.7, due to Sciortino (2002), which well illustrates the salient features of attractive colloidal systems.

Higher-order singularities

The significant agreement between experiments, simulations and theoretical MCT predictions has prompted researcher to test, via specific accurate simulations, the most striking and unusual predictions of the theory. Numerical simulations have been designed to study numerically both the glass–glass transition and the

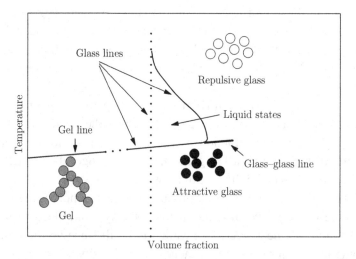

Figure 4.7 Schematic phase diagram of a colloid with hard-sphere interactions and a short-range attractive part. Two intersecting glass lines are observed, separating the metastable supercooled liquid by the repulsive and attractive glasses. At high temperatures the repulsive glass line approaches the hard-sphere one. At lower temperatures the repulsive glass line moves to higher ϕ and gives rise to a pocket of liquid states. At low temperatures there is an attractive glass line. The attractive glass line generates at high ϕ a glass–glass line along which the elastic properties of the glass change discontinuously. At much lower ϕ, a gel line is experimentally observed (Sciortino, 2002). Reprinted by permission from Macmillan Publishers Ltd: Nature Materials, copyright 2002.

dynamics close to an A_4 singularity (Götze and Sperl, 2004). These numerical studies have been performed using the square-well binary mixture model discussed above. An important prediction of the theory regards indeed the presence of a kinetic (as opposed to thermodynamic) glass–glass transition, which should take place in the glass phase on crossing a critical temperature. Heating a short-range attractive glass should produce a sudden variation of all dynamical features, without significant structural changes. For example, at the transition temperature, the value of the long-time limit of the density–density correlation function, the non-ergodicity factor f_k, should jump from the value characteristic of the short-range attractive glass to the significantly smaller value characteristic of the HS glass. The numerical work suggests that the ideal attractive glass line in short-range attractive colloids has to be considered as a crossover line between a region where ideal-MCT predictions are extremely good (in agreement with the previous calculation in the fluid phase) and an activated-dynamics region, where ideal-MCT predictions apply in a limited time window. The anomalous dynamics, which stems from the presence of a higher-order singularity in the MCT equations (Sciortino *et al.*, 2003), still affects the dynamical processes in the fluid and in the glass, even if activated processes preempt the possibility of fully observing the glass–glass transition phenomenon, at least in the square-well case (Zaccarelli *et al.*, 2003). It is important to note that systems in which the short-range inter-particle potential has a shape enhancing the bond lifetime, could produce dynamics which are less affected by hopping processes, favouring the observation of the glass–glass phenomenon.

Simulations have also been performed to investigate the dynamics close to the A_4 point (Sciortino *et al.*, 2003). In the SW case, the systems are characterised by three control parameters, the packing fraction ϕ, the ratio of the thermal energy to the typical well depth u_0, and the range Δ of the attractive potential. Within MCT, the phase diagram of this three-dimensional control parameter space is organised around a critical point (T_4, ϕ_4, Δ_4), referred to as a type A_4 higher-order glass-transition singularity in the MCT classification. A_4 is the end point of a line of higher-order singularities (of type A_3). From a physical point of view, A_4 is characterised for being the simplest higher-order singularity on a type B transition line accessible from the liquid phase. For $\Delta > \Delta_4$ no singular points are predicted by the theory, while for $\Delta < \Delta_4$ the A_3 singularity points are buried in the glass phase, and their presence can be observed only indirectly.

Near the A_4 singularity, MCT predicts a structural relaxation dynamics, which is utterly different from that known for conventional glass-forming liquids. It is ruled by logarithmic variations of correlators as a function of time and sub-diffusive increase of the mean square displacement. Theory makes precise predictions for the time interval where this unusual dynamics is expected, as well as for its variation with changes of the T, Δ and ϕ control parameters. To show that the above described features are consistent with the ones predicted by MCT, we cite the general asymptotic decay law for the density correlators, near the higher-order singularity:

$$\Phi_q(t) = f_q - h_q \left[B_q^{(1)} ln(t/\tau) + B_q^{(2)} ln^2(t/\tau) \right], \tag{4.13}$$

where τ is a time scale which diverges if the state approaches the singularity. The formula is obtained by asymptotic solution of the MCT equations, considering small parameter differences $\phi^* - \phi$, $\theta^* - \theta$, $\Delta^* - \Delta$ of order ϵ. The coefficient $B_q^{(1)}$ is of order $\sqrt{\epsilon}$, while $B_q^{(2)}$ is of order ϵ. The amplitude h_q is independent of ϵ. The first term f_q is the sum of the non-ergodicity parameter at the singularity and a correction of order ϵ. Terms of order $\epsilon^{1/2}$ are neglected.

In order to test the theoretical predictions, a simulation was performed in a binary mixture of hard spheres of equal mass, with slightly different diameters σ_{AA} and σ_{BB}, $\sigma_{AA} = \sigma_{BB} = 1.2$ sufficient to prevent crystallisation at high values of volume fractions. The hard-core potential is complemented by an attractive square-well potential of depth u_0, independent of the particle type.

In order to locate the A_4 point for the binary mixture square-well model and select the parameters to be used in the simulations, the procedure is as follows:

(i) First a simulation is made of a specific mixture, labelled S_1, with width of the square well $\Delta_{ij} = 0.031 \, \sigma_{ij}$.
(ii) The glass line is calculated for various temperatures, using diffusivity data extrapolated according to the power law predicted by MCT.
(iii) Then one calculates the MCT ideal glass–transition line using partial structure factors calculated with the Percus–Yevick approximation.
(iv) The theoretical glass line is mapped on the simulation line. Since MCT cannot reproduce the appropriate value of the critical parameters, one has to use a linear transformation of variables (ϕ, T) in order to superimpose the MCT result to the simulation one.
(v) The theoretical MCT A_4 is then mapped, using the same variable transformation, to the simulation variables $k_B T / u_0 = 0.4$, $\phi = 0.6075$ and $\Delta_{ij} = 0.043 \, \sigma_{ij}$, a state point labelled as S_2.

Figure 4.8 shows the mean square displacement for the larger species for both systems S_1 and S_2. Notice the presence of a subdiffusive regime at intermediate times for about three decades. The logarithmic regime extends by more than a decade if the attraction range is increased to that for system S_2.

Figure 4.9 shows the density-fluctuation autocorrelation functions for a representative set of wave vectors q. The decay curves do not show the two-step scenario with a plateau characteristic of conventional glass-forming liquids. It is also impossible to fit these curves for large q with the standard stretched exponential function.

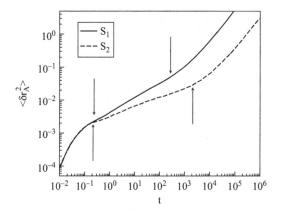

Figure 4.8 Mean square displacement of the large particles, showing subdiffusive behaviour. At long times, the diffusive behaviour is recovered (Sciortino *et al.*, 2003). Copyright 2003 by The American Physical Society.

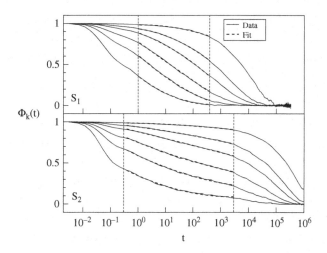

Figure 4.9 Numerical study of the density-fluctuation correlators in a 50–50 binary mixture of particles of size ratio $\sigma_{BB}/\sigma_{AA} = 1.2$ at $T/u_0 = 0.4$ and $\phi = 0.6075$ for two different values of the well width, close to the estimated position of the A_4 singularity of the SW ($T/u_0 = 0.416$; $\Delta_{ij} = 0.043\sigma_{ij}$; $\phi = 0.6075$). Six different wave vectors (from top to bottom $k\sigma_{BB} = 6.7$, 11.7, 16.8, 23.5, 33.5 and 50.3) are shown. The long dashed lines are fits according to the MCT predictions for systems close to the A_4 singularity. The vertical dashed lines indicate the fitting interval. The top and bottom panel refer to the states S_1 and S_2 respectively. Note the large window in which an essentially logarithmic decay in time is observed for a selected value of k (Sciortino *et al.*, 2003). Copyright 2003 by The American Physical Society.

There are instead regions of clear logarithmic decay over time intervals of similar size as those for the mean square displacement.

To estimate the possibility of describing the time dependence of $\Phi_q(t)$ according to Eq. (4.13), the density autocorrelation functions are fitted to a quadratic

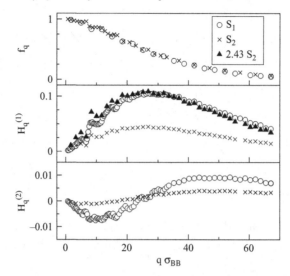

Figure 4.10   The fitting parameters of the asymptotic logarithmic law, for the states S_1 and S_2. The central panel also shows the parameter $H_q^{(1)} = h_q \, B_q^{(1)}$. At the wave vector where $H_q^{(2)} = h_q \, B_q^{(2)} = 0$ the correlation functions display a pure logarithmic decay (Sciortino *et al.*, 2003). Copyright 2003 by The American Physical Society.

polynomial in the logarithm of time for different q values, the fitting parameters being reported in Figure 4.10. The fitting parameter f_q provides an estimate of the non-ergodicity parameter at the A_4 point and does not depend on the state point. This confirms the preceding conclusion that the studied state points are very close to the singularity and that corrections of order ϵ in f_q cannot be detected.

These results have an intrinsic value associated with the observation, in a particularly simple system, a square-well potential of a particularly complex dynamics over more than four decades in time.

They show that MCT, a theory essentially developed to address the problem of the excluded volume glass transition is able, without any modification, to handle the logarithmic dynamics and to provide an interpretative scheme in terms of the A_4 point. The simulation data – with parameters explicitly chosen close to the A_4 point – exhibit the mentioned laws over time intervals up to four orders of magnitude, as shown in Figure 4.9. More importantly, the decay patterns vary with changes of the control parameters and wave vectors as expected (Götze and Sperl, 2004), properly testing the theoretical predictions.

Mechanical properties

The dynamic differences between the attractive and repulsive glasses discussed above generate differences in the mechanical properties of the two glasses. In fact one expects that attractive glasses have a stronger rigidity under shear than those produced simply by packing forces.

This phenomenon originates from the changes in the non-ergodicity factor associated to the glass–glass transition, which lead the system to soften with respect to shear on approaching the repulsive glass. In other words, the short-range bonding characteristic of the low-temperature attractive glass produces a much stiffer material than the repulsive glass formed at high temperature.

As we reported above, by tuning the volume fraction and the attraction range, it should also be possible to design a material where the transition between the attractive and repulsive glass takes place without any intermediate liquid phase, i.e. a glass-to-glass transition. In these specific conditions, a small change in the external parameters would produce a remarkable change in the elastic properties of the material with no significant structural change. Theoretically, more than one order of magnitude change in the stiffness is expected. For material scientists, design production and technological exploitation of such a material is one of the most fascinating challenges opened by the study of the dynamic properties of short-range attractive colloids. In this respect, MCT calculations, whose predictive power has been significantly strengthened by the close agreement between predictions and experimental observation, might become a valuable instrument for guiding the design of novel soft materials with specific properties.

The richness of the dynamics in the short-range attractive colloids can provide a valuable test case for comparing theoretical predictions and numerical data of the viscoelastic behaviour of the fluid phase, close to dynamic arrest. Viscoelastic properties of the square-well models were studied numerically, computing the shear viscosity η as the integral of the correlation function of the non-diagonal terms of the microscopic stress tensor,

$$\sigma^{\alpha\beta} = \sum_{i=1}^{N} m v_{i\alpha} v_{i\beta} - \sum_{i<j}^{N} \frac{r_{ij\alpha} r_{ij\beta}}{r_{ij}} V'(r_{ij}),$$

where $v_{i\alpha}$ is the α-th component of the velocity of particle i, and V' is the derivative of the total potential with respect to its argument. Then

$$\eta \equiv \int_0^\infty dt \, C_{\sigma\sigma}(t) = \frac{\beta}{3L^3} \int_0^\infty dt \sum_{\alpha<\beta} \langle \sigma^{\alpha\beta}(t) \sigma^{\alpha\beta}(0) \rangle, \qquad (4.14)$$

where L^3 is the volume of the simulation box. It has been found that η diverges following a power law as the transition points (attractive and repulsive) are approached, with the same exponent as the time scale of the density fluctuations. On approaching the glass transition, the slow-decaying part of the $C_{\sigma\sigma}(t)$ can be time rescaled into a master function which has the same shape as the square of the density–density correlation function at a specific wave-vector k. For the case of the repulsive glass the inverse k value corresponds to the nearest neighbour distance, while in the case of the attractive glass it corresponds to distances comparable to the (short) range of attraction. This interesting feature can be related, within MCT, to the fact that $C_{\sigma\sigma}(t)$ can be expressed as (Bengtzelius *et al.*, 1984)

$$C_{\sigma\sigma}(t) = \frac{k_B T}{60\pi^2} \int_0^\infty dk\, k^4 \left[\frac{d\,\ln S_k}{dk} \Phi_k(t) \right]^2 .$$

Using these formulae a particularly interesting calculation of the elastic shear modulus G' was performed in the region where the attractive glass line extends into the glass regions, separating two different type of glass, the repulsive and the attractive one, and ending in the higher-order singularity A_3 (Zaccarelli *et al.*, 2001). Figure 4.11 shows the discontinuity in G' when crossing the glass line, a signal of the difference between the two types of glass, which tends to disappear on

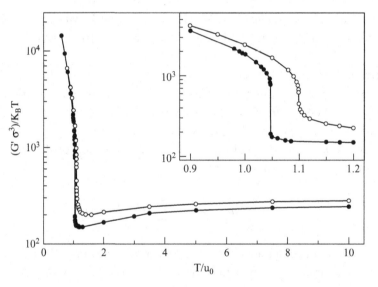

Figure 4.11 Mode-coupling predictions for the temperature dependence of the elastic shear modulus G' for a short-range square-well system with range of interaction equal to 0.031 of the hard core σ. $\phi = 0.5397$ (full circles) and $\phi = 0.5441$ (empty circles). The first ϕ value corresponding to crossing the glass–glass coexistence line. At this ϕ value a discontinuous change in G' is observed. The second corresponds to crossing the glass–glass line close to the end point A_3. Here the width of the jump in G' approaches zero. Note also that G' in the attractive glass can be almost two orders of magnitude larger than the repulsive one. The inset shows details of the transition region (Zaccarelli *et al.*, 2001). Copyright 2001 by The American Physical Society.

approaching the singularity A_3. In fact, for the smaller of these two volume fractions, it is evident from the figure that upon crossing the transition, there appears a sharp discontinuity in the shear modulus. For the larger one, which is very close to the value of the packing fraction of the A_3 end-point, crossing the curve, there is a negligible difference between the shear moduli of the two glasses. Evidently, it is of interest to define $\Delta G'$, the difference in G' found in the two glasses at the transition and to examine this as a function of ϕ and T, respectively, the differences in volume fraction and temperature from their end point values ($\phi_{A_3} = 0.5441$ and $T_{A_3} = 1.09975$). Over the whole range of the glass–glass transition one finds the following power laws in terms of volume fraction or temperature distance from A_3:

$$\Delta G' \sim (\Delta \phi)^p, \tag{4.15}$$

$$\Delta G' \sim (\Delta T)^q, \tag{4.16}$$

with $p \sim 0.33 \pm 0.03$ and $q \sim 0.32 \pm 0.08$. These exponents can be explained by a simple argument. Near an A_2 singularity f_q is a solution of a second-order polynomial, while near an A_3 point it is solution of a cubic one. This implies Δf_q is a power law with exponent $1/3$ in $\Delta \phi$ or ΔT near the A_3 singularity. Thus, evaluating Eq. (4.3) for G' gives, to leading order, the same exponent $1/3$.

It is clear from these results is that the repulsive glass stiffness with respect to shear is much smaller than the attractive glass one, as indicated by the large vertical discontinuity shown in the inset of Figure 4.11. This reflects the fact that particles in the attractive glass are bonded by the stickiness of the potential, whilst in the repulsive one there is no real bonding between them, thus implying that they are more easily broken apart under shear. Also, as expected, the attractive glass shear modulus increases considerably with decreasing temperature, the attractions between particles becoming more relevant, whilst for the repulsive glass there are no significant changes with temperature, since there is no energy scale involved in its formation.

4.4 Module – Clustering in systems with competing interactions

According to the van der Waals theory, any inter-particle potential which involves rapidly increasing repulsion at small separations and slowly varying attraction at larger separations will produce a gas–liquid transition. It is also known that this transition is suppressed in favour of a gas–solid transition when the attraction is sufficiently short range. When a further repulsion of range longer than the attraction is added, new phenomena appear, such as microphase separation, also referred to as formation of patterned or modulated phases, characterised by inhomogeneous densities and cluster formation. The phenomenon is especially important in complex liquids, where micellar solutions, microemulsions, and block copolymer systems, amongst others, are all examples of microphase separated systems with periodic density modulations. Examples of microphase phases are clusters and stripes in two dimensions, or spheres, columns or lamellae in three dimensions.

In several colloidal systems the excluded volume and the attractive component are complemented by screened electrostatic interactions which can have quite different physical origins. The classical example goes back to the celebrated Derjaguin, Landau, Vervey and Overbek (DLVO) form of the inter-particle potential, which has contributions from the hard core of the particles, the dipolar van der Waals attraction and the screened Coulomb electrostatic interaction. The final form of the potential depends parametrically on various physical quantities relating to the state of the colloidal dispersion, e.g. the charge and concentration of the counterions, the co-ions and the temperature. Quite often, the range of the attractive interaction is significantly smaller than the range of the repulsive electrostatic interaction. In these cases, the competition between the two length scales gives rise to a wide variety of phenomena. We consider in more detail cluster phases in equilibrium, as a possible path to spontaneous pattern formation in kinetically arrested states.

We report here the results of the application of MCT to a model system in which particles interact via short-range attractive and long-range repulsive Yukawa potentials. The input to the MCT equations is the structure factor calculated from the mean-spherical approximation for the equilibrium properties. The glass–glass reentrance phenomenon in the attractive colloidal case is observed in the presence of the long-range repulsive barrier, which results in a situation similar to the attractive and repulsive glasses reported above. Competition between the short-range attraction and the long-range repulsion produces characteristic behaviours of the structure factor and the non-ergodicity factor. A careful analysis gives rise to new regimes which appear in the attractive glass, and are tentatively associated with a cluster liquid (called static cluster glass) and heterogeneous structures (dynamic cluster glass). Crossover points separating the different glass states are also identified along the liquid–glass transition line between the liquid and the attractive glass.

A physically more direct theoretical approach to describe equilibrium cluster phases consists in using concepts typical of the glass formation considered as a more general kinetic arrest phenomenon. The inter-particle potential used is composed by a long-range Yukawa repulsive interaction

$$V_Y(r) = Ae^{-r/\xi}/(r/\xi), \tag{4.17}$$

and by a short-range potential V_{SR} of the form proposed by Vliegenthart and coworkers, which is a generalisation of the Lennard-Jones potential, and that mimics a hard core followed by an attractive well:

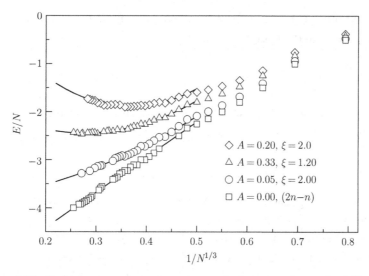

Figure 4.12 Ground state energy for connected clusters composed of N particles for different choices of A and ξ. While in the $A = 0$ case, a monotonous linear decrease of the energy in $N^{-1/3}$ is found, as expected from the balance between bulk and surface effects, for large A the lowest energy state is reached for a finite N^* value and provides an upper limit to the size of thermodynamically stable clusters (Sciortino *et al.*, 2004). Copyright 2004 by The American Physical Society.

$$V_{SR} = 4\epsilon \left[(\sigma/r)^{2\alpha} - (\sigma/r)^{\alpha} \right],$$

with $\alpha = 100$, which corresponds to an attractive range of about 0.03σ, a situation well studied when no long-range repulsion is present. The cluster ground state energy, shown in Figure 4.12 for particles interacting with the potential $V_{SR} + V_Y$ for appropriate values of A and ξ, has a minimum at a finite size N, indicating that clusters of size larger than N^* are energetically disfavoured (and, hence, that liquid condensation is inhibited).

Since entropic contributions to the free energy will always favour small clusters, aggregation will never proceed beyond N^*. The existence of an optimal cluster size (which will of course depend on T and ϕ) suggests that, at low T, clusters act as building blocks of the supramolecular cluster fluid.

A numerical study confirmed the possibility of generating both equilibrium cluster phases and kinetically arrested phases, and identified the structural signature in the low k pre-peak of the structure factor. The final state results from a subtle competition between aggregation (triggered by T) and repulsion. We will show later that dynamical arrest can be interpreted as a Wigner glass transition of the clusters fluid. Indeed, under the hypothesis of spherical homogeneous clusters of radius R, the cluster–cluster interaction is found to be of the Yukawa form, with the same screening length ξ as in the monomer–monomer interaction, but with a renormalised amplitude given by the expression

$$A(R) = A \left\{ 2\pi \xi^3 n e^{-R/\xi} \left[1 + R/\xi + (R/\xi - 1) e^{2R/\xi} \right] \right\}^2 ,$$

where n is an effective monomer number density in the cluster. Following a T-jump, the aggregation process can be thus visualised as a flow on the Yukawa phase diagram, due to the simultaneous change of the cluster number density and T^*

$$T^* \equiv k_B T / V_Y(\lambda \xi) = k_B T \lambda / A e^{-\lambda}. \tag{4.18}$$

If the melting or the glass line of the Yukawa model are crossed during the cluster growth, the system becomes solid, i.e. a cluster crystal or, due to polydispersity of the aggregates, a cluster glass. With this mechanism arrested disordered states with small ϕ can be generated, so that this route to the gel state demonstrates, in the case of weakly charged short-range attractive colloids, the identity of the gel and the glass states of matter.

The evolution of the structure with T is studied analysing the static structure factor $S(q)$. As shown in Figure 4.13, $S(q)$ progressively develops a peak at wave vectors q, associated with distances in real space of the order of several particle diameters, reflecting the nearest neighbour cluster–cluster distance. The amplitude of this novel peak changes significantly in a small temperature range around the

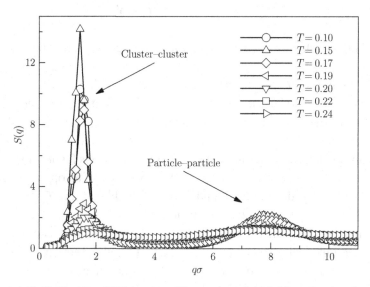

Figure 4.13 Evolution of $S(q)$ on cooling at $\phi = 0.125$. Note the development of a peak at small q, reflecting the formation of clusters in the system (Sciortino *et al.*, 2004). Copyright 2004 by The American Physical Society.

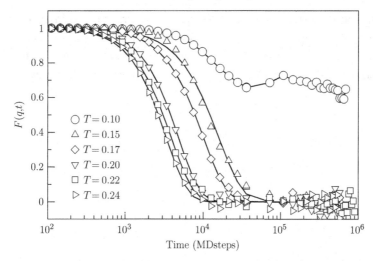

Figure 4.14 $F(q, t)$ at $q\sigma = 2.7$. At $k_B T = 0.1$ the autocorrelation function no longer decays to zero, highlighting the formation of a structurally arrested phase with a small packing fraction. Curves are stretched exponential fits with exponent $\beta > 1$, except for the gel state (Sciortino *et al.*, 2004). Copyright 2004 by The American Physical Society.

T-region where the flow diagram predicts the crossing of the crystallisation line. Therefore, the system progressively changes from a solution of interacting particles toward a system of interacting clusters. It is interesting to note that both the location and the amplitude of the cluster–cluster peak are in agreement with the experimental data reported for colloidal gels.

To confirm that the aggregation process indeed generates a disordered arrested state, Figure 4.14 shows the coherent scattering function $F(q, t)$. It displays a (one-step) decay, with a characteristic time which increases on cooling. At the lowest studied T, the correlation function does not decay to zero, signalling that the system is trapped in a non-ergodic state. The transition from fluid to glass is very sharp, again in full agreement with results for colloidal gels.

In these systems, the gel phase is not stabilised by the short-range attraction and hence its formation is not connected to the attractive glass transition observed at high packing fractions. The competition between isotropic short-range attraction and long-range repulsion is capable of generating a very rich range of phenomena.

Within the present model, where a short-range attractive interaction is complemented by a weak electrostatic repulsion, as in colloidal systems and in protein solution, the formation of low-density arrested states can be modelled as a glass transition, where clusters (as opposed to particles) are trapped in cages generated by the long-range repulsions. An interesting question is to study if the arrested disordered state is metastable with respect to fcc or bcc crystals of spherical clusters (a Wigner crystal) or with respect to lamellar phases, as observed in other models.

5

Experiments on structural arrest

5.1 Introduction

Following the pioneering experiments on the glass transition in PMMA, which gave a good approximation to monodisperse hard-sphere systems, Pusey and van Megen (1989) and coworkers started a series of experiments with the aim of characterising the transition. Most of the experiments were performed through laser light scattered by density fluctuations characterising the transition from liquid to supercooled liquid and eventually to an amorphous solid. An efficient method was also used to measure the time average in a non-ergodic system using averages over different scattering volumes and wave vectors. The comparison with the predictions of MCT was performed in an extended fashion, showing a relatively good agreement with the experimental findings, to a 20% level of accuracy. The most relevant result of this important set of measurements was the detection of the structural arrest point, a result that is not easy to obtain in normal liquids due to the existence of activated dynamics or hopping effects. The latter are supposedly responsible for the crossing of the barriers that confine the system in a potential well.

MCT was subsequently applied to potentials with an attractive tail following the short-range repulsion, and lead to the behaviour described in the previous chapter on the the theory of supercooled liquids. The most relevant finding was the evidence of the existence of higher-order singularities, which were already defined and studied within MCT, in systems with short-range attractive interactions. Shortly after the predictions of MCT on the consequences of an attractive interaction in hard-sphere systems obtained, many attempts were made to experimentally demonstrate their validity. In particular, first the re-entrant glass line was detected, then the effect of the A_3 singularity was shown and finally the higher singularity of type A_4 was identified. The experiments on these various aspects of the behaviour of supercooled liquids are illustrated in the following sections.

5.2 Module – Experiments on the glass transition in PMMA

In this section we introduce the basic aspects of the experiments performed in supercooled liquids, with special emphasis on the pioneering work of Pusey and van Megen, and leaving the more recent experiments on the glass transition in attractive colloids to the next sections.

As we mentioned in the previous chapter, Pusey and van Megen (1989) have shown not only the ability of the colloidal hard-sphere model systems to form an arrested glass, but also performed an extensive set of measurements, using mainly dynamic light scattering, in order to characterise the supercooled liquid close to the glass transition. An interesting procedure was introduced by them in order to measure the intermediate scattering function in a system which is not ergodic, such as a supercooled liquid. The electric field scattered at an angle defined by the momentum transfer q by a system of N particles is

$$E(q, t) = \sum_{j=1}^{N} e^{iq \cdot r_j(t)}, \tag{5.1}$$

and in ergodic systems it can be considered a Gaussian random variable with zero average, so that in this case the time average

$$\langle E(q, t) \rangle_T = \lim_{T \to \infty} \frac{1}{T} \int_0^T dt\, E(q, t), \tag{5.2}$$

and the ensemble average $\langle E(q, t) \rangle$ coincide. In a dynamic light-scattering experiment one measures the correlation function of the scattered intensity $I(q, t) = E(q, t)E^*(q, t)$ and because of the Gaussian behaviour of $E(q, t)$ the Siegert relation can be easily derived:

$$\frac{\langle I(q, \tau)\, I(q, 0) \rangle}{\langle I(q, 0) \rangle^2} = 1 + c \left[\frac{\langle E(q, \tau)E^*(q, 0) \rangle}{\langle |E(q, 0)|^2 \rangle} \right]^2. \tag{5.3}$$

Using this relation it is possible to obtain the time correlation function of the electric field fluctuations using the measured scattered radiation intensity correlations. The proportionality constant c takes into account the geometrical arrangement of the scattering experiment and is essentially related to the ratio of the coherence area observed and the area of the detector, a quantity typically of the order 0.8.

In the case of a non-ergodic samples, such as a supercooled fluid in the vicinity of a glass transition, the electric field is no longer a Gaussian variable and the Siegert relation loses its validity. Pusey and van Megen introduced a simple procedure which allows the evaluation of the time correlation function. When illuminated by coherent laser light, the scattered intensity constitutes a random speckle pattern which, when close to an arrest transition, is strongly space modulated but essentially constant in time. Each speckle, which is also a coherence area accepted by the detector, represents a single spatial Fourier component of the density. An ensemble average can still be made through the acquisition of data from many independent scattering volumes within the medium, each of which gives rise to speckles of different intensities. In fact, in a glass at non-zero temperature the particles are allowed some freedom for local motion and small intensity fluctuations are superimposed on an otherwise constant speckle pattern. In this case a single experiment samples only a restricted region of phase space associated with the local particle motions. Therefore measurements on many independent volumes are still needed to provide a full ensemble average, i.e. an average that also includes the constant components of the speckle pattern that results from the frozen density fluctuations, whose amplitude may vary significantly from speckle to speckle. In order to study non-ergodic samples, a variant of the technique of performing the measurements on a very large number of independent scattering volumes is to use an enlarged scattering volume obtained by using an unfocused laser beam and an enlarged detection slit, so that the coherence area factor is $c \approx 0.3$. With this arrangement the detector aperture accepted about ten coherence areas in the scattered light field, corresponding to ten independent spatial Fourier components of the particle density. Measurements were made for ten spatially separated scattering volumes obtained by moving the sample, so that in the limit of infinitely many volumes the procedure gives the full ensemble averages. The effective sampling was performed experimentally on the time evolution of roughly 100 different Fourier components.

A more efficient procedure was later introduced (Pusey and van Megen, 1989) in terms of partially fluctuating speckle patterns, where the instantaneous position of a particle is written as the sum of a fixed average position and a fluctuating displacement relative to this position, as expected in deeply supercooled liquids. A detector area much smaller than one coherence area or speckle is also assumed. As a result one gets the following expression for the normalised ensemble-averaged coherent intermediate scattering function:

$$\frac{\langle E(q,\tau)E^*(q,0)\rangle}{\langle |E(q,0)|^2\rangle} = 1 + \frac{\langle I(q,0)\rangle_T}{\langle I(q,0)\rangle} \tag{5.4}$$
$$\times \left\{ \left[\frac{\langle I(q,\tau)I(q,0)\rangle_T}{\langle I(q,0)\rangle_T^2} - \frac{\langle I^2(q,0)\rangle_T}{\langle I(q,0)\rangle_T^2} + 1 \right]^{1/2} - 1 \right\},$$

which again connects the field correlation function to time-averaged correlations of the scattered intensities $\langle I(q,0)\rangle_T$, $\langle I(q,0)\rangle_T^2$ and $\langle I(q,\tau)I(q,0)\rangle_T$, and the intensity ensemble average $\langle I(q,0)\rangle$. In the ergodic case the time and ensemble averages are equal, the field $E(q, t)$ is a complex Gaussian random variable, so that

$$\frac{\langle I^2(q, 0)\rangle_T}{\langle I(q, 0)\rangle_T^2} = 2, \tag{5.5}$$

and Eq. (5.4) reduces to the Siegert relation Eq. (5.3).

Pusey and van Megen gave the general prescriptions for the determination of the intermediate scattering function from dynamic light scattering from non-ergodic samples, in this case supercooled liquids and glasses, considering the slowest structural relaxation time τ_s and the measurement time T:

(i) $T \gg \tau_s$. The duration of the experiment significantly exceeds the relaxation time, the system is ergodic and a single measurement of the time-averaged intensity correlation, together with the Siegert relation, allows the calculation of the intermediate scattering function.

(ii) $T \approx \tau_s$. The system intrinsic relaxation time and the experiment duration are comparable. In this case a large number of independent measurements of the mean intensities and their correlations can be summed to give the normalised ensemble intermediate scattering function through the use, in this case too, of the Siegert relation.

(iii) $T \ll \tau_s$. When the relaxation time is much larger than the measuring time, i.e. the concentration fluctuations are largely arrested during time T, a single measurement of the time-averaged intensity autocorrelation function is performed, the ensemble-averaged intensity is evaluated through the intensity accumulated while the sample is moved in the laser beam. Use is then made of Eq. (5.4) to get the intermediate scattering function.

The systems studied were the original hard sphere-like PMMA samples described in the previous chapter. Typical results are reported in Figure 5.1, which shows the progressive loss of ergodicity on approaching the glass transition and the way in which MCT is capable of describing the experimental data.

Figure 5.2 reports the Debye–Waller factor for PMMA hard spheres as a function of the momentum transfer q, in units of the inverse particle radius, for various volume fractions.

5.3 Module – Copolymer solutions and higher-order singularities

The prediction of a new phenomenology, in comparison with the simple case of hard sphere, appearing when considering attractive interactions between colloidal particles, gave rise to an interesting experimental activity concerning the glass transition in colloidal systems. The attention was in particular directed

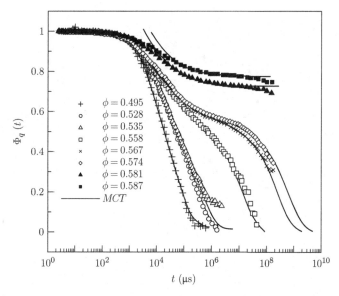

Figure 5.1 Experiments on hard spheres close to the glass transition. Normalised density fluctuations time correlation functions $\Phi_q(t)$ measured by dynamic light scattering using photon correlation spectroscopy. The particle radius is $R = 205$ nm and the wave vector $qR = 4.10$. The samples at various volume fractions are represented by the symbols, while the solid lines refer to MCT. Reprinted from Pusey and van Megen (1989), Copyright 1989, with permission from Elsevier.

toward the study of higher-order singularities and the accompanying enhanced slowing down of the relaxational properties of these systems. The investigations were performed mainly with radiation scattering experiments, using both light and neutrons.

In one of these studies (Mallamace *et al.*, 2000) the colloidal system consisted of a non-ionic triblock copolymer called L64, belonging to the Pluronic family. Pluronic is made of polyethylene oxide (PEO) and poly-propilene oxide (PPO), with the two PEO chains placed symmetrically on each end of the PPO chain. The chemical structure of L64 is [PEO$_{13}$ – PPO$_{30}$ – PEO$_{13}$], with a molecular weight of 2900 u and is comprised of 60 wt% PPO. PPO tends to become less hydrophilic than PEO at higher temperatures. Thus the copolymers acquire a surfactant property and self-assemble, spontaneously forming spherical micelles in an aqueous solution at sufficiently high temperatures and above certain concentrations. The micelles exist in a wide range of copolymer concentrations from a few wt% up to more than 50 wt% for $T > 30\,°$C. The behaviour of the disordered micellar phase, the microstructure of the micelles, and their mutual interactions were investigated by small angle neutron and light scattering. The micelles turn out to be essentially monodispersed, with polydispersity below 3%.

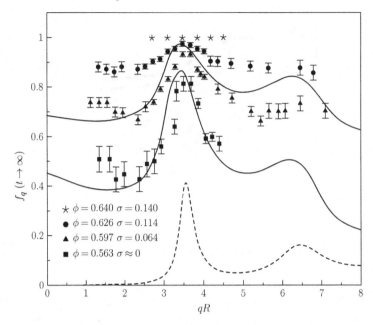

Figure 5.2 The non-ergodicity parameter of hard-sphere colloidal glass as function of volume fraction and qR, the product of scattering vector and particles radius. The solid curves are the MCT predictions for $\sigma = 0$ (lower curve) and $\sigma = 0.066$ (upper curve). The dashed curve is the Percus–Yevick structure factor for hard spheres at $\phi = 0.562$ reduced in magnitude by a factor of 10. Reprinted from Pusey and van Megen (1989), Copyright 1989, with permission from Elsevier.

The phase diagram, shown in Figure 5.3, is characterised by an inverted binodal curve with a critical point at $c_c = 5$ wt% and $T_c = 330.3$ K and a percolation line depending on c and T cutting across the phase diagram starting from near the critical point to at least 40 wt%. The percolation locus is determined by the observation that the shear moduli jump by more than two orders of magnitude across it, and by a power-law behaviour of the frequency-dependent viscosity. It can also be determined by the change of slope of the relaxation time measured by photon correlation spectroscopy as a function of T at a given c.

When the attractive potential has a range that is short compared to the particle size, it is usually represented by an exponential or a square-well interaction. Typical examples are a Yukawa potential or the sticky sphere potential introduced by Baxter and discussed in previous chapters. In both cases the Ornstein–Zernike equation for the pair correlation function can be solved in the mean spherical approximation or in the Percus–Yevick approximation respectively. This type of approach has been rather successful in interpreting percolation phenomena in microemulsions and in the L64/water system. In particular, the phase diagram

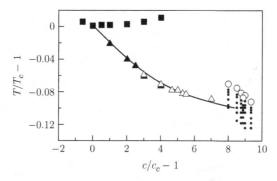

Figure 5.3 The phase diagram of Pluronic, showing the binodal line (squares), the percolation line (triangles), the glass line (open circle) and the measured states (small closed circles) (Mallamace *et al.*, 2000). Copyright 2000 by The American Physical Society.

can be fairly well described in terms of the Baxter sticky hard-sphere model, i.e. a square-well potential of zero range and infinite depth. In order to relate the model to the thermodynamic properties of the solvent, one has to assume a functional relation between the measured temperature T and the parameter $1/\tau$ which characterises the stickiness in the Baxter model. In the present case a simple linear relation $T \approx 1/\tau$ is sufficient to give a fairly good representation of the percolation and binodal lines, once the critical point given by the theory is assumed to coincide with the experimental one. The good agreement with the measured percolation line is clearly seen from Figure 5.3. The percolation points, measured through shear modulus and viscosity, and the points marking the structural arrest, measured by dynamic light scattering, cannot easily be explained in terms of other physical phenomena.

The photon correlation spectroscopy data have been taken using a digital correlator with a logarithmic sampling time scale, which allows an accurate description of both the short and long time regions of the intermediate scattering function (ISF), up to times of the order of seconds. Following the method used for colloidal hard spheres described earlier, since the system shows a structural arrest and therefore a non-ergodic behaviour, particular care has been taken in detecting the correlation function by averaging over many different sample positions over long time regions of the time correlation function. For each sample and each temperature from 100 to 200 measurements were performed, observing a larger scattering area corresponding to three or more independent Fourier components and changing the position of the sample in order to observe different scattering volumes. The measurements were taken for $c > 40$ wt%, the concentration below which no structural arrest has been observed, although the ISF tends to develop

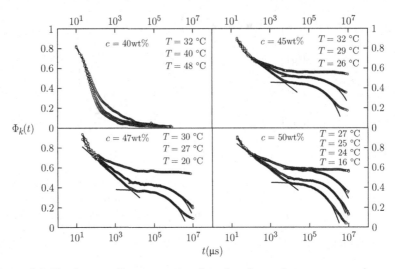

Figure 5.4 The intermediate scattering function for various concentrations and temperatures in the glass region. The lines are fits to both the von Schweidler law and the logarithmic decay (Mallamace *et al.*, 2000). Copyright 2000 by The American Physical Society.

a long time tail which can be described by means of a stretched exponential. For $c > 40$ wt%, moving at constant c and increasing T above the percolation line, one observes a progressive slowing down of the decay of the ISF, up to a point where it becomes flat, indicating structural arrest. The structural arrest is experimentally identified by the breakdown of the Siegert relation and the impossibility to obtain reproducible spectra. A typical behaviour is shown in Figure 5.4 at various concentrations and temperatures. The ISF for $c = 40$ wt% clearly shows that the structural arrest appears only above that threshold. In fact, below it the usual stretched exponential behaviour of dense micellar systems is observed, but no non-ergodicity sets in.

The points in the c–T plane where measurements of the ISF were performed are indicated in the phase diagram of Figure 5.3. In what follows MCT is used to describe the structural arrest in model systems with a potential characterised by a hard core and a short-range attractive tail. As was reported in details in the previous chapter, the MCT equations for the normalised ISF $f_q(t) = F_q(t)/S_q$ predict the existence, for a given c, of a critical temperature T_{MCT} where the non-ergodic transition takes place. It can be characterised by the fact that $f_q(t)$ tends to a finite plateau $f_q(t \to \infty) = f_q^c > 0$. The predictions of MCT is reported can be briefly summarised as follows.

The parameter driving the transition is the separation parameter $\sigma = 1 - T/T_{MCT}$. MCT introduces various time regimes in the decay of $f_q(t)$. The short time region $t < t_0$ is dominated by microscopic dynamics and is followed by the β relaxation region close to the plateau f_q^c, characterised by the time scale $t_\sigma = t_0/|\sigma|^{1/2a}$, whose initial part gives the approach to f_q^c while the final one follows von Schweidler's law,

$$f_q(t) = f_q^c - h_q(t/t_\sigma')^b, \tag{5.6}$$

which includes the second characteristic time scale $t_\sigma' = t_0/|\sigma|^\gamma$ with $\gamma = (2a)^{-1} + (2b)^{-1}$ and h_q a constant amplitude.

The two quantities a and b, with $0 < a < 0.5$ and $0 < b < 1$, are non-universal exponents determined solely by the so-called exponent parameter $\lambda = \Gamma^2(1 - a)/\Gamma(1 - 2a) = \Gamma^2(1 + b)/\Gamma(1 + 2b)$, Γ being the Euler gamma function, and λ is in turn determined by the static structure factor S_q. Analyses of the experimental data give the parameters $b = 0.6$ characterising Eq. (5.6) and the exponent parameter $\lambda = 0.7$.

The universal plot of the von Schweidler law is reported in Figure 5.5, while the scaling time is shown in Figure 5.6 and is described by the exponent $\gamma = 2.3$. Note that data taken at low T tend to deviate from the power law, indicating that corrections to scaling might be important. The situation is less satisfactory when considering the data above the plateau of the ISF, which should follow a power law with an exponent a. In that region, in fact, the data show a tendency to a logarithmic dependence on time. This logarithmic behaviour is related to the existence of higher-order singularities, as described earlier, which are in turn related to the existence of short-range attractive interactions. In these cases the glass transition line in the c–T plane shows two branches, one essentially corresponding to the glass line of a hard-sphere system at high c, the other extending to much lower values of c. The former is attributed, with the usual mechanism related to excluded volume, to the repulsive part of the interaction, and the latter to the attractive part of the potential.

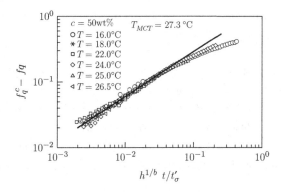

Figure 5.5 Experiments on A_3 (Mallamace *et al.*, 2000). Copyright 2000 by The American Physical Society.

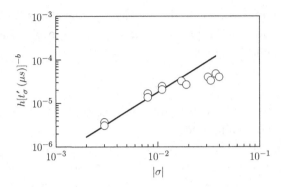

Figure 5.6 Scaling time of the von Schweidler law, showing the fitted exponent $\gamma = 2.3$ (Mallamace *et al.*, 2000). Copyright 2000 by The American Physical Society.

The non-ergodicity parameters have a qualitatively different behaviour on approaching the two lines. In the case of the sticky hard-sphere system a new phenomenon arises, namely, a glass-to-glass transition due to the crossing of the two glass lines. This transition is associated with a higher-order glass transition of type A_3. The experimental ISF decay also shows a time region with a logarithmic time dependence. This feature is characteristic of a cusp-like singularity of the A_3 type, in the vicinity of which a very slow logarithmic relaxation sets in and influences a large time domain. The observed logarithmic behaviour in the ISF is a manifestation of such an effect.

5.4 Module – The glass–glass transition

As we have emphasised many times, the study of the kinetic glass transition due to structural arrest, in systems with short-range attraction, has attracted increasing attention because it may lead, among other things, to a deeper understanding of gel states. These states are commonly observed in soft matter, particularly in colloidal systems and protein solutions (Russel, 1996), but are still not completely characterised from the physical point of view. From the theoretical point of view mode-coupling theory (Götze, 1991; Götze and Sjögren, 1992), which gives a fairly accurate quantitative description of the dynamic slowing-down process near the transition from a supercooled liquid to a solid-like amorphous glassy state, also made rather detailed predictions on the behaviour of colloidal suspensions where attractive interactions are important.

The recent MCT calculations that used more accurate descriptions of the particle interactions (Bergenholtz and Fuchs, 1999a,b; Fabbian *et al.*, 1999; Dawson *et al.*,

2000; Zaccarelli *et al.*, 2001) showed that when the system is characterised by a hard core plus an additional short-range attractive interaction (such as an adhesive hard-sphere system, AHS), a different dynamical arrest scenario emerges, as we described in detail earlier. In an AHS, in addition to the volume fraction, temperature is introduced as a second external control parameter in order to describe the phase behaviour of the system. It is possible to induce loss of ergodicity not only by excluded volume effects, by increasing the volume fraction, but also by changing the temperature which introduces bonding between particles with attractive interactions. At high temperatures and at sufficiently high volume fractions, the system evolves into the well-known glassy state due to excluded volume effects, called a repulsive glass. However, at relatively low temperatures, an attractive glass can form where the particle motion is hindered by cluster formation with neighbouring particles. In an AHS, aside from the hard-core diameter, an additional length scale, the range of the attractive well, should come into play. Making this distinction, one may divide spherical colloidal systems into two categories: the one-length scale hard-sphere system, in which the glass formation is dictated by the cage effect, and the two-length scale AHS, in which the two glass-forming mechanisms may coexist and compete with each other. Furthermore, MCT shows that in an AHS with sufficiently small ratio of the range of the attractive interaction to the hard-core diameter, variation of the control parameters allows to control the transition between these two distinct forms of glass (see Figure 5.13(b) for the predicted phase diagram). Of particular interest is the occurrence of an A_3 singularity at which point the glass-to-glass transition line terminates. MCT suggests that the two distinct dynamically arrested states become identical at and beyond this point.

Several experimental investigations have confirmed the theoretical predictions, such as the re-entrant glass transition phenomenon and the logarithmic relaxation of the glassy dynamics. They are considered to be signatures of the glassy dynamics in the two-length scale system (Mallamace *et al.*, 2000; Segré *et al.*, 2001; Bartsch *et al.*, 2002; Pham, 2002).

A particularly extensive set of measurements was performed (Chen *et al.*, 2002, 2003) on the triblock copolymer L64 described above, one member of Pluronic (BASF AG, Ludwigshafen, Germany) family system. The T–ϕ phase diagram of L64/D_2O system in the volume fraction temperature plane was reproduced in Figure 5.3 and discussed in the previous section. The aim of these studies was to characterise the re-entrant region of the repulsive-attractive glass, the higher-order singularities and the glass–glass transition, the most relevant features of colloids with attractive interactions. The micellar system was made by dissolving the triblock copolymer L64 described above, one member of Pluronic (BASF AG, Ludwigshafen, Germany) family, into D_2O at different weight fractions c. The amorphous states in the region of phase space under investigation were metastable states of the system, such as in supercooled liquids. Crystallisation does not happen unless the system is perturbed by an applied shear stress or placed very near a surface. Thus, the glass transition is observed in a metastable state of the micellar solution which lasts during the period of observation, even for days.

Three different experimental methods and set-ups were used to investigate various expected properties of the colloid, namely:

(i) Photon correlation spectroscopy (PCS) for the study of the re-entrant repulsive-attractive kinetic glass transition line in the colloid.
(ii) Shear viscosity to determine the location of the liquid–glass and the glass–liquid transitions in the re-entrant zone of the phase digram.
(iii) Small angle neutron scattering (SANS) experiments to trace the details of the liquid–glass transition line and the glass–glass transition.

Photon correlation spectroscopy measurements were made at a scattering angle of 90°, using a continuous wave solid state laser (Verdi-Coherent) operating at λ 5120 Å with 50 mW and an optical scattering cell of 1 cm diameter in a refractive index matching thermostatic bath. The intensity data were also corrected for turbidity and multiple scattering effects. The PCS data were taken using a digital correlator with a logarithmic sampling time scale. The latter feature allows us to describe accurately both the short-time and the very-long-time regions, up to times of the order of seconds. From the measured photon correlation function it is possible to calculate the intermediate scattering function using the usual procedures for non-ergodic systems.

The measurements correspond to a concentration range $0.41 < c < 0.51$ and various temperatures and are reported in Figure 5.7. Starting from a liquid state at a lower temperature, where the ISF decays to zero at long time, on increasing the temperature, it approaches a kinetic transition line characterised by a diverging relaxation time and tends to a finite plateau at long time. The ISF shows an initial Gaussian-like short-time relaxation and, immediately before the non-ergodicity transition, exhibits a time interval with a logarithmic decay before entering into the plateau region. The important prediction of the MCT calculation is that for an attractive colloid system there exists an attractive branch of the transition line near the singularity separating two different glass phases. In order to check this prediction the study of the micellar system was extended to higher temperatures. Starting from a glass state, upon further increasing the temperature, the measured ISF reveals a marked re-entry from glass to liquid state, as seen by the ISF again decaying to zero at higher temperatures.

Viscosity measurements (Chen *et al.*, 2002) were made with a strain controlled rheometer, using a double wall Couette geometry. To ensure a linear response the applied strains was fixed at $\gamma = 0.04$. In Figure 5.8 the low shear viscosity is

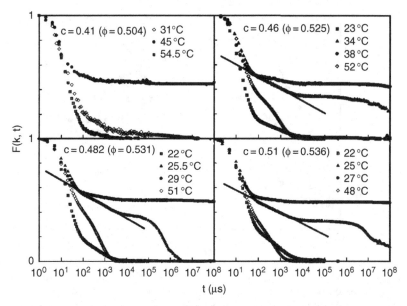

Figure 5.7 Typical examples of the intermediate scattering functions as measured by PCS, at different concentrations, as a function of temperature, showing the ergodic–non-ergodic–ergodic transition. It is possible to observe the occurrence of a region of a logarithmic time dependence preceding the plateau region for the system in the ergodic state just before the transition. The temperature dependence also indicates the existence of a re-entrant region. From Chen *et al.* (2002). Copyright 2002 by The American Physical Society.

reported at five different concentrations, measured as a function of the temperature. The viscosity at all concentrations are characterised by an initial increase, starting from viscosity values typical of a liquid, $\eta \approx 10^{-2}$ Pa, followed by a flattening at the highest temperatures. For the concentration $c = 0.35$, the viscosity shows only a single step increase that covers three orders of magnitude. For the remaining concentrations, there is an additional step increase located in a different temperature range. Thus for concentrations $c > 0.35$ the overall viscosity increase is about five orders of magnitude. The first step viscosity increase in the L64/D_2O micellar system is due to the percolation process that is characteristic of a colloid with a short-range attractive interaction.

Using the power-law frequency dependences of the complex viscosity and elastic and storage components of the shear moduli G' and G'' as a function of temperature one can determine the percolation locus in the composition temperature plane. The second step increase in viscosity in which the viscosity reaches very large absolute values (larger than 10^3 Pa), is located in the region of the glass transition. The elastic component G' becomes larger than the viscous component G'', whereas the frequency dependence of the G' develops a plateau and that of G'' a minimum.

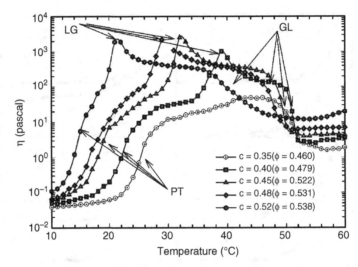

Figure 5.8   The shear viscosity L64/D_2O at five different concentrations as a function of temperature. Starting from low temperatures, the shear viscosity increases steeply, first going through a percolation transition indicated as PT, then a liquid–glass transition LG, and finally a glass–liquid transition. Note that, for $c = 0.35$, only the percolation transition is present. From Chen *et al.* (2002). Copyright 2002 by The American Physical Society.

Finally, the sharp reduction in viscosity and its eventual flattening off at high temperatures, with the final viscosity value of the order of a few pascals, can be related to the re-entry of the system into the liquid state. Figure 5.9, which is more accurate than the corresponding Figure 5.3, shows the Pluronic phase diagram including the measured percolation points.

SANS measurements were made at NG7, the 40-m SANS spectrometer in the Center for Neutron Research NIST, and at the SAND station of the Intense Pulse Neutron Source IPNS at the Argonne National Laboratory. At NG7, incident monochromatic neutrons of wavelength $\lambda = 5\,\text{Å}$ with $\Delta\lambda/\lambda = 11\%$ were used. The sample to detector distance was fixed at 6 m, covering the magnitude of wave vector transfer range $k = 0.008 - 0.3\,\text{Å}^{-1}$. IPNS generates high-energy spallation neutrons by bombarding a heavy metal target with repetitive pulses of 500 MeV protons. After moderation of these high-energy neutrons, it is possible to select a pulse of white neutrons with an effective wavelength range from 1.5 Å to 14 Å. At SAND all these neutrons were utilised by encoding their individual time of flight and the scattering angles determined by their detected position at a two-dimensional area detector. The two-dimensional area detector had an active area of $40 \times 40\,\text{cm}^2$ and the sample to detector distance was 2 m. This configuration allowed a maximum scattering angle of about 9°. The reliable Q range covered in the measurements were from $0.004\,\text{Å}^{-1}$ to $0.6\,\text{Å}^{-1}$. Q-resolution functions of both

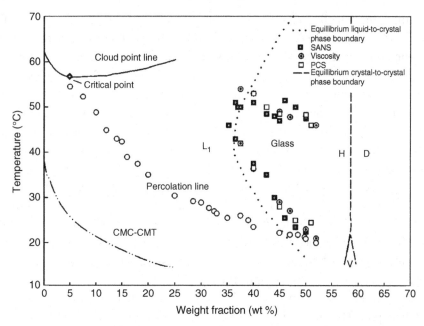

Figure 5.9 The temperature-concentration phase diagram of the Pluronic L64/D$_2$O solution. The phase diagram contains the CMC-CMT line, the cloud point line, the percolation line determined by viscosity measurements, the equilibrium liquid–crystal phase boundary (dotted line), the equilibrium crystal–crystal phase boundary (dashed line), and the re-entrant kinetic glass transition lines determined by PCS, SANS, and zero shear viscosity measurements. From Chen *et al.* (2002). Copyright 2002 by The American Physical Society.

of these SANS spectrometers are Gaussian and well characterised. It is essential to apply these resolution broadenings to the theoretical cross section when fitting the intensity data. The sample liquid was contained in a flat quartz cell with a 1 mm path length. The measured intensity was corrected for background and empty cell contributions and normalised using a reference scattering intensity of a polymer sample of known cross-section.

The absolute intensity of neutron scattering from a system of monodisperse micelles can be expressed by

$$I(k) = cN \left[\sum_i b_i - \rho_w \, v_p \right]^2 \bar{P}(k) \, S(k), \tag{5.7}$$

where c is the number density of polymer, N is the aggregation number of polymers in a micelle, the sum refers to the coherent scattering lengths of atoms including a polymer molecule, ρ_w the scattering length density of D$_2$O, v_p the molecular volume of the polymer, $\bar{P}(k)$ is the normalised intra-particle structure factor, and $S(k)$ is the inter-micellar structure factor.

Assuming the micelle to have a compact spherical hydrophobic core of a radius a, consisting of all the PPO segments with a dry core and a diffuse corona region consisting of PEO segments and the solvent molecules, it is possible to apply an original model called the cap-and-gown model (Liao *et al.*, 2000a). This model allows the calculation of the normalised form factor $\bar{F}(k)$, which defines the normalised intra-particle structure factor $\bar{P}(k) = |\bar{F}(k)|^2$ (Chen *et al.*, 2002). A square-well potential with a hard sphere with an adhesive surface layer can be introduced to describe the inter-micellar attractive interaction. The Ornstein–Zernike equation in the Percus–Yevick approximation for this potential can be solved analytically to first order in a small expansion parameter, the square-well width parameter $\epsilon = (R - R')/R$, where R and R' are the hard-core diameter and the full diameter of the colloidal particles respectively (Liu *et al.*, 1996). To summarise the main points concerning the analysis of the absolute SANS intensity distribution, it can be fitted uniquely with four parameters: the aggregation number N, the volume fraction, the fractional well width parameter ϵ, and the effective temperature $T^* = k_B T/u$, where u is the square-well potential depth. The normalised intra-particle structure factor $P(k)$ is function of N only. The inter-micellar structure factor $S(k)$ is function of all the four parameters.

An example of SANS intensity distribution for a 44.0 wt% micellar solution at 53 °C in an absolute scale is shown together with its model fit, taking into account the effect of the resolution function in Figure 5.10. It is apparent that the resolution correction is essential to fit the experimental data properly. The lower panel gives the normalised intra-particle structure factor $P(k)$ and the inter-particle structure factor $S(k)$ for this case. The observed SANS data is proportional to the product of these two functions. The fact that the first diffraction peak of $S(k)$ occurs at relatively smooth tail part of $P(k)$ implies that the interaction peak in the SANS intensity distribution is primarily due to the first diffraction peak in the inter-micellar structure factor $S(k)$.

A series of SANS measurements were performed on the micellar system within the appropriate volume fraction and temperature ranges, in order to examine the local structure of the glassy states, both in the region of the re-entrant glass transition (Chen *et al.*, 2002) and in the vicinity of the A_3 point (Chen *et al.*, 2003). A detailed analysis was performed in the original literature, here we discuss in particular the glass–glass transition and the higher-order singularity.

In the amorphous states, a SANS intensity distribution, which reflects the local structure, is characterised by a single peak located at k_{max} or equivalently by a unique length scale $\Lambda = 1/k_{max}$, which is the mean distance between micelles. It is well known that SANS absolute intensity distribution of a two-phase system (the

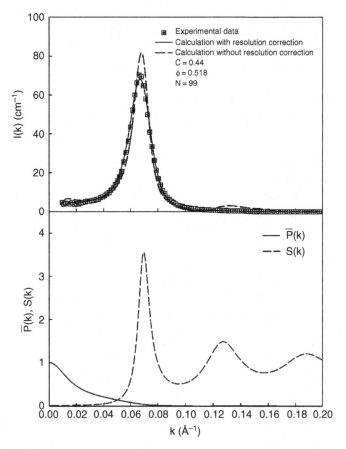

Figure 5.10 The SANS intensity distribution in an absolute scale and its model fit taking into account the effect of resolution function (top panel). The dashed line represents the theory while the solid line is the theory properly convoluted with the resolution function. The lower panel gives the normalised intra-particle structure factor $P(k)$, dotted line, and the inter-particle structure factor $S(k)$, solid line, for a 44.0 wt% Pluronic L64 solution in D_2O at 53 °C. The observed SANS data is the product $P(k)S(k)$ and clearly shows that the interaction peak observed in the SANS data is due essentially to the first diffraction peak in the inter-micellar structure factor. From Chen *et al.* (2002). Copyright 2002 by The American Physical Society.

micelles and the solvent) is proportional to a three-dimensional Fourier transform of the Debye correlation function, which in this case must be of the form $\Gamma(r/\Lambda)$. Therefore, by a simple transformation of variables, the dimensionless, scaled intensity distribution can be expressed as a unique function of a scaled scattering wave vector y in the following form:

$$\frac{k_{max}^3 I(k)}{<\eta^2>} = \int_0^\infty dx \, 4\pi x^2 j_0(xy)\Gamma(x), \qquad (5.8)$$

where $x = k_{max}r$, $y = k/k_{max}$, and $<\eta^2>$ is the so-called invariant of the scattering.

Thus, by plotting the scaled intensity as a function of the scaled variable y, all the scattering intensity distributions at different temperatures within a single phase region should collapse into a single master curve. In this way, the distinct local structures associated with different phases occurring at different temperature ranges can be identified.

Four sets of scaling plots as a function of temperature are shown in Figure 5.11 at the four different volume fractions indicated in the phase diagram. In Figure 5.11(a), the scaling plots of SANS intensities at $\phi = 0.535$ are shown at

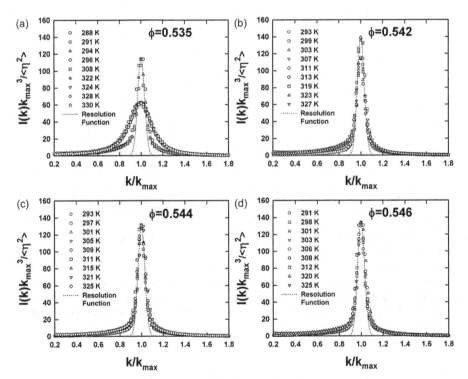

Figure 5.11 (a) The scaling plots of SANS intensities at $\phi = 0.535$ at different temperatures. (b) The scaling plots of SANS intensities at $\phi = 0.542$ at different temperatures. A temperature-dependent degree of disorder characterises the system. (c) The scaling plots of SANS intensities at temperatures ranging from 293 to 325 K at $\phi = 0.544$, which is predicted to be the volume fraction in which the A_3 singularity point is located. (d) The scaling plots of SANS intensities at $\phi = 0.546$, a volume fraction beyond the A_3 point. From Chen *et al.* (2003). Reprinted with permission from AAAS.

different temperatures. The two-dimensional SANS pattern showing a single uniform ring implies that the sample is amorphous in the entire temperature range at this concentration. From the width of the peak it is possible to infer that there are two distinct degrees of disorder depending on temperature. Whereas the narrower peak, which is resolution-limited, represents the glassy state, the broader peak, which is much broader than the resolution, represents the liquid state with a wider distribution of the inter-particle distances. The figure indicates that the system shows a re-entrant liquid to attractive glass to liquid transition. Therefore the transition between the ergodic liquid state and the non-ergodic attractive glass state can be driven by varying the temperature, in agreement with the phase diagram. It shows a substantial difference between the local structure peaks of the resolution-limited non-ergodic state and a much broader ergodic state peak (Chen *et al.*, 2002). Figure 5.11(b) shows the scaling plots of SANS intensities at $\phi = 0.542$ at different temperatures. A temperature-dependent degree of disorder characterises the system. Whereas the narrower peak is resolution limited, the slightly broader peak is also nearly resolution-limited but lower in intensity. It can be interpreted as showing a glass-to-glass transition, i.e. a transition between two amorphous solid states with different degrees of disorder. The higher peak corresponds to the repulsive glass, and the lower peak to the attractive glass. Thus, it is possible to conclude that the repulsive glass is better ordered locally than the attractive glass. Therefore, when increasing ϕ one drives the system into the phase region where the two glass-forming mechanisms nearly balance each other ($T^* \approx 1$). Changing T in turn affects the well depth and thus T^*, and a transition from the repulsive glass (lower temperatures) to the attractive glass (higher temperatures) is triggered. In the scaling plots of SANS intensities at temperatures ranging from 293 K to 325 K at $\phi = 0.544$, Figure 5.13(c), which is predicted to be the volume fraction within which the A_3 singularity point is located, one sees that the two sharp peaks merge into one sharp peak independent of temperature, signifying that the local structures of the two glasses become identical in their local structure at this volume fraction. Increasing ϕ further to 0.546, Figure 5.13(d), a volume fraction beyond the A_3 point, the scaling plots of SANS intensities remain the same as in the previous volume fraction. All the scaled intensities are again characterised by a unique length scale, and collapse into one single master curve independent of temperature, showing the identical local structure of the two glasses. This constitutes a proof that the MCT predictions are accurate.

Photon correlation spectroscopy was extended also in this case in order to investigate the predicted dynamical singularity at the A_3 point. In Figure 5.12, intermediate scattering functions measured at $k = 0.00222 \, \text{Å}^{-1}$ and at the same four different volume fractions as before are shown as a function of temperature. As shown in Figure 5.12(a), in the liquid state, the long-time limits of the ISFs,

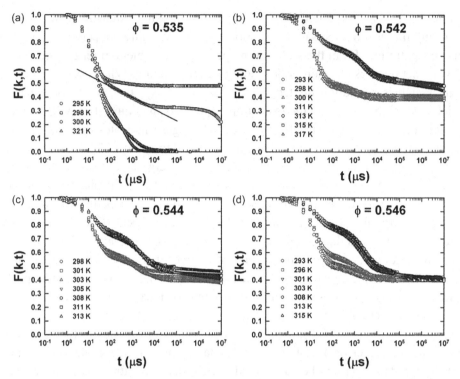

Figure 5.12 (a) The ISFs measured at $\phi = 0.535$, where the liquid-to-attractive glass transition is predicted, as a function of temperature. (b) The ISFs measured at $\phi = 0.542$. According to MCT, there is a possibility to observe glass-to-glass transition by varying T^*. (c) The ISFs measured at $\phi = 0.544$, where, according to MCT, the long-time limit of ISF of the two glasses become identical. (d) The ISFs measured at $\phi = 0.546$. This figure shores the same features as (c). It hints at critical point-like characteristics of the A_3 point. From Chen *et al.* (2003). Reprinted with permission from AAAS.

f_q values, are zero, whereas in the attractive glass state f_q is about 0.5, which is the Debye–Waller factor of the attractive glass state, i.e. the height of the plateau. The structural arrest transition is thus characterised by a bifurcation transition, i.e. a discontinuous change of f_q.

The ISF measured in an ergodic state exhibits a logarithmic relaxation (indicated by a straight-line fit) at intermediate times followed by a power-law decay before the final α relaxation, in agreement with MCT. The non-ergodic state ($T = 300\,\text{K}$), which has a finite plateau for long times, as indicated by the ISF, represents the attractive glass for which the Debye–Waller factor is about 0.5. Upon increasing ϕ to 0.542, Figure 5.12(b), by comparing the long-time limit of the ISFs with the predictions of MCT, one can identify the two different types of the glasses by the respective DWFs $f_{qA} \approx 0.5$ and $f_{qR} \approx 0.4$. The reason for observing two different values of DWFs can be interpreted as the different degrees of localisation of density fluctuation in the two glasses. This figure combined with Figure 5.11(b) gives firm evidence of the repulsive glass ($f_{qR} \approx 0.4$) to attractive glass ($f_{qA} \approx 0.5$) transition. The ISFs in Figure 5.12(c), $\phi = 0.544$ and Figure 5.12(d), $\phi = 0.546$, indicate that at A_3 and beyond, the DWFs of the two glassy states become identical, with a value of $f_q \approx 0.45$ for the two glassy states.

This is proof of the existence of the A_3 singularity in the phase diagram occurring at exactly the predicted volume fraction. It is, however, worth noting that at A_3

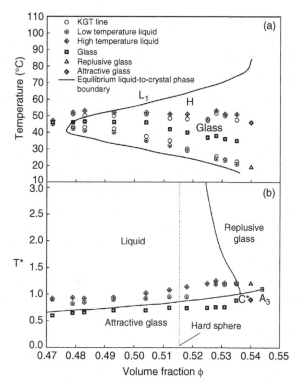

Figure 5.13 Mapping of the experimental phase diagram of the micellar solution on the theoretical phase diagram determined by MCT for a colloidal system with a short-range attraction. In (a), the solid line is the equilibrium liquid–crystal transition line, empty symbols are the re-entrant kinetic glass transition lines, and filled symbols are the phase points where the experimental data are analysed. (b) Shows the theoretical phase diagram of a colloidal system interacting by a short-range ($\epsilon = 0.03$) attractive square-well potential predicted by MCT. The phase points shown in (a) are then mapped into the corresponding symbols in (b) using the results of SANS data analyses. The results of the mapping is seen to confirm the existence of the attractive branch in the predicted glass transition boundary and the glass–glass transition (Chen *et al.*, 2002). Copyright 2002 by The American Physical Society.

and beyond, the intermediate-time relaxations (the β relaxation region) of the two glassy states are clearly different in spite of the fact that the long-time relaxation becomes identical. In summary, photon correlation spectroscopy was employed to verify that the L64/D_2O micellar system follows the phase behaviour predicted by MCT with the use of a square-well potential with a short-range attraction of size characterised by the parameter $\epsilon = 0.03$. In particular, the existence of the A_3 singularity was experimentally detected in the phase diagram predicted by MCT. The SANS experiment further showed that although the local structures of the attractive and repulsive glasses are in general different, they become identical at the A_3 singularity.

In order to make a detailed comparison of the experimental results with the MCT predictions for colloids with attractive interactions, we summarise in Figure 5.13 the essential results of the extensive SANS data analysis, complemented by photon correlation measurements. Figure 5.13(a) gives the experimental phase diagram replotted in the temperature-volume fraction plane. It contains the equilibrium liquid–crystal phase boundary (solid line), the locus of experimentally determined kinetic glass transition temperatures (open circles), the points of low-temperature liquid locating approximately $1\,°C$ below the lower transition line, the points of high-temperature liquid locating approximately $1\,°C$ above the upper line, and the points in the glass region between the lower and upper lines. The upper transition line and the lower one merge at volume fraction $\phi = 0.472$ and at a temperature of $46\,°C$. Figure 5.13(b) shows the theoretical phase diagram predicted by mode-coupling theory in the effective temperature and the volume fraction plane of a colloid system with a short-range attraction, for the specific case of $\epsilon = 0.03$. The solid line gives the liquid–glass phase boundaries and the symbols represent the experimental data points mapped from Figure 5.13(a). It is important to note that fitted parameters ϕ and T^* obtained by analysing SANS intensity distributions enable us to perform this mapping and thus confirm the existence of the liquid-to-attractive-glass transition line and the glass–glass transition.

6

Models of gel-forming colloids

6.1 Introduction

As we mentioned in previous chapters, structural arrest refers in general to various phenomena which have in common a marked slowing down of the dynamics, including aggregation and cluster formation, percolation and glass transition. In the last few years the study of the glass transition of colloidal systems in the super-cooled region revealed a series of new phenomena which were interpreted as the possibility of including in a single framework glass and gel transition in liquids (Sciortino and Tartaglia, 2005).

Summarising what we described at length in previous chapters, the study of the glass transition in supercooled liquids was initiated in hard-sphere systems, where the cage effect is the relevant physical mechanism for the structural arrest. In the case of colloids an interaction potential with a hard core and an attractive tail brings in the usual phase separation with a critical point, but also a new mechanism of structural arrest, predicted theoretically. Besides the repulsive glass transition due to excluded volume effects, a line of attractive liquid-glass appears, the driving mechanism of which is the bonding between the particles. Since the attractive glass line extends to a wide range of volume fractions of the dispersed phase, the colloid could also give rise to a low-density gel, similar to the gelling due to cluster formation and aggregation at very low densities.

In order to clarify the mutual relation between the coexistence curve and the glass lines, a detailed study was performed (Tartaglia, 2007) in a model binary system with hard-core repulsion followed by a short-range square-well attractive potential. The ideal glass transition line has been studied with a simulation of the dynamics of the colloidal system and by evaluating the diffusion coefficient D for long times. The loci of constant D, the iso-diffusivity lines, are used and by extrapolating to the limit $D \to 0$ one gets an indication of the position of the arrest line. Mathematically the extrapolation, according to MCT, follows a power law in

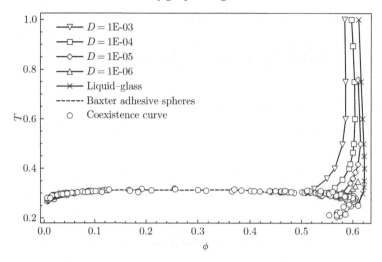

Figure 6.1 Phase diagram of a short-range square-well binary system. Points on the coexistence curve are shown, together with the iso-diffusivity lines for various values of D. The extrapolated liquid–glass line intersects the binodal near $\phi \approx 0.6$ and $T \approx 0.26$. The dashed line refers to a computation of the Baxter adhesive spheres model (Tartaglia, 2007). Copyright 2008, AIP Publishing LLC.

temperature or volume fraction of the dispersed phase. The generic result is that in the case of colloids the glass line intersects the binodal line below the critical point. A direct evaluation of the binodal line, using the distribution of the particles in sub-volumes of the the total volume confirms the previous results. Figure 6.1 shows the relative positions of the binodal and glass–liquid transition lines.

The interesting consequence of the situation described above is that it is possible, with a deep quench in the two-phase region, to generate a glassy phase, as opposed to the dense liquid one gets when quenching to a temperature above the intersection between binodal and glassy lines. This has been interpreted as a non-equilibrium route to the structural arrest characteristic of a glass phase, which was identified with a gel phase of the colloidal suspension.

The formation of an equilibrium gel in colloids is not possible, unless one is able to extend the glass line to lower values of the volume fraction. Since crossing the binodal line seems to be unavoidable, the only possibility remains to reduce the size and extension of the two-phase region. The efforts made in this direction using experiments, theory and simulations are the topics of the rest of the chapter.

From the experimental point of view the competition between spinodal decomposition and dynamical arrest was studied in an aqueous solution of the globular protein lysozyme, which is close to the colloidal system with short-range attractive interactions we considered earlier (Cardinaux *et al.*, 2007). The aim of the experiment was to investigate whether dynamical arrest can only occur via an arrested spinodal decomposition, or whether there exist also an equilibrium route

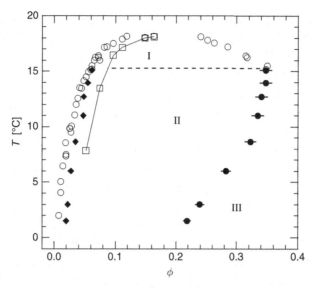

Figure 6.2 Kinetic phase diagram of aqueous lysozyme solutions showing regions of complete demixing (I), gel formation (II) and glasses (III). Full symbols stand for the results of the centrifugation experiments: circles, arrested dense phase, diamonds, dilute phase (Cardinaux *et al.*, 2007). Copyright 2007 by The American Physical Society.

to gelation, where the gel line becomes more stable than the coexistence curve. The experimentally determined phase diagram was obtained using various measuring techniques, essentially rheology and centrifugation, to separate the glassy phase and quantitatively determine the location of the arrest line in the unstable region of the phase diagram. It is reported in Figure 6.2.

The local arrest line formed by the volume fractions of the glassy phase crosses the coexistence curve at $\phi = 0.34$ and $T = 15.0 \pm 0.3$ and extends deep into the unstable region. It delimits a region where homogeneous attractive glasses can be reached by quenches at sufficiently high volume fractions and low T (III) from a large region (II) where gels are formed via an arrested spinodal decomposition. Microscopically, these gels correspond to a coexistence of a dilute fluid with a dense percolated glass phase. Finally, shallow quenches into region I lead to complete demixing. The situation is quite similar to the one observed in the numerical simulation reported previously.

Once the interactions between coexistence of phases and dynamical arrest are clarified, the way of separating their effects suggests the possibility of changing the attractive interactions among the elementary constituents, something that is in fact more easily attainable in the case of colloidal suspensions. The most promising route adopted recently is the use of colloidal particles where the surface structure is chemically modified in order to modulate the attractive interactions, the so-called patchy particles. In these systems the valence can be modified in a controlled way, and the solutions become similar to associated liquids, but on a much larger length scale.

In Section 6.2 simple models are treated which allow the numerical investigation of the effects of limiting the surface valence to the colloidal particles, a model which shows the expected effects on the shape and extension of the coexistence curve.

Section 6.3 is devoted to a much more physical system of patchy particles, a type of colloidal system that was recently manufactured in a rather efficient way. In this case the analysis of the phase coexistence and criticality can be brought much further and leads also to some new and interesting effects, among them the possibility of defining limiting structures, such as the so-called empty liquids.

Section 6.4 is more specifically devoted to the analogy between the artificially manufactured patchy particles and associated liquids, with particular emphasis on tethraedal ordering which leads to systems undergoing glass formation in molecular liquids and gellification in colloidal suspensions.

6.2 Module – Limited valence models

A first attempt to reduce the effect of the attractive interactions between two colloidal particles, and therefore to decrease the size of the two-phase region, is to consider limited valence models (Zaccarelli *et al.*, 2005). The interaction considered has a hard-sphere component followed by a square-well attraction if the number n of particles surrounding a given one is less than a predetermined number n_{max}, $n < n_{max}$, but only hard core when $n > n_{max}$. This gives rise to an effective many-body interaction which limits the number of neighbours of a particle when n_{max} is assumed to be less than 12, the maximum number of neighbours allowed by geometrical packing. As a result, the driving force for phase separation is significantly reduced, and open structures are favoured. We find that, for $n_{max} < 6$, the system can access regions usually dominated by phase separation and experience a dynamical slowing down by several orders of magnitude, thus entering the gel regime. It is interesting to note that the system undergoes the fluid–gel transition in equilibrium, so that the process is fully reversible.

Event-driven molecular dynamics simulations were performed of particles interacting via a square-well potential of the type described above, with the additional constraint that particles can form a maximum number of bonds. In Figure 6.3 the evolution of the spinodal line for different values of n_{max} in the (ϕ, T) plane is reported. The location of the spinodal line is estimated by bracketing it with the last stable state point and the first phase separating state point along each isochore. The unstable area in the phase diagram shrinks on decreasing n_{max}, showing that

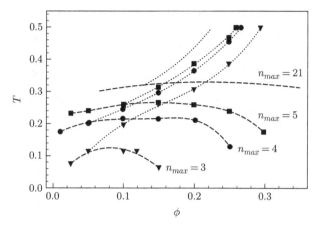

Figure 6.3 Phase diagram of the limited valence model for various values of n_{max}, showing phase coexistence (dashed lines) and corresponding percolation (dotted lines) loci. Lines are only for guidance (Tartaglia, 2007). Copyright 2008, AIP Publishing LLC.

the additional constraint opens up a significant portion of phase space, where the system can be studied in equilibrium one-phase conditions. In fact it is possible to study the dynamics of the model at very low temperatures, where the lifetime of the inter-particle bond increases, stabilising for longer and longer time intervals the percolating network. When the bond lifetime becomes of the same order as the observation time, the system will behave as a disordered solid. It is worth stressing that in the present model there is no thermodynamic transition associated with the onset of a gel phase.

To quantify these statements, the dynamics for all temperatures in the region where the system is in a single phase was studied. There are two distinct arrest lines for the system, a glass line at high packing fraction, and a gel line at low ϕ and T. In the (ϕ, T) plane the former is rather vertical and controlled by the volume fraction, while the latter is rather horizontal and controlled by temperature. Dynamics on approaching the glass line along isotherms exhibit a power-law dependence on ϕ, while dynamics along isochores follow an activated Arrhenius dependence. The gel has clearly distinct properties from those of both a repulsive and an attractive glasses, and a gel to glass crossover occurs in a fairly narrow range in ϕ along low-T isotherms. Figure 6.4 shows the logarithm of the diffusivity D as a function of $1/T$, for $n_{max} = 4$. Arrhenius behaviour of D is observed at all ϕ at low T. Since at the lowest studied temperatures the structure of the system is already essentially independent of T, there is no reason to expect a change in the functional law describing the $T \rightarrow 0$ dynamics. In this respect, the true arrest of the dynamics is located along the $T = 0$ line, limited at low ϕ by the spinodal and

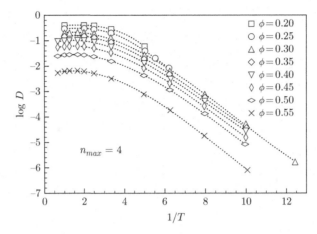

Figure 6.4 Arrhenius plot for the diffusivity for the $n_{max} = 4$ model along lines at constant volume fraction, showing activated behaviour. Reprinted with permission from Tartaglia (2007). Copyright 2008, AIP Publishing LLC.

at high ϕ by crossing of the repulsive glass transition line. This peculiar behaviour is possible only in the presence of limited valence, since when such a constraint is not present, phase separation preempts the possibility of accessing the $T \to 0$ Arrhenius window.

6.3 Module – Patchy colloids and Wertheim theory

The limited valence model introduced earlier has a precise experimental realisation in the so-called patchy colloids. In fact the possibilities offered by the physico-chemical manipulation of colloidal particles, due to the precise control of the interparticle potential, has made it possible to design colloidal molecules, particles decorated on their surface by a predefined number of attractive sticky spots, i.e., particles with specifically designed shapes and interaction sites (Manoharan *et al.*, 2003). A model version used for simulation purposes, but close to the real colloid molecules, is presented in Figure 6.5. To characterise patchy colloids it is necessary to be able to determine their phase diagram, and in particular to predict the regions in which clustering, phase separation, or even gelation are expected.

The theory of the physical properties of these systems was formulated around 1980 in the context of the physics of associated liquids. In an attempt to derive the essential features of association, the molecules were treated as hard-core particles with attractive spots on the surface, a realistic description close to the recently manufactured patchy colloidal particles. A thermodynamic perturbation theory appropriate for these models was introduced by Wertheim (1984) to describe association under the hypothesis of absence of closed ring configurations, and that

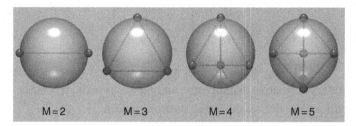

Figure 6.5 Patchy colloids with different number M of sticky spots. Schematic representation of the location of the square-well interaction sites, centres of the small spheres on the surface of the hard-core particle (Bianchi *et al.*, 2006). Copyright 2006 by The American Physical Society.

a sticky site on a particle cannot bind simultaneously to two (or more) sites on another particle. Such a condition can be naturally implemented in colloids, using an appropriate choice of the range of attraction, due to the relative large size of the particle compared to the range of the sticky interaction. Wertheim thermodynamic perturbation theory (WTPT) provides an important starting point to study the phase diagram of this new class of colloids, and in particular the role of the patches number.

The WTPT describes the free energy of associating liquids made by molecules with fixed valence, i.e. a small number of patchy interacting sites. The main assumption in the theory is that molecules form open clusters without closed bond loops, a condition that can be realised with patchy particles when the average functionality is small. The free energy due to bonding, to be added to the hard-sphere contribution, is given by

$$\frac{\beta A_{bond}}{N} = M \ \ln(1 - p_b) - \frac{M}{2} p_b. \tag{6.1}$$

The theory predicts that the bond probability p_b can be calculated from the chemical equilibrium between two non-bonded sites forming a bonded pair. For the present model, such a relation reads

$$\frac{p_b}{(1 - p_b)^2} = M \rho \Delta, \tag{6.2}$$

where

$$\Delta = 4\pi \int \ dr_{12} \ r_{12}^2 \ g_{HS}(r_{12}) \ \langle f_{ij}(12) \rangle_{\omega_1, \omega_2}. \tag{6.3}$$

The Mayer $f_{ij}(12)$-function between two arbitrary sticky sites i and j – respectively on particles 1 and 2, separated by $\mathbf{r}_{12}^{ij}$ and interacting through the square-well potential V_{SW} – is

$$f_{ij}(12) = \exp\left[-\frac{V_{SW}(\mathbf{r}_{12}^{ij})}{k_B T}\right] - 1. \tag{6.4}$$

In Eq. (6.4), $\langle f_{ij}(12)\rangle_{\omega_1,\omega_2}$ represents an angle average over all orientations of the two particles at fixed relative distance r_{12}, and $g_{HS}(r_{12})$ is the reference hard-sphere radial distribution function in the linear approximation

$$g_{HS}(r_{12}) = \frac{1 - 0.5\phi}{(1-\phi)^3} - \frac{9\phi(1+\phi)}{2(1-\phi)^3}(r_{12} - 1). \tag{6.5}$$

The theory was applied to various cases with different values of the number M of sticky spots, in order to show how the use of patchy colloids gives rise to diverse physical situations that can be satisfactorily described by the theoretical approach.

Bianchi *et al.* (2006) studied the general properties of a system of hard-sphere particles with a small number M of identical short-range, square-well attraction sites per particle, distributed on the surface. The number of possible bonds per particle is the key parameter controlling the location of the critical point, as opposed to the fraction of surface covered by attractive patches. We summarise here results of extensive numerical simulations of this model in the grand-canonical ensemble to evaluate the location of the critical point of the system in the (ϕ, T) plane as a function of M. The simulation results are complemented with the evaluation of the region of thermodynamic instability according to the WTPT reported above. Both theory and simulation confirm that, on decreasing the number of sticky sites, the critical point moves toward smaller volume fraction and temperature values. Note that while adding to hard spheres a spherically symmetric attraction creates a liquid–gas critical point which shifts toward larger ϕ; on decreasing the range of interaction the opposite trend is present when the number of interacting sites is decreased. Simulations and theory also provide evidence that for binary mixtures of particles with two and three sticky spots (where $\langle M \rangle$, the average M per particle can be varied continuously down to two by changing the relative concentration of the two species) the critical point shifts continuously toward vanishing. This makes it possible to realise equilibrium liquid states with arbitrary small ϕ (which have been given the name *empty liquids*), a case which cannot be realised using spherical potentials.

To locate the critical point, grand-canonical Monte Carlo (MC) simulations were performed and histogram reweighing for $M = 5, 4$ and 3 and for binary mixtures of particles with $M = 3$ (fraction α) and $M = 2$ (fraction $1 - \alpha$) at five different compositions, down to $< M > \equiv 3\alpha + 2(1 - \alpha) = 2.05$. MC steps involved 500 random attempts to rotate and translate a random particle and one attempt to insert or delete a particle. On decreasing $< M >$, numerical simulations become particularly time-consuming, since the probability of breaking a bond $\sim e^{1/T_c}$ becomes progressively small. To improve statistics, averages over 15–20 independent MC realisations were performed. After choosing the box size, the T and the chemical potential μ of the particle(s), the grand-canonical simulation evolves the system toward the corresponding equilibrium density. If T and μ correspond to the critical point values, the number of particles N and the potential energy E of the simulated system show ample fluctuations between two different values. The linear combination $x \sim N + sE$ (where s is named field mixing parameter) plays the role of the order parameter of the transition. At the critical point, its fluctuations are found to follow a known universal distribution, i.e. (apart from a scaling factor) the same that characterises the fluctuation of magnetisation in the Ising model.

The upper panel of Figure 6.6 shows the resulting volume fraction fluctuations distribution $P(\phi)$ at the estimated critical temperature T_c and critical chemical potentials for several M values. The distributions, whose average is the critical packing fraction ϕ_c, shift to the left on decreasing M and become more asymmetric, signalling the progressive increasing role of the mixing field. In the lower panel, the calculated fluctuations of x, $P(x)$, are compared with the expected fluctuations for systems in the Ising universality class. They provide evidence that the critical point has been properly located and the transition belongs to the Ising class. Data show a clear monotonic trend toward decreasing T_c and ϕ_c on decreasing M.

Figure 6.7 shows a quantitative comparison of the numerical and theoretical estimates for the critical parameters T_c and ϕ_c performed using WTPT. Theory predicts quite accurately T_c but underestimates ϕ_c, nevertheless clearly confirming the M dependence of the two quantities. The overall agreement between Wertheim theory and simulations reinforces our confidence in the theoretical predictions and supports the possibility that, on further decreasing $< M >$, a critical point at vanishing ϕ can be generated.

WTPT allows also to evaluate the locus of points where $(\partial P/\partial V)_T = 0$, which provides, at the mean field level, the spinodal locus. Figure 6.8 shows the predicted spinodal lines in the (ϕ, T) plane for several M values. On decreasing M also the liquid spinodal boundary moves to lower values, suggesting that the region of stability of the liquid phase is progressively enhanced. It will be desirable to investigate the structural and dynamical properties of such empty liquids by experimental and numerical work on patchy colloidal particles.

One can therefore conclude that for particles interacting with attractive spherical potentials, phase separation always destabilises the formation of a homogeneous arrested system at low temperature. On the contrary, with patchy particles and small $< M >$, disordered states in which particles are interconnected in a persistent gel network can be reached at low T, without encountering phase separation. Indeed, at such low T, the bond-lifetime will become comparable to the experimental observation time. Under these conditions, a dynamic arrest phenomenon at

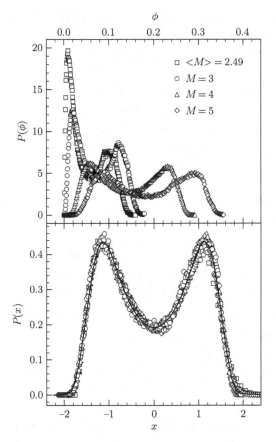

Figure 6.6 Order parameter distributions of patchy colloids. The upper panel shows the volume fraction fluctuations distribution $P(\phi)$ at the critical point for various M values. The lower panel shows the order parameter x fluctuations distribution $P(x)$ for all studied cases, compared with the expected distribution (full line) for systems at the critical point of the Ising universality class (Bianchi *et al.*, 2006). Copyright 2006 by The American Physical Society.

small ϕ will take place. It will be thus possible to approach dynamic arrest continuously from equilibrium and to generate a state of matter as close as possible to an ideal gel. Patchy colloids made of particles with two sticky spots self-assemble only in chains, while systems with $M > 2$ form ramified clusters and show a phase separation at low volume fractions. It is also interesting to study the process of formation of clusters or, in other words, the mechanism of self-assembly of the colloidal particles.

It is well known that one of the most peculiar aspects of the physics of colloids is their ability to organise spatially in self-assembling structures. For systems with a small average functionality it is possible to provide a parameter-free full

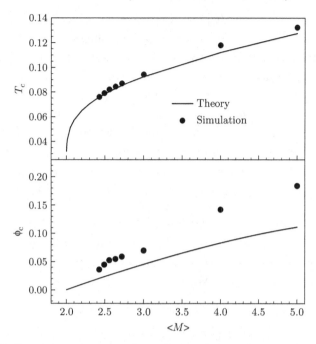

Figure 6.7 Comparison between theoretical and numerical results for patchy particles with different number of sticky spots. Upper and lower panels show respectively the location of the critical values T_c and ϕ_c (Bianchi *et al.*, 2006). Copyright 2006 by The American Physical Society.

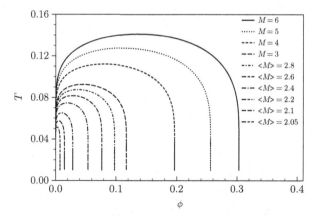

Figure 6.8 Spinodal curves calculated according to WTPT for the patchy particles models for several values of $< M >$ (Bianchi *et al.*, 2006). Copyright 2006 by The American Physical Society.

description of the self-assembly process (Bianchi *et al.*, 2007). One of the simplest, but not trivial self-assembly processes, namely a binary mixture of particles with two and three attractive sites, was studied theoretically and numerically. The presence of three-functional particles – which act as branching points in the self-assembled clusters – introduces two important phenomena which are missing in equilibrium chain polymerisation: a percolation transition, where a spanning cluster appears; and a region of thermodynamic instability, where phase separation between colloid-poor and colloid-rich regions appears.

The investigation considers a binary mixture composed of $N_2 = 5670$ bi-functional particles and $N_3 = 330$ three-functional ones. The resulting average number of sticky spots per particle, i.e. the average functionality, is $\langle M \rangle = (2N_2 + 3N_3)/(N_2 + N_3)$. Particles are modelled as hard spheres of diameter σ, whose surface is decorated by two or three bonding sites at fixed locations. Sites on different particles interact via a square-well potential V_{SW} of depth u_0 and attraction range $\delta = 0.119\sigma$. More precisely, the interaction potential $V(1, 2)$ between particles **1** and **2** is

$$V(\mathbf{1}, \mathbf{2}) = V_{HS}(\mathbf{r_{12}}) + \sum_{i=1,n_1} \sum_{j=1,n_2} V_{SW}(\mathbf{r}_{12}^{ij}), \tag{6.6}$$

where V_{HS} is the hard-sphere potential and $\mathbf{r_{12}}$ and $\mathbf{r}_{12}^{ij}$ are respectively the vectors joining the particle–particle centres and the site–site (on different particles) locations; n_i indicates the number of sites of particle i. The well width δ is chosen to ensure that each site, due to the steric effect, is engaged at most in one interaction. Distances are measured in units of σ. Temperature is measured in units of the potential depth (i.e. the Boltzmann constant $k_B = 1$). In the studied model, bonding is properly defined: two particles are bonded when their pair interaction energy is $-u_0$. This means that the potential energy of the system is proportional to the number of bonds. The lowest energy state of the system (the ground state energy) coincides with configurations in which all bonds are formed, i.e $E_{gs} = -u_0 (2N_2 + 3N_3)/2$. As a result, the bond probability can be calculated as the ratio of the potential energy E and E_{gs}, $p_b = E/E_{gs}$. Pairs of bonded particles are assumed to belong to the same cluster. Standard Monte Carlo Metropolis simulations were performed at several temperatures and densities. An MC step was defined as an attempted move per particle. A move was defined as a displacement of a randomly selected particle in each direction of a random quantity distributed uniformly between $\pm 0.05\,\sigma$ and a rotation around a random axis of a random angle uniformly distributed between ± 0.1 radiant.

In the following pages the comparison between the simulation and the WTPT is briefly described, highlighting the salient results.

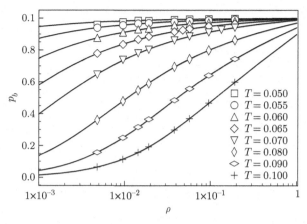

Figure 6.9 Temperature and volume fraction dependence of the bond probability p_b. Points are simulation results based on Monte Carlo simulations. Lines are parameter-free predictions based on the WTPT. Note that at low T the system reaches a fully bonded configuration. Reprinted with permission from Bianchi *et al.* (2007). Copyright 2007 American Chemical Society.

Bond probability

The WTPT theory describes the free energy of associating liquids made by molecules with fixed valence, i.e. a small number of patchy interacting sites, and forming open cluster without closed bond loops. The comparison between the theoretical predictions and the numerical data for the T and ρ dependence of p_b is shown in Figure 6.9. Data show clearly that the theory is able to predict precisely p_b (or equivalently the system potential energy) in a wide T and ρ range. At low T, $p_b \to 1$, and the system approaches a fully bonded disordered ground state configuration.

Combining Wertheim and Flory–Stockmayer theories

To derive information on the structure of the system and the connectivity of the aggregates we combine the Wertheim and Flory–Stockmayer (FS) theories (Flory, 1953; Stockmayer, 1943), both of which rely on the absence of closed bonding loops. The Wertheim prediction for p_b can thus be consistently used in connection with the FS approach to predict the T and ρ dependence of the cluster size distributions.

The number of clusters (per unit volume) containing l bi-functional particles and n three-functional ones can be written

$$\rho_{nl} = \rho_3 \frac{(1 - p_b)^2}{p_3 p_b} [p_3 p_b (1 - p_b)]^n [p_2 p_b]^l w_{nl},$$

$$w_{nl} = 3 \frac{(l + 3n - n)!}{l! n! (n + 2)!}, \tag{6.7}$$

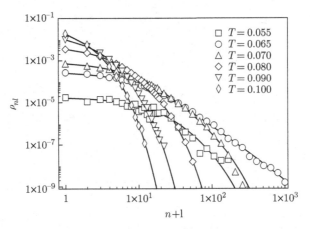

Figure 6.10 Cluster size distribution for $\rho\sigma^3 = 0.04$ for some of the investigated T. Points are simulation data and lines are the corresponding theoretical curves. The number of finite clusters (per unit volume) containing l bi-functional units plus n three-functional ones. Reprinted with permission from (Bianchi *et al.*, 2007). Copyright 2007 American Chemical Society.

where $p_3 \equiv 3N_3/(2N_2 + 3N_3)$ and $p_2 = 1 - p_3$ are the probabilities that a randomly chosen site belongs to a three-functional or to a two-functional particle, p_b is given by Eq. (6.2), and w_{nl} is a combinatorial contribution. Distributions are normalised in such a way that $\sum_{ln,l+n>0}(l+n)\rho_{nl} = p_2 + p_3$. As shown in Figure 6.10, on decreasing T, the ρ_{nl} distribution becomes wider and wider and develops a power-law tail with exponent -2.5, characteristic of loopless percolation (Rubinstein and Colby, 2003). On further decreasing T, the distribution of finite size clusters progressively shrinks, since most of the particles attach themselves to the infinite cluster. Data show that Eq. (6.7), with no fitting parameters, predicts extremely well the numerical distributions at all state points, both above and below percolation.

In the framework of the FS approach it is also possible to evaluate the number of clusters $\rho_c \equiv \sum_{nl} n_{nl}$ as a function of p_b, irrespectively of the cluster size. Below percolation, in the absence of bonding loops, the relation between ρ_c and p_b is linear, since each added bond decreases the number of clusters by one. Above percolation the relation crosses to a non-linear behaviour, so that the number of clusters becomes one when $p_b = 1$. As shown in Figure 6.11, the simulation data conform perfectly to the theoretical expectation both below and above percolation. This suggests that, when the average functionality is small, bonding loops in finite size clusters can always be neglected. This agreement, which covers the entire range of p_b values, implies that closed loops of bonds are statistically less favoured than the corresponding open structure.

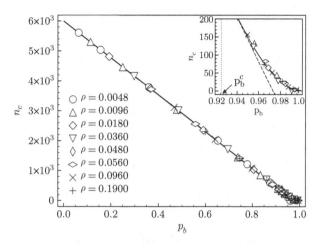

Figure 6.11 The number of finite-size clusters n_c (irrespective of their size) as a function of the bond probability. Symbols are simulation result and solid lines FS predictions. Below percolation, in the absence of bonding loops, n_c is given by the difference between the total number of simulated particles N and the number of bonds N_b, since each added bond decreases n_c by one, i.e. $n_c = N - N_b = N - p_b(2N_2 + 3N_3/2)$. This linear relation (dashed line) can be extended above percolation (dotted line), but never beyond the point where $n_c < 1$. The inset enlarges the region of large p_b values, to provide evidence that the FS approach is valid over the entire p_b range. Reprinted with permission from Bianchi *et al.* (2007). Copyright 2007, American Chemical Society.

Percolation

The three-functional particles act as branching points of the network formed by long chains of two-functional ones. It is possible to predict the number of finite size clusters composed of n three-functional units, irrespective of the number of bi-functional units. The system can thus be considered as a one-component fluid of three-functional particles forming clusters, in which the bonding distance between the three-functional particles is given by the length of the chains formed by the bi-functional units. Following again FS, it is possible to predict the p_b value at which the systems develops a percolating structure: when $p_b \geq p_b^c \equiv 1/(1+p_3) = 0.9256$, an infinite cluster is present in the system. The percolation line is thus the locus of points in the phase diagram such that $p_b(T, \rho) = p_b^c$, with $p_b(T, \rho)$ given in Eq. (6.2).

Liquid–gas phase separation

The WTPT predicts a phase separation into two phases of different density and connectivity at small ρ for any non-vanishing amount of branching point (Bianchi *et al.*, 2006). The theoretical spinodal curve, the line separating the stable (or

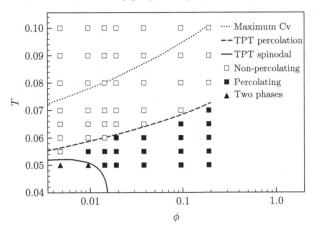

Figure 6.12 Phase diagram of the 2–3 mixture. Lines are theoretical predictions the continuous line is the spinodal curve obtained from the WTPT equation of state; the dotted line shows the locus of maximum specific heat C_V; the dashed line shows the percolation locus. Points are simulation results, white and black squares are the non-percolating and the percolating equilibrium states. Triangles are non-equilibrium states in the two-phase region. Reprinted with permission from Bianchi *et al.* (2007). Copyright 2007, American Chemical Society.

metastable) state points from the unstable ones, is defined as the locus of points such that the volume V derivative of the pressure P vanishes, i.e. $(\partial P / \partial V)_T = 0$. It is located below the percolation line and the two lines merge asymptotically for $T \to 0$ and $\rho \to 0$ as shown in Figure 6.12. The configurations for the two investigated state points located inside the spinodal are characterised by a bimodal distribution of the density fluctuations and a very large value of the small-angle structure factor, indicating a phase-separated structure.

Maximum of the specific heat C_V

As shown in the case of chain polymerisation (Sciortino *et al.*, 2007), the theory predicts a line of constant volume specific heat C_V maxima, calculated from the inflection point in the p_b as a function of T curves, see Figure 6.9, which also agrees very well with the simulation results. The line of C_V extrema is also shown in Figure 6.12. The presence of a maximum in C_V is a characteristic of bond-driven assembly, and the locus of maxima in the phase diagram is one of the precursors of the self-assembly process for low-functionality particles.

6.4 Module – Gel-forming colloids and network glass formers

As stressed in the preceding chapters, colloidal solutions have often been considered as mesoscopic molecular systems, i.e. systems on expanded length and time scales, which show many of the typical behaviour of molecular systems, notwithstanding the wide difference between the respective characteristic scales. A particularly interesting case is the one of colloids with chemically treated surfaces, such as the ones manufactured with different sticky spots, so that they show a specific valence. They form colloidal gels which are an arrested state of matter at low density or packing fraction, the gel being formed via a long lifetime bonded-particle percolating structure. In the gel, particles are tightly bonded to each other, so that the thermal energy is significantly smaller than the bonding energy, and the lifetime of the thermo-reversible bonds is comparable or longer than the experimental observation time. This type of colloids show a pronounced analogy with the behaviour of associated liquids. One of these analogies between patchy colloidal particles and associated liquid was considered in detail by Sciortino (2008), suggesting that network strong glass-forming liquids are the atomic and molecular counterpart of patchy thermoreversible colloidal gels.

The connection between patchy colloidal gels and network glass-forming liquids is provided by the limited valence of the inter-particle interactions, i.e. by the presence of a limit in the number of bonded nearest neighbours. The structure of systems with limited valence is not determined by short-range repulsion, like in hard-sphere systems, but rather by strong directional interactions, such as hydrogen bond in molecular liquids or three-body interactions, as in silicon.

In the previous sections it was shown that the field of stability of fluid phases in colloidal systems is limited by phase separation and the coexistence curve, but also from glass formation and the glass line, which meet and allow only the formation of a non-equilibrium arrested phase.

The way to avoid this situation is to include directional interactions, the effect of which is to shrink the extent of the phase-separated region and opening a gap between phase separation and glass formation. Indeed, it becomes now possible to reach low temperatures without encountering phase separation in conditions such that packing is not an effective arrest mechanism. The sketch in Figure 6.13 gives a schematic representation of the salient features, both static and dynamic, of the phase diagram of network-forming limited valency systems (Largo *et al.*, 2007).

As noted earlier, the thermodynamic of systems with limited valence is characterised by the presence of a temperature at which the constant volume specific heat C_V has a maximum. The locus of C_V extrema in the phase diagram is a clear experimentally detectable evidence of the leading role of the bonding interaction in the thermodynamic of the system, since it can be interpreted as arising from the equilibrium between broken and formed bonds. Moreover, bonds are very often due to well defined interactions that can be detected experimentally or in numerical simulations. We also stress that the study of the connectivity properties of limited valence systems shows the presence of a bond percolation line, passing always

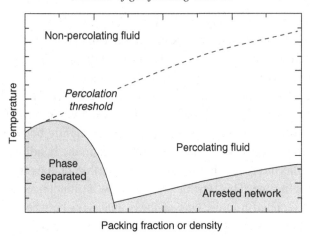

Figure 6.13   Schematic representation of the relevant thermodynamic and dynamic properties of network-forming limited valency systems. It applies to tetrahedral network-forming systems, such as silica and water, described in this section. Reprinted with permission from Largo *et al.* (2007). Copyright 2007, American Chemical Society.

slightly above the critical point. At small valence, the C_V-max line lies above the percolation line. The opposite behaviour is observed at large valence.

The limited valence is introduced through strong directional interactions such as the hydrogen bond in molecular systems, or via interplay of isotropic competing interactions or due to specific three-body interactions, as in silicon. Therefore it is interesting to discuss the properties of simple models with valence four, in order to make contact with the most common network-forming glass formers. Several simple patchy models of four coordinated particles have been recently studied, with the aim of finding out the universal features related to the valence. These models include the ones contained in the following list together with a brief description. All the models have been investigated essentially using numerical simulations, a good account is also given in Zaccarelli (2007).

(i) **Limited valence square-well model (NMAX),** described at length in a previous section.

(ii) **Primitive model for water (PMW)** This model was introduced several years ago (Kolafa and Nezbeda, 1987) as a model for the water molecule, described as a hard sphere decorated with four interaction sites, arranged on a tetrahedral geometry, which are meant to mimic the two hydrogens and the two oxygen lone pairs of the water molecule (Figure 6.14). The PMW has been studied in detail since it is both a valid candidate for testing theories of bond association and a model capable of reproducing the thermodynamic anomalies of water. It is representative of the larger class of

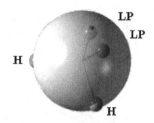

Figure 6.14 In the primitive model for water (PMW) each particle is modelled as a hard-core sphere of diameter σ. The four interaction sites are located in a tetrahedral arrangement. Two of the sites (the H-sites) are located on the surface whereas the remaining two sites (LP) are located inside the sphere, at distance 0.45σ from the centre. Only H-sites and LP-sites on different particles interact with a square-well interaction (Romano *et al.*, 2007). ©IOP Publishing. Reproduced with permission. All rights reserved.

particles interacting via localised and directional interactions, a class of systems that includes, besides network-forming molecular systems, also proteins and colloidal particles, the systems we are treating here.

Numerical studies of the static properties of the model (Romano *et al.*, 2007), show that, similarly to what happens in the case of particles interacting via short-range attractive spherical potentials, the gas–liquid phase separation is metastable with respect to crystal formation. In fact it appears in the region of the phase diagram where the crystal phase is thermodynamically favoured.

In this system there is a window of packing fraction values in which it is possible to avoid phase separation, reaching very low temperatures where states with extremely long bond lifetimes exist which favour the formation of spanning networks of long-lived bonds. The network provides indication of gel formation for colloidal systems or glass formation for network-forming liquids. The region of gas–liquid instability of the PMW model is significantly reduced as compared to that from equivalent models of spherically interacting particles, showing the possibility to detect kinetic arrest in a homogeneous sample driven by bonding as opposed to packing interactions. The connectivity properties of the equilibrium configurations allow also the study of percolation in PMW. Similar to what happens in short-range square-well potentials and in other simple models, the percolation line intersects the spinodal curve close to the critical point, on the left side of it. The equilibrium gas phase at volume fractions smaller than the crossing point is always non-percolating. Percolation at very small volume fractions can only be achieved as a result of an out-of-equilibrium process, quenching the system inside the spinodal curve. As far as the dynamical behaviour is concerned (De Michele *et al.*, 2006a), one of the interesting features of PMW is that, unlike models of spherically symmetric interacting particles, the liquid can be supercooled

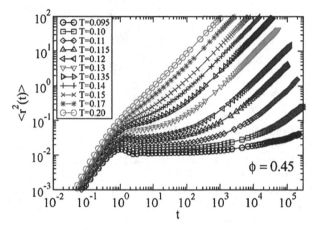

Figure 6.15 Mean square displacement in the MPW for various temperatures at a volume fraction $\phi = 0.45$. Reprinted with permission from De Michele *et al.* (2006a). Copyright 2006, American Chemical Society.

without crossing the gas–liquid spinodal in a wide region of packing fractions. Around an optimal density, a stable fully connected tetrahedral network of bonds develops. By analysing the dynamics of the model one finds evidence of anomalous behaviour since around the optimal packing, dynamics accelerate on both increasing and decreasing it. The shape of the lines of constant diffusivity in the volume fraction and temperature plane allows to trace the shape of the dynamic arrest line in the phase diagram of the model.

An example of the mean shared displacement used to evaluate diffusivity is shown in Figure 6.15 for a selected volume fraction $\phi = 0.45$ and different temperatures (De Michele *et al.*, 2006a). It is worth observing that the static percolation curve has no effect on dynamics since there is no dynamic arrest at the percolation transition. When a well-developed tetrahedral network can form, it is possible to cool the system to temperatures at which arrest is observed, without any phase separation. The mean square displacement develops a clear intermediate region where only the dynamic inside the cage is left, so that caging is not associated with excluded volume interactions, but with the formation of energetic bonds.

(iii) **Primitive model for silica (PMS)**   Recently a simple model based on low coordination and strong association has been introduced for silica (Ford *et al.*, 2004), one of the most important network-forming materials. PMS is a rigid site model that describes a silicon atom as a hard sphere, whose surface is decorated by four sites, arranged according to a tetrahedral geometry (Figure 6.16). The oxygen atom is also modelled as a hard sphere, but with only two additional sites. The only well attraction takes place between distinct

Figure 6.16 Three-dimensional view of the PMS molecular model of silica. Left: O atom modelled as a hard-core sphere plus two additional bonding sites on the surface. The bond angle is fixed to 145.8. Right: Si atom modelled as a hard-core particle, plus four additional bonding sites located along the direction of a tetrahedral geometry on the surface of the hard sphere. The O and Si sites interact with a spherical square-well potential. Reprinted with permission from De Michele *et al.* (2006b). Copyright 2006, AIP Publishing LLC.

sites of Si and O atoms. Despite these simple assumptions, the resulting phase diagram, which includes three solid phases, corresponding to cristobalite, quartz and coesite, and a gas and a fluid phase, compares very favourably with the experimental one.

A molecular dynamics simulation (De Michele *et al.*, 2006b) allows to make a comparison of the PMS model with more realistic potentials that are available, in particular detailed information on the structure of the system on cooling are contained in the particle–particle radial distribution function $g_{i,j}(r)$ where i and j label the different particle species. Figure 6.17 shows the three partial radial distribution functions and compares them with data from the BKS (Bohr–Kramers–Slater) silica model (van Beest *et al.*, 1990).

From the dynamical point of view the mean square displacement was carefully studied, since its long time limit gives a measure of the diffusion coefficient D. In Figure 6.18 the slowing down of the dynamics of the PMS model is studied over more than six orders of magnitude. While at high temperature diffusion becomes T independent (an evidence that density is the only relevant variable at high T), on cooling the diffusion coefficient crosses to a T dependence consistent with an activation law.

The linearity of log D vs. $1/T$ is very striking in the region of the optimal number densities, where an unconstrained tetrahedral network can form at low T. The diffusion of oxygen particles is faster than that of Si particles and follows the same qualitative behaviour, crossing to an Arrhenius law at low T, but with an activation energy smaller compared to Si particles.

(iv) **Floating bond model (FB)** The models reported in this section for network-forming liquid systems are based either on many-body interactions (NMAX, DNA dendrimers) or angular constraints (patchy colloids). We now describe a simple model characterised by both pairwise additivity and

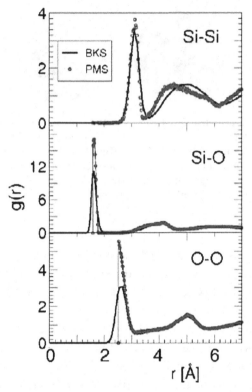

Figure 6.17 Radial distribution functions for Si–Si, O–O and Si–O at density $n =$ 0.30 and temperature $T = 0.07$ (symbols, in reduced units). Lines are results for the BKS silica (van Beest *et al.*, 1990) at $T = 2750$ K. Reprinted with permission from De Michele *et al.* (2006b). Copyright 2006, AIP Publishing LLC.

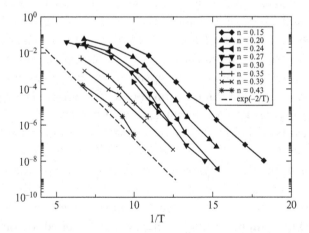

Figure 6.18 Temperature dependence of the diffusion coefficient of Si particles for various densities, showing Arrhenius behaviour at low temperatures. Reprinted with permission from De Michele *et al.* (2006b). Copyright 2006, AIP Publishing LLC.

spherical interactions, but capable of producing geometrical arrangement into a locally ordered network structure (Zaccarelli *et al.*, 2007).

The spherical model with directional interactions is obtained through the introduction of a second species in the mixture that represents a sort of floating bond. In this way, an effective one-component directional potential for the colloidal particles is obtained. The system is composed of a binary system of large hard-sphere colloidal particles which interact with small hard-sphere particles, which are the floating bonds in the original nomenclature. The interaction is a non-additive, short-range, attractive square-well potential. The role of the floating bonds is to link particles, providing a connection between them. The potential parameters are chosen in such a way that each floating bond connects no more than two particles, and the maximum number of floating bonds binding to a single particle is fixed. With this choice of the interactions and of the relative concentration of the two species it is possible to control the effective valence of the system.

The static phase diagram of the system is very similar to the one already described for patchy models, as well as for the other models of network-forming liquids. In fact the phase-separating region of the system, which is unstable, is confined to low packing fractions and small temperatures. The phase separation region is followed at larger volume fractions by a window where the mixture can be equilibrated to very low temperatures with a progressive formation of a tetrahedral network of bonds, an optimal network region where the system almost reaches the fully bonded configuration, the disordered ground state of the system, and hence a further lowering of the temperature would not produce any significant structural change. Destruction of the tetrahedral bonding is observed at high densities, induced by the packing constraints due to the competition between excluded volume and attractive interactions, so that the system forms a defective network.

From the dynamical point of view it is possible to discuss the relative location of the percolation locus, the locus of specific heat maxima and the liquid–gas spinodal. In the fluctuating bonds model, the percolation line provides the first indication of the clustering process upon lowering the temperature. Successive cooling brings to the presence of a line of specific heat maxima, and finally of the spinodal line. The constant-volume specific heat maxima line appears to be a characteristic of all bonded systems, since it is observed also in the case of valence two where no percolation is present.

(v) **Four-armed DNA dendrimers model**   One of the possibilities to control bond formation between mesoscopic particles is one of the pathways to the construction of complex materials from the molecular point of view. One of the most popular choices is offered by biological molecules such as DNA,

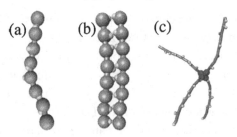

Figure 6.19 (a) Sketch of the model for a single DNA strand. The large spheres represent the repulsive cores along the chain, while the small ones are the attractive sticky spots carrying information on the base type. (b) Two isolated strands at low temperature spontaneously aggregate to form a double-stranded unit. (c) Single DNA tetramer, designed by joining four strands to a central point. Arms of different tetramers can bind in the same fashion as the two individual strands. Reprinted with permission from Largo *et al.* (2007). Copyright 2007, American Chemical Society.

the specific interactions of which are specially important. In particular, DNA single strands grafted on a core particle have been used, since controlling the length and sequence of the DNA strand makes it possible to modulate the interaction and alter the temperature or concentration at which assembly by the formation of double strands occurs. A specific minimal model we refer to represents particles composed of a central core, a molecule or a colloidal particle, to which several single-stranded DNA arms are attached (Largo *et al.*, 2007). This model was designed to mimic the general features of a four-armed DNA dendrimer complexes that have been synthesised experimentally. Each arm consists of a single DNA strand with an even number of bases; the sequencing of the bases is chosen such that two identical strands will bond to each other in head-to-tail order. As a result, these dendrimers have the potential to naturally assemble into higher-order structures. The head-to-tail bonding is achieved by choosing a sequence where the bases of the second half of a strand are complementary to the bases of the first half, but in reverse order.

As far as the structure of the fluid is concerned, when base pairing is favoured, in a small temperature interval, the system passes from a fluid state made of monomers to a fully connected network of four-coordinated particles. When the volume fraction is small the bonding interaction produces a phase coexistence of two phases, characterised by growing density fluctuations. When the density grows the phase separation disappears for every temperature in favour of the formation of a network of four-coordinated particles. The behaviour of the model is analogous to the one observed in the other models of four-coordinated patchy particles characterised by reduced valency, where a region of densities exists where the system remains stable against phase separation up to low temperatures. The system can access equilibrium states up to volume fractions where the formation of a glass due to packing constraints prevents equilibrium. In this window of intermediate densities, kinetic arrest is driven by the fact that the bond probability follows an Arrhenius law at low temperatures.

The process of network formation can be described by random percolation theory, the cluster size distribution at percolation being described by the characteristic power law with exponent 2.2. On decreasing the temperature a spanning cluster appears, but the dynamics do not arrest, since the bond lifetime is finite, showing that the percolating structure is transient.

All the models described above satisfy the single-bond per patch condition. Defining the distance between a pair of bonded particles as unit of length and the critical temperature as unit of temperature, all these models are characterised by phase diagrams which show extended similarities, as shown in Figure 6.20.

The gas–liquid unstable region is limited to scaled densities lower than ~ 0.6. Above that density, the liquid is stable and the structure of the liquid is characterised by an extended network of four coordinated particles (being the liquid state well within the percolation region). In all the single-bond per patch models, a well-defined ground state exists, provided by the geometrically fully bonded configuration. In all the models reported, the fully bonded state can be accessed in a region of densities whose width is controlled by the angular constraints. The approach to the ground state energy along isochoric paths is well described by an Arrhenius law in temperature. Universal features also control the particle dynamics in these systems. At low T, the temperature dependence of the diffusion coefficient and the viscosity approaches an Arrhenius law, the characteristic behaviour of so-called strong glass-formers.

Having examined the colloidal model systems, we turn to the case of molecular liquids in order to draw the analogy between the two classes. The class of network-forming liquids, i.e. systems with valence four, includes two very important liquids, namely water and silica. For these materials, several continuous potentials have been developed and extensively studied, providing a reasonable reproduction of several thermodynamic and dynamic properties of the real substances.

As far as silica is concerned, van Beest *et al.* (1990) proposed inter-atomic potentials of the force field consisting of a Coulomb term and a covalent short-range contribution (BKS). They assume that the electronic structure of silica is well described as a system of small rigid spherically symmetric ions which are undistorted by their environment. There is no significant difference in the electronic structure of the Si and O ions in different phases; the electronic structure

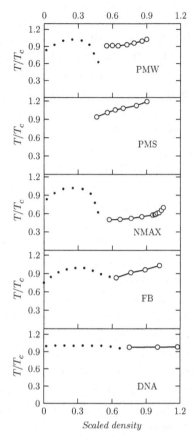

Figure 6.20 Gas–liquid spinodal line and iso-diffusivity line for five different models of patchy particles, in the scaled variables T/T_c and scaled number density. In all models, the gas–liquid unstable region extends for scaled number density < 0.6. The iso-diffusivity lines, providing an estimate of the shape of the glass line, are all mostly parallel to the temperature axis, bending up only on approaching large densities. Scaled densities are defined as number of particles divided by the volume, measured in units of the average distance between two bonded particles (Sciortino, 2008). With kind permission from Springer Science and Business Media.

of ions in bulk silica is well described by that in small clusters. The parameters have been chosen to reproduce the energetics of small hydrogen terminated silica clusters as well as the experimental structural parameters and elastic constants of quartz. The resulting potential has been extensively applied to study a large range of dynamic and thermodynamic properties of silica in many different phases and under many different thermodynamic conditions.

In the case of water there were many attempts to produce interaction potentials capable of describing the anomalous static and dynamic properties of water. In the SPC/E model (Berendsen *et al.*, 1987) of the water molecule each of the three atoms, arranged in a rigid geometry, is assigned a point charge, while the oxygen atom also has Lennard–Jones interactions. This type of model is widely used for molecular dynamics simulations because of its simplicity and computational efficiency. The ST2 model (Stillinger and Rahman, 1974) is a five-site model, which places the negative charge on dummy atoms representing the lone pairs of the oxygen atom, with a tetrahedral-like geometry. These classical potentials, based on electrostatic and Lennard–Jones type of interactions, are able to reproduce the tetrahedral structure characteristic of water and silica under the appropriate density and temperature conditions. At low temperature, the diffusion coefficient shows an Arrhenius dependence.

It is thus worth comparing the behaviour of these more complicated and accurate potentials with the generic features observed previously in the simpler one-bond per patch models. For all models, the generic features observed in the patchy colloidal particles models are confirmed. Even the structure of the tetrahedral network generated by the ionic model for silica is astonishingly similar to the one generated by the short-range patchy model for silica. The gas–liquid unstable region is located in the same region of scaled densities and the iso-diffusivity lines are essentially controlled by temperature (parallel to the T-axis), as shown in the scaled plot of Figure 6.21.

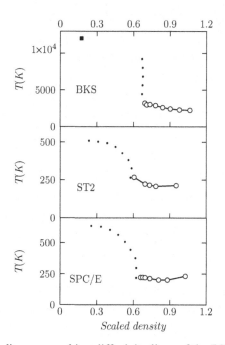

Figure 6.21 Phase diagrams and iso-diffusivity lines of the BKS model for silica, and the ST2 and the SPC/E models for water. Dotted lines are the gas–liquid spinodals, circles indicate the slowest available iso-diffusivity lines, the square is the critical point. From Sciortino (2008), Figure 3. With kind permission from Springer Science and Business Media.

In the figure, water distances are measured in units of 2.8 Å (the HB distance), while in the case of silica distances are measured in units of 3.1 Å, a typical value for the Si–Si distance.

Several issues arise when considering in detail the proposed analogy between network forming liquids and patchy particles:

(i) The reduction of the valence appears to be critical in opening up an intermediate region of densities where bonding, as opposed to packing, becomes the leading interaction.

(ii) The slowing down of the dynamics in this region is found to be Arrhenius and hence typical of the so-called strong glass formers.

(iii) Arrest is driven by the formation of a network of long-lived bonds, i.e. by the formation of energetic cages, as opposed to the more familiar excluded volume cages characteristic of fragile glasses.

(iv) In the models reported, the slowing down of the dynamics is triggered by the formation of bonds and the Arrhenius dynamics arises when most of the possible bonds are formed. The simplicity of the Arrhenius dynamics appears to be connected to the elementary local independent process of bond-breaking.

(v) In the simple models investigated, the diffusion process is controlled by the very small concentration of free unbounded particles diffusing in an empty matrix of fully bonded and hence not reactive particles.

(vi) In these models the ground state is continuously approached on lowering the temperature, proving the absence of any thermodynamic transition associated to the dynamic arrest. No finite Kauzmann temperature appears.

In fragile glass-formers, homogeneous arrested states are observed only at very high densities, when excluded volume caging becomes dominant. At low density, phase separation prevents the possibility of homogeneous arrest. It is thus tempting to speculate that dynamics can be mostly interpreted in term of hard-sphere dynamics, similarly to the role played by the hard-sphere model as reference system for structural and thermodynamic quantities. In this picture, energy and temperature play a role mainly via the change in the parameters of the reference state. If this is the case, fragile and strong liquids are two distinct limiting classes of materials. A final comment regards the behaviour of liquid water. As discussed, the gas–liquid unstable region is analogous to the one observed in tetrahedral coordinated patchy models. Interestingly, several water models are characterised in the region of intermediate densities by an additional region of instability (the second liquid–liquid critical point) separating two disordered liquid structures with different densities. The low-density liquid has indeed the properties of an empty liquid structure while the high density one appear to be characterised by a much larger density. It is likely that the dynamics in the low-density liquid is Arrhenius (strong liquid) while it is super-Arrhenius (fragile liquid) in the high-density one.

Part III
Water

7

Dynamic crossover phenomena in confined water

7.1 Introductory remarks on confined water[1]

Water affects all aspects of life (Ball, 2000; Franks, 2000). Hence, it comes as no surprise that water has tremendous political, cultural and historical significance. The role that water plays in different branches of the natural sciences is varied, but always influential. Understanding the physical characteristics of water has the potential to elucidate how it affects countless systems that are of fundamental importance to biology, chemistry, geology, meteorology, etc. While much more is known about the properties of water than those of most other substances, there are still many open questions and substantial gaps in our understanding. Thus, despite systematic investigation throughout the history of scientific inquiry, the behaviour of water remains an active and exciting area of research.

The properties of water are anomalous. The density maximum of liquid water observed at 4 °C under ambient pressure is probably the most widely known anomaly, but water's thermodynamic response functions and transport coefficients also behave in seemingly counterintuitive ways. On the molecular level, water's ability to form hydrogen bonds that induce an open tetrahedral inter-molecular structure underlies its unique qualities. These hydrogen bonds, which have energy intermediate to van der Waals interactions and covalent bonds, are continuously breaking and reforming while maintaining some average degree of association. At lower temperatures, the hydrogen bonding becomes increasingly prominent and the average degree of association in the fluid increases. The macroscopic properties resulting from this microscopic picture can be conceptualised in terms of a competition between *density* ordering, associated with the normal liquid state behaviour found in van der Waals' view of the liquid state (Chandler *et al.*, 1983) and *bond-orientational* ordering associated with the formation of hydrogen bonds (Tanaka, 1998). This antagonism is captured by models of water used in molecular

[1] Bertrand *et al.* (2013a).

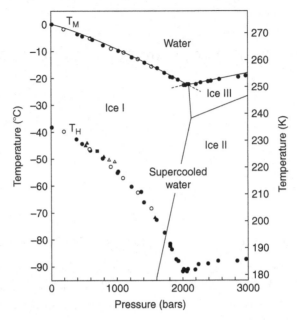

Figure 7.1 Phase diagram of supercooled water indicating experimentally determined melting temperatures T_M and homogeneous nucleation temperatures T_H. Also shown are the phase boundaries between the stable crystalline phases: Ice I (also called hexagonal ice, Ih), Ice II (rhombohedral), and Ice III (tetragonal). From Kanno *et al.* (1975). Reprinted with permission from AAAS.

simulations that treat the central oxygen atom with a Lennard–Jones potential and incorporate the preference for bond-orientational ordering through the addition of decentralised hydrogen atoms, virtual sites or three-body interactions.

It has been known for over a century that water's low-temperature behaviour can be qualitatively understood in terms of these two competing tendencies (Röntgen, 1892). However, a comprehensive theory of water still remains elusive. With some care, water can be supercooled below the bulk melting temperature T_M (shown in Figure 7.1 as a function of pressure) without crystallising. The homogeneous nucleation temperature $T_H \approx 235$K (at 1 atm) defines the experimental lower limit of how deeply bulk water can be supercooled. At this temperature liquid water becomes kinetically unstable with respect to the formation of ice nuclei. Experimental values of T_H are plotted in Figure 7.1 as a function of pressure. Like T_M and the temperature of maximum density, T_H decreases with increasing pressure until about 200 MPa, above which *normal* liquid behaviour is restored. The region bounded by the melting line and the homogeneous nucleation line defines the (experimentally relevant) supercooled region of liquid water's phase diagram. The strength of water's anomalies continues to increase until T_H is reached and further measurements of equilibrium liquid state properties are rendered impossible below this temperature. However, the freezing transition in water can be suppressed by confining water to various substrates with small characteristic dimensions such as the mesoporous silica materials MCM-41. Under *strong confinement* (in this case, it means MCM-41 with cylindrical pore diameter of less than 16 Å) with hydrophilic surface, the freezing of water is completely suppressed and the properties of confined water can be measured well below the bulk T_H. It should be noted here that while bulk supercooled water is metastable with respect to ice; strongly confined water is not in a metastable state. Hence, we will refer to strongly confined water as *deeply-cooled* for any state points below the bulk freezing temperature. Understanding the novel properties of *deeply cooled* water under strong confinement and their relation to the anomalies observed in bulk water has been a focal point of water research in recent years.

This chapter presents recent experimental results mostly from neutron scattering studies of water under strong confinement. In this introductory section, the so-called *dynamic crossover* phenomenon observed in water under confinement in MCM-41 is reviewed.

Ito and co-authors (Ito *et al.*, 1999) have noted that, even though water behaves as a fragile glass former at high temperatures, there are experimental indications that water should behave as a strong glass former close to the calorimetric glass transition temperature T_g. A *strong* glass former is one for which the temperature dependence of the viscosity or the structural relaxation time τ (the two quantities are proportional to each other) follows the Arrhenius law:

$$\tau = \tau_0 \exp\left(\frac{E_a}{k_B T}\right), \tag{7.1}$$

where τ_0 is a characteristic time scale, E_a is a temperature-independent activation energy, k_B is Boltzmann's constant, and T is temperature. A glass former with a super-Arrhenius structural relaxation time is deemed as *fragile*. To reconcile the apparent contradiction of water's fragility, these authors proposed that water undergoes a fragile-to-strong crossover (FSC) at a temperature of roughly 228 K under ambient pressure. Homogeneous nucleation prohibits the measurement of bulk-water dynamics in the deeply supercooled regime where the crossover was conjectured to occur. However, the properties of non-crystalline water can be measured in this temperature range when water is strongly confined in MCM-41. Using incoherent quasi-elastic neutron scattering (QENS), Faraone *et al.* (2003b) observed a dynamic crossover in the temperature dependence of the single-particle translational relaxation time of strongly confined water. The crossover is from super-Arrhenius behaviour at high temperatures to Arrhenius behaviour at low temperatures and has consequently been termed a FSC. In this section we review recent

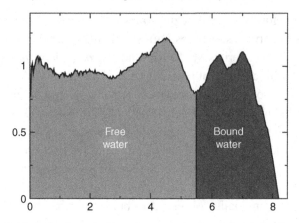

Figure 7.2 Radial density profile of water confined in an MCM-41 pore as function of radial distance from the center of the pore. Water in the surface layer is labelled *bound water* and has a density roughly 10% greater than the density of the *free water*. The profile was calculated from molecular dynamics simulations of SPC/E water. Reprinted (adapted) with permission from Gallo *et al.* (2010b). Copyright 2010, American Chemical Society.

experimental measurements of the dynamics of water under strong confinement in MCM-41 and the FSC phenomenon.

Just as with the thermodynamic measurements of confined water, the distinction between surface water and free water is important for interpreting the dynamic phenomena observed in strongly confined water. Molecular dynamics simulations of strongly confined water have proven particularly valuable in illuminating this distinction. Simulations of SPC/E water in an MCM-41 pore of 15 Å diameter have been performed by Gallo *et al.* (2000), see Figures 7.2 and 7.3 (Gallo *et al.*, 2010a,b, 2012). These works extended earlier simulations of water in a larger diameter silica pore (40 Å) intended to model Vycor glass (Rovere *et al.*, 1998; Gallo *et al.*, 2000).

More recently, Limmer and Chandler (2012) performed simulations of mW water in MCM-41 pores of varying diameter. One advantage of these simulations is that the dynamics of surface water, and the so-called *free water* which behaves dynamically more or less like bulk water, can be determined separately. It has been observed by Gallo *et al.* (2010a) that, at low hydration levels, the water inside the pore is found to coat the surface of the pore leaving the centre vacant. In the fully hydrated case, the surface water relaxes significantly more slowly than the *free (core) water*. More importantly, the surface water relaxes so slowly that, to a good approximation, it primarily contributes to the elastic scattering component of the measured QENS spectra, $S_H(Q, E)$, for the typical experimental energy resolution of a backscattering spectrometer (Gallo *et al.*, 2010b). Figure 7.3 compares the self-intermediate scattering function (SISF) of simulated water confined in

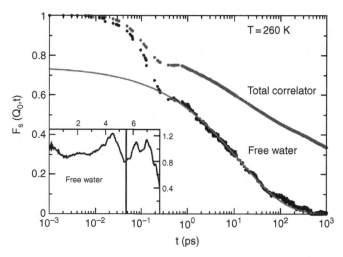

Figure 7.3　Self-intermediate scattering function (SISF) of the oxygen atoms of water confined in MCM-41 (green points, labeled *Total correlator*). Also shown is the SISF of the *free water* located in the centre of the pore (black points). The free water exhibits stretched exponential relaxation (grey curve) that becomes increasingly glassy as the temperature is lowered. The SISFs are evaluated at $Q_0 = 2.25\,\text{Å}^{-1}$, where Q_0 corresponds to the peak of the static structure factor at $T = 260\,\text{K}$ (Gallo *et al.*, 2012). ©IOP Publishing. Reproduced with permission. All rights reserved.

MCM-41 with the SISF of the *free water*. The incoherent scattering cross-section of hydrogen is about 80 barns, which is 20 times larger than its coherent scattering cross-section. But for oxygen and silicon, the incoherent scattering cross-sections are negligible in comparison to their coherent ones, which is only a few barns.

Since the oxygen and silicon just mentioned above are the primary constituents of the silica confining medium described in this section, incoherent QENS is ideally suited for studying the single-particle dynamics of the hydrogen atoms in confined water. Figure 7.4 presents examples of the dynamic structure factor $S_H(Q, E)$ measured by the High-Flux Backscattering Spectrometer at the NIST Center for Neutron Research (NCNR).

Early measurements of the dynamics of H_2O confined in MCM-41 were made by Takahara *et al.* (1999), who found that confined water is generally less mobile than bulk water. Subsequently, Faraone *et al.* (2003a,b) studied the glassy dynamics of moderately supercooled water confined in the mesoporous silica matrices MCM-41 and MCM-48 via QENS. The synthesised material MCM-48 differs from MCM-41 in its pore geometry. By making a decoupling approximation these authors differentiated between rotational and translational relaxation times, and found that both slow down significantly upon cooling. These results have been reviewed and elaborated by Liu *et al.* (2004a), who justified the assumptions of

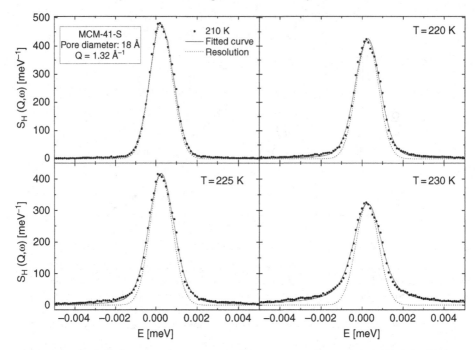

Figure 7.4   QENS spectra from water under strong confinement in MCM-41. The energy transfer E is related to the angular frequency ω by $E = \hbar\omega$. The black dots are the experimentally measured spectra. Solid curves are results of RCM model fitting. The dashed curve is the Gaussian resolution function (Liu *et al.*, 2006). ©IOP Publishing. Reproduced with permission. All rights reserved.

the analysis model by making comparisons with molecular dynamics simulations. Measurements of the single-particle dynamics of water confined in MCM-41 were extended to lower temperatures by Faraone *et al.* (2004), who found a FSC in the (Q-independent) average single-particle translational relaxation time, τ_T, of water in MCM-41 at a temperature of $T_L \approx 225$ K. In Figure 7.5, τ_T is shown for water confined in pores of four different diameters.

The data were analysed with the model used by Faraone *et al.* (2003b) previously. Above T_L, the temperature dependence of τ_T can be described by the phenomenological Vogel–Fulcher–Tammann (VFT) law:

$$\langle \tau_T \rangle = \tau_0 \exp\left(\frac{DT_0}{T - T_0}\right), \tag{7.2}$$

where τ_0 is a characteristic time scale, which is generally different from τ_0 in Eq. (7.1), D is a constant, and T_0 is a temperature that characterises the apparent divergence of the relaxation time. The VFT law is frequently used to describe the behaviour of fragile glass formers. Larger values of D correspond to greater fragility. The apparent divergence in τ_T is averted at T_L below that the average

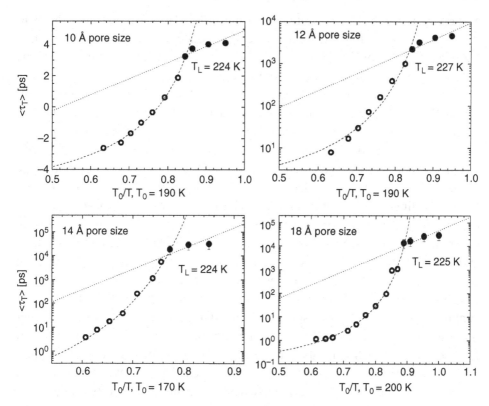

Figure 7.5 Temperature dependence of the average translational relaxation time of water under strong confinement in MCM-41. A fragile-to-strong crossover is observed at $T_L \approx 225$ K for all pore sizes (10, 12, 14 and 18 Å). Solid symbols are experimental data. The relaxation times were extracted from QENS. Dashed and dotted lines denote the Vogel–Fulcher–Tammann and Arrhenius fits, respectively (Liu *et al.*, 2006). ©IOP Publishing. Reproduced with permission. All rights reserved.

translational relaxation time tends towards the Arrhenius law, Eq. (7.1). The intersection of the VFT law and the Arrhenius law defines T_L. It is notable that T_L is close to other characteristic temperatures that have been found in thermodynamic measurements of water's properties, in particular the extremum of the thermal expansivity α_P.

Takahara *et al.* (1999) have explored the possibility of using different models to describe QENS spectra of H_2O confined in MCM-41 with nominal pore diameters of 21.4 Å and 28.4 Å. Their analyses compared three different methods of extracting the results. The models included simple exponential decay, stretched exponential decay, and stretched exponential decay with an additional elastic contribution intended to model immobile surface water. The last model, which also has the greatest number of fit parameters, was found to describe the data most accurately. While the exact quantitative results differed to a limited extent, the qualitative findings were found to be the same. Consequently, while the models used to analyse the scattering data may not be exact, the observed phenomena, such as the FSC, are expected to be robust and model-independent. The pressure dependence of the FSC has been studied by Liu *et al.* (2005, 2006) for MCM-41 confined water using QENS spectroscopy. The temperature dependence of the measured average translational relaxation times is plotted in Figure 7.6 for six different pressures.

These investigators found that the temperature of the crossover T_L is lowered as the pressure is increased, so that the line of T_L has a negative slope in the temperature–pressure plane, as do the melting line and the homogeneous nucleation limit line. At the highest experimental pressures, above 1600 bars, the change in the relaxation time was not marked by a sharp transition from super-Arrhenius to Arrhenius behaviour. If the line of FSC is extrapolated in the temperature–pressure plane, the smoothing of the crossover appears to coincide with the crossing of the bulk homogeneous nucleation temperature.

The results presented thus far have focused on the single-particle dynamics of confined H_2O. The FSC has also been observed in the collective dynamics of confined D_2O. Yoshida *et al.* (2008, 2012) made neutron spin echo measurements of D_2O confined in MCM-41 with a nominal pore diameter of 20.4 Å. The full intermediate scattering functions were fit with a stretched exponential function at all temperatures. The absence of crystallisation was confirmed by differential scanning calorimetry and neutron diffraction measurements. As seen in Figure 7.7, the extracted relaxation times indeed show a FSC at $T_L \approx 220$ K.

Yoshida *et al.* (2012) have also recently studied the hydration level dependence of the FSC. The experiment described in the preceding paragraph was repeated for monolayer hydration in addition to full hydration. The relaxation time of the monolayer did not exhibit a FSC and could be well described by Arrhenius temperature dependence. It is notable that the extracted monolayer relaxation times are much shorter than those found in the case of full hydration. These authors concluded that the dynamic crossover is associated with the *free water* in the core region of the pore (cf. Figure 7.2) and its ability to form a tetrahedrally bonded structure. Recently Bertrand *et al.* (2013b) performed a QENS experiment to study the hydration-dependent single-particle dynamics of deeply cooled water under strong confinement. They concluded that the monolayer relaxation times are longer than the full hydration times at high temperatures, but become shorter at lower temperatures.

There has been some debate in the literature surrounding the FSC in the single-particle dynamics of water confined in MCM-41. Swenson *et al.* (2006) argued that the concept of fragility only applies to the *glass-transition related α relaxation* in supercooled liquids. Since the glass transition temperature found by extrapolating

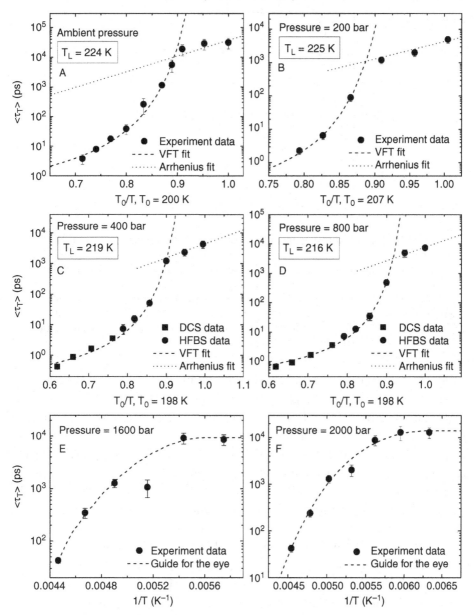

Figure 7.6 Average translational relaxation times of water under strong confinement in MCM-41 at elevated pressures. A distinct FSC was observed at 1, 400, 800 and 1200 bars, but not at 1600 and 2000 bars. Translational motion was isolated by considering $Q < 1.1\,\text{Å}^{-1}$, so that only the translational portion was used in the analysis (Liu *et al.*, 2005). Copyright 2005 by The American Physical Society.

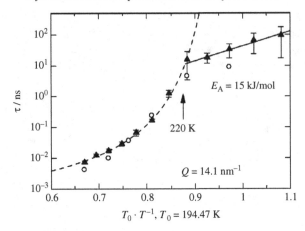

Figure 7.7 Temperature dependence of the collective relaxation time of heavy water under strong confinement in MCM-41 (triangles). A FSC is observed at $T \approx 220\,\mathrm{K}$. Relaxation times were extracted from stretched exponential fits to neutron spin echo data. The stretching exponent was fixed at $\beta = 0.5$. Open circles are data from a previous measurement (Yoshida *et al.*, 2008). Dashed and solid lines denote the Vogel–Fulcher–Tammann and Arrhenius fits, respectively Yoshida *et al.* (2012). Reprinted with permission from Yoshida *et al.* (2008). Copyright 2008, AIP Publishing LLC.

the low temperature Arrhenius behaviour of confined water is unrealistically low ($T_g \approx 50\,\mathrm{K}$), it was concluded by Swenson *et al.* (2006) that the translational relaxation time observed by Liu *et al.* does not correspond to the α-relaxation of the confined water, but instead results from a vanishing of the α-relaxation. Chen *et al.* (2006d) concede that the dynamic crossover observed in QENS scattering experiments may not match the strict definition of the fragile-to-strong crossover, as introduced by Ito *et al.* (1999). However, they emphasise that the QENS spectra are consistent with the RCM model for all experimental temperatures and that the long-time behaviour is well described by a stretched exponential function. That the perceived disagreement may be semantic in origin, is emphasised by Sjöström *et al.* (2008), who asserted that 'there is no doubt that a crossover of the H diffusivity occurs around 220 K in QENS, NMR, and also neutron spin echo experiments' while reporting the absence of a FSC in dielectric relaxation spectroscopy measurements. It has been suggested by Cerveny *et al.* (2006) that the dynamic crossover observed in confinement may be a finite-size effect and therefore depends sensitively on the details of the confinement. This does not appear to be the case for confinement in MCM-41 silica materials. As shown by Chen *et al.* (2006d) the FSC is robust to systematic changes in the diameter of the confining pores (cf. Figure 7.5).

As reported by Mallamace *et al.* (2006) and Chen *et al.* (2006d), the FSC has been confirmed via nuclear magnetic resonance experiments. In this case,

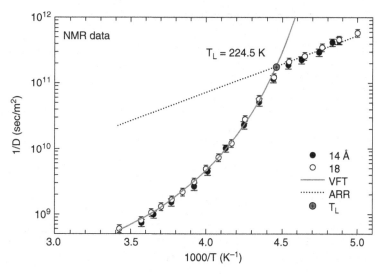

Figure 7.8 Arrhenius plot of the inverse self-diffusion coefficient of water under strong confinement in MCM-41. A FSC is observed at $T_L \approx 225$ K. As previously noted (cf. Figure 7.5), T_L is independent of the pore diameter (14 and 18 Å). Solid and dotted lines denote the Vogel–Fulcher–Tammann and Arrhenius fits, respectively (Chen *et al.*, 2006d). Copyright 2006, National Academy of Sciences, USA.

the crossover is inferred from the temperature dependence of the self-diffusion coefficient D. The inverse self-diffusion coefficient $1/D$, shown in Figure 7.8 was found to exhibit a crossover from non-Arrhenius to Arrhenius behaviour at $T_L \approx 220$ K. In a simple fluid, the self-diffusion coefficient of a spherical particle of radius R is related to the shear viscosity by the Stokes–Einstein relationship (SER):

$$D = \frac{k_B T}{6\pi \eta R}. \tag{7.3}$$

The average translational relaxation time, τ_T, measured by incoherent QENS is assumed to be proportional to the shear viscosity η. On the basis of Eq. (7.3), one expects that the product $D\langle \tau_T \rangle / T$ is temperature independent. By taking the results from two separate experiments, Chen *et al.* (2006d) and Mallamace *et al.* (2006) have calculated the product $D\tau_T / T$ as a function of temperature for H_2O confined in MCM-41. It was found that this product begins to increase sharply below the FSC temperature in the regime when the fluid behaves as a strong liquid. This violation of the SER is illustrated in Figure 7.9. This finding implies the existence of dynamic heterogeneities in confined supercooled water (Kumar, 2006).

The single-particle dynamics of water confined in (hydrophobic) carbon nanotubes have also been investigated in Mamontov *et al.* (2006), Chu *et al.* (2007, 2010) and Faraone *et al.* (2009). Mamontov *et al.* (2006) measured the single

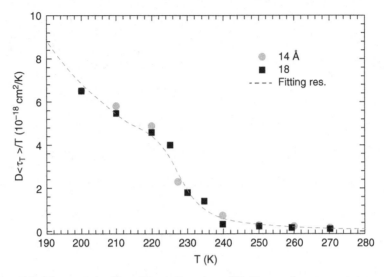

Figure 7.9 The quantity $D\tau_T/T$ as a function of T. Dots and squares represent its values coming from the experimental data of D and τ_T in samples with diameters of 14 and 18 Å, respectively. The dashed line represents same quantity obtained by using the fitting values obtained from the data reported (Chen *et al.*, 2006d). Copyright 2006, National Academy of Sciences, USA.

particle relaxation times of water confined in single- and double-walled carbon nanotubes with inner diameters 14 and 16 Å respectively. They analysed their data using a simple jump diffusion model. A FSC was observed in the α-relaxation time of water confined in the single-walled nanotube at $T_L \approx 218$ K, which is slightly lower than observed in hydrophilic confinement. While a crossover was not observed for the double-walled nanotube in this experiment (due to a limited experimental temperature range), Chu *et al.* (2007) subsequently found a dynamic crossover in a double-walled carbon nanotube of inner diameter 16 Å at $T_L \approx 190$ K, roughly 35 K below the crossover temperature observed in hydrophilic confinement. The average translational relaxation times found by Chu *et al.* (2007) for the double-walled nanotube are shown in Figure 7.10.

The single-particle dynamics of water confined in other hydrophobic materials have been measured as well. Faraone *et al.* (2009) made QENS and NSE measurements of strongly confined water in hydrophobically modified MCM-41 (18 Å), which was formed by replacing a fraction of the silanol-OH groups that line on surface of the pore with methyl groups.

The relaxation times were well described by the VFT law in the entire experimental temperature range of $T = 210$–300 K. Based on the behaviour of water in double-walled carbon nanotubes it was suggested that the experimental temperature range may not have been low enough to observe a dynamic crossover.

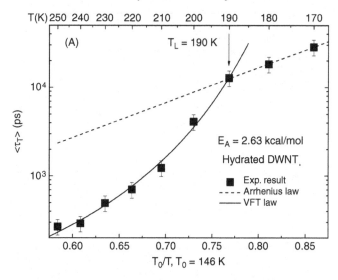

Figure 7.10 Average translational relaxation time of water confined in a double-walled carbon nanotube (inner diameter 16 Å). The relaxation times show a well-defined FSC at $T_L = 190$ K. Relaxation times were extracted by fitting the RCM model to the QENS spectra. Solid and dashed lines denote the Vogel–Fulcher–Tammann and Arrhenius fits, respectively (Chu *et al.*, 2007). Copyright 2007 by The American Physical Society.

However, Chu *et al.* (2010) subsequently found a FSC in hydrophobic CMK-1. This experiment was conducted in the temperature range $T = 170$–250 K and the crossover temperature were found to be $T_L \approx 225$ K. These authors located the crossover temperature using a method that places the position of the dynamic transition 10–20 K higher than found by the *traditional* method (fitting an Arrhenius plot of the translational relaxation times with a super-Arrhenius curve at high temperatures and an Arrhenius curve at low temperatures). Therefore, the data from CMK-1 is consistent with the crossover temperature being lower in hydrophobic confinement. However, the shift does not appear to be as great as observed in double-walled carbon nanotubes. It is notable that although a density minimum is not observed for water confined in CMK-1, the dynamic crossover is still observed. It has been conjectured that the dynamic crossover phenomenon is associated with an inflection point in the density curve (or peak in the thermal expansivity) and does not require an absolute density minimum. Chen *et al.* (2009b) analysed the relaxation times of several glass forming liquids and found that experimental relaxation times, as characterised by the viscosity, are consistent with a FSC. Notably this crossover occurs at temperatures above T_g. The relaxation time of the glass former α-phenyl-o-cresol is plotted in Figure 7.11.

The crossover temperature in the relaxation times was found to coincide with the maximum in the heat capacity associated with the glass transition. This analysis

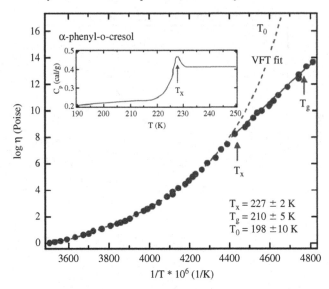

Figure 7.11 Arrhenius plot of the viscosity of the glass former α-phenyl-o-cresol. The temperature of the proposed FSC, $T_x = 227$ K, is coincident with a peak in the isobaric heat capacity (inset). Dashed and solid lines denote the Vogel–Fulcher–Tammann and Arrhenius fits, respectively (Chen *et al.*, 2009b).

supports the suggestion of Mallamace *et al.* (2010) that the FSC may be a general phenomenon in glassy systems. The existence of a FSC above T_g is supported by an extension of mode-coupling theory that incorporates ergodicity-restoring activated hopping processes (Chong, 2008; Chong *et al.*, 2009).

7.2 Module – Model for water's single-particle dynamics–relaxing cage model (RCM)

Due to the large incoherent scattering length of hydrogen, quasi-elastic neutron scattering (QENS) essentially measures the self-dynamic structure factor $S_H(Q, E)$ of the hydrogen atoms in water as a function of the magnitude of the wave-vector transfer Q and energy transfer E. The dynamic structure factor $S_H(Q, E)$ is related to the self intermediate scattering function (SISF) of the hydrogen atoms $F_H(Q, t)$ by the following Fourier transform:

$$S_H(Q, E) = \frac{1}{2\pi\hbar} \int_{-\infty}^{\infty} dt\, F_H(Q, t) \exp(iEt/\hbar), \qquad (7.4)$$

where t is time, and $\hbar$ is the reduced Planck constant. For a collection of N hydrogen atoms, i.e. $N/2$ water molecules, with the j-th hydrogen atom located at position $x_j(t)$ at time t, the hydrogen SISF is defined by

$$F_H(Q, t) = \left\langle \frac{1}{N} \sum_{j=1}^{N} e^{i Q \cdot [x_j(t) - x_j(0)]} \right\rangle, \tag{7.5}$$

where the brackets denote an average over the time origins. Typically, a model of the SISF is first developed for data analysis and then the Fourier transform of this function is using to fit experimental data.

The single-particle SISF of bulk supercooled H_2O can be decomposed into a product of a translational and a rotational part, both of which contribute to the quasi-elastic scattering peak measured by QENS (Dore and Teixeira, 1991). To model the effects of the combined translational and rotational it is useful to define the translational SISF, $F_T(Q, t)$, which is essentially the SISF of the centre mass (oxygen atom) of a water molecule, and the rotational SISF, $F_R(Q, t)$, which comes from consideration of the motion of the hydrogen atoms attached to the oxygen atom. The oxygen atom is essentially the centre of mass of the water molecule. The SISF of the hydrogen atom can then be written as a product of the translational SISF of the centre mass, $F_T(Q, t)$, and the rotational SISF of the hydrogen atoms around the centre of mass, $F_R(Q, t)$, namely

$$F_H(Q, t) = F_T(Q, t) \cdot F_R(Q, t). \tag{7.6}$$

This product form of $F_H(Q, t)$ is called the decoupling approximation, which is a fairly good approximation for a spherical molecule, such as a water molecule. The relaxing cage model (RCM) was then introduced to calculate the translational SISF, $F_T(Q, t)$, for many of the experimental results described in this chapter. The RCM was developed by Chen *et al.* (1999) on the basis of molecular dynamics simulations of SPC/E water. Initially it was used to describe the dynamics of water confined in Vycor glass (Zanotti *et al.*, 1999; Bellissent-Funel *et al.*, 2000) and is an extension of previous work on the relaxation processes in supercooled water (Gallo *et al.*, 1996; Sciortino *et al.*, 1996, 1997). The model is inspired by ideas from mode coupling theory that account for the increase in local structure surrounding a tagged particle as a fluid is supercooled or as the density is increased through the application of pressure. On short time scales, the particle is envisaged to be rattling around in a *cage* formed by the surrounding particles. Diffusion of the particle out of the cage requires significant structural rearrangement, which implies a strong coupling between single particle dynamics and density fluctuations. These two characteristics of the motion are accounted for by a self-intermediate scattering function of the form

$$F_T(Q, t) = F_{vib}(Q, t) \cdot F_\alpha(Q, t), \tag{7.7}$$

where $F_{vib}(Q, t)$ represents in-cage vibrational motions of the centre mass of the water molecule. And $F_\alpha(Q, t)$ represents the long-time motions of the water molecule migrating away from the relaxing cage. In RCM, $F_{vib}(Q, t)$ is modelled as

$$F_{vib}(Q, t) = \exp\left\{-Q^2 v_0^2 \left[\frac{1-C}{\omega_1^2}\left(1 - e^{-\omega_1^2 t^2/2}\right) + \frac{C}{\omega_2^2}\left(1 - e^{-\omega_2^2 t^2/2}\right)\right]\right\}.$$
(7.8)

In a QENS experiment, we are measuring the motion in the time scale of $F_\alpha(Q, t)$, which is in the nanosecond range. So in this long time limit, $F_{vin}(Q, t)$ is approaching its infinite-time limit form, $A(Q)$,

$$A(Q) = \lim_{t \to \infty} F_{vib}(Q, t) = \exp\left\{-Q^2 v_0^2 \left[\frac{1-C}{\omega_1^2} + \frac{C}{\omega_2^2}\right]\right\}$$

$$= \exp\left(-\frac{1}{3}Q^2 a^2\right)$$
(7.9)

where a is the root mean-squared radial displacement of the water molecules in the cage. For supercooled water, a has been found to be roughly constant in temperature and has a value of $a \approx 0.5\,\text{Å}$ (Chen *et al.*, 1999; Liu *et al.*, 2002). The long-time α-relaxation, in the sense of the mode-coupling theory, is described by $F_\alpha(Q, t)$. Its form is given by the stretched-exponential Kohlrausch–Williams–Watts function

$$F_\alpha(Q, t) = \exp\left[-\left(\frac{t}{\tau_T(Q, T)}\right)^\beta\right],$$
(7.10)

where $\tau_T(Q, t)$, is the Q- and temperature T-dependent characteristic relaxation time and β is a stretching exponent ($\beta < 1$). Stretched exponential relaxation can be interpreted as resulting from a particular distribution of exponential relaxation processes. Both β and γ are temperature-dependent parameters.

In general the average translational relaxation times extracted from QENS spectra vary with temperature and Q. Faraone *et al.* (2004) and subsequent investigators have found that the Q dependence of $\tau_T(Q, T)$ of MCM-41 confined water is consistent with the following power-law behaviour:

$$\tau_T(Q, T) = \tau_0(T)(aQ)^{-\gamma(T)},$$
(7.11)

where $\tau_0(T)$ is the Q-independent pre-factor, $a = 0.5\,\text{Å}$, and $\gamma(T)$ is the temperature-dependent power law exponent. This power-law dependence of the measured relaxation time on Q was later shown to be consistent with the extended mode coupling theory of Chong (2008) by Chen *et al.* (2009b). If we accept this power-law Q dependence, then the Q-independent average relaxation time $\tau_T(T)$ was shown to be given by

$$\langle \tau_T(T) \rangle = \frac{\tau_0(T)}{\beta}\Gamma\left(\frac{1}{\beta}\right),$$
(7.12)

where $\Gamma(x)$ is the gamma function. An Arrhenius plot of this Q-independent average relaxation time has been used extensively in all publications of Chen's group to show the existence of the dynamic crossover temperature, T_L, in confined water under pressure (Liu *et al.*, 2002).

A model for the rotational intermediate scattering function $F_R(Q, t)$ of super-cooled water was developed by Liu *et al.* (2002). For small values of Q ($Q <$ 1 Å^{-1}), the rotational ISF is approximately equal to unity, and the translational ISF makes the dominant contribution to $F_H(Q, t)$, (Dore and Teixeira, 1991). For larger values of Q, an accurate model of the rotational dynamics is required to extract meaningful rotational relaxation times and to separate the translational and rotational dynamics. To this end, a Sears (1967) expansion of the rotational ISF is made in terms of the (l-th)-order rotational correlation functions $C_l(t)$. It is assumed that $C_l(t)$ which makes the dominant contribution can be separated into short-time $C_l^S(t)$ and long-time $C_l^L(t)$ parts such that $C_l(t)$ is written as the product

$$C_1(t) = C_1^S(t) \cdot C_1^L(t). \tag{7.13}$$

The whole picture resembles the relaxing cage model of the translational dynamics. At short times, the orientation of the central water molecule is fixed by the H-bonds with its neighbours. It performs nearly harmonic oscillations around the hydrogen-bond direction. This dynamics is described by $C_l^S(t)$. At longer times, the bonds break and the cage begins to relax. So the particle can reorient itself, losing memory of its initial orientation. Thus the first-order rotational correlation function eventually decays to zero by a stretched exponential relaxation.

This function $C_l^S(t)$ describes the short-time behaviour of the first-order rotational correlation function. It starts from unity at $t = 0$, exhibits an oscillation at time 0.05 ps and then decays to a flat plateau determined by $\exp\left(-4\langle\omega^2\rangle/15\omega_3^2\right)$ for times longer than 0.1 ps (Liu *et al.*, 2002). For time longer than 0.1 ps, the function $C_l(t)$ follows an α-relaxation model given by $C_l^L(t)$:

$$C_1^L(t) = \exp\left[-(t/\tau_R)^{\beta_R}\right], \tag{7.14}$$

where τ_R is the rotational relaxation time and β_R is the rotational stretching exponent. In the RCM, for the treatment of the rotational ISF, $F_R(t)$, one makes a model explicitly for the function $C_l(t)$ and shall generate the other higher-order rotational correlation functions approximately using the maximum entropy method of Berne *et al.* (1968). The two higher-order rotational correlation functions retained in the expansion, $C_2(t)$ and $C_3(t)$, are then expressed in terms of $C_l(t)$. In many ways this description of the rotational relaxation resembles the treatment of the translational part of ISF, $F_T(t)$, by the RCM. Fits to the MD calculated rotational correlation functions (Liu *et al.*, 2002) are shown in Figure 7.12.

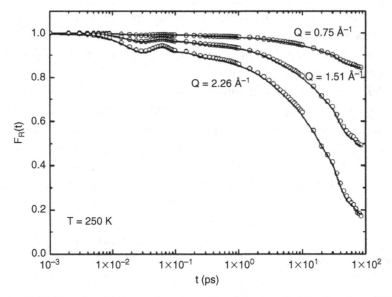

Figure 7.12 Rotational intermediate scattering function $F_R(Q, t)$ at three differ-ent Q values. The circles are calculated from molecular dynamics simulations of SPC/E water. The curves are fits from the Sears expansion up to the fourth-order term (Liu *et al.*, 2002). Copyright 2002 by The American Physical Society.

Although the RCM and the rotational correlation function expansion discussed above were developed for application to bulk water, Faraone *et al.* (2004) have argued that the physical principles embodied by these models are general enough to successfully represent the dynamics of confined water as well. In particular, the RCM emphasises the separation of short and long time scales and the non-exponential nature of the long time decay, features which are also observed in simulations of confined water (Gallo *et al.*, 2010b).

7.3 Module – Dynamic crossover in hydration water of biomaterials

Water is an active constituent in cell biology (Ball, 2008). This section describes a series of neutron scattering experiments made during 2005–2009 by Chen *et al.* for the understanding of the single-particle (hydrogen atom) dynamics of a protein (and other bio-macromolecules) and their hydration water and the strong coupling between them. They found that the key to this strong coupling is the existence of a fragile-to-strong dynamic crossover (FSC) phenomenon occurring at around $T_L = 225\pm5$ K in the hydration water. On lowering of the temperature toward FSC, the structure of hydration water makes a transition from predominantly the high-density form (HDL), a more fluid state, to predominantly the low density form (LDL), a less fluid state, derived from the existence of a liquid–liquid critical point at an elevated pressure. They show experimentally that this sudden switch in the mobility of hydration water on lysozyme, B-DNA and RNA triggers the dynamic transition, at a universal temperature $T_D = 220$ K, for these biopolymers. In the glassy state, below T_D, the biopolymers lose their vital conformational flexibility resulting in a substantial diminishing of their biological functions. They also performed molecular dynamics (MD) simulations on a realistic model of hydrated lysozyme powder, which confirms the existence of the FSC and the hydration level dependence of the FSC temperature.

7.3.1 Introduction to the importance of the concept of hydration water in proteins

One of the most striking examples of the importance of water in bio-macromolecules is that, for example, proteins cannot perform their functions if they are not covered by a minimum amount of hydration water. Hydration can be considered as a process of adding water incrementally to a dry protein, until a level of hydration is reached beyond which further addition of water produces no change of the essential properties of the protein but only dilutes the protein (Rupley and Careri, 1991).

In the case of a well-known enzyme called lysozyme, Rupley *et al.* (1980) measured its reaction with the hexasaccharide of N-acetylglucosamine over a full hydration range. The threshold hydration level was determined as $h = 0.2$, where h is the ratio between grams of water and grams of dry protein. They showed that enzymatic activity closely parallels the development of surface motion, which is thus responsible for the functionality of the protein. At low hydration levels ($h \leq 0.6$ g H_2O/g protein), the space between globular proteins in a powder sample serves as a confining medium for hydration water. The confined protein hydration water can be supercooled below the bulk homogeneous nucleation temperature of water without crystallising (Chen *et al.*, 2006b,c; Chu *et al.*, 2008). Chen *et al.* (2006c) have studied the self-dynamics of the hydration water in a system of hen egg-white lysozyme powder at a hydration level of $h \approx 0.3$. This hydration level corresponds to that of having enough water to coat the surface of an isolated globular protein with roughly a monolayer of water. When the protein globules are packed together in a powder sample, the distribution of water is non-uniform. Protein hydration water can be classified as either bound internal water, or surface water, or inter-protein free water. The bound internal water does not contribute significantly to the quasi-elastic broadening in the incoherent scattering process. The distinction between surface water and inter-protein free water is very similar to what is seen in water confined in MCM-41. Just as in MCM-41, it is believed that the relaxation of the surface water, which is much slower than that of the free water, essentially contributes to the elastic scattering only. The quasi-elastic scattering component is then primarily due to the motion of the free water found in the inter-protein space beyond the first hydration layer.

7.3.2 The elastic scan and the mean square displacement (MSD) of hydrogen atoms in protein and its hydration water

It is customary that a quasi-elastic neutron scattering (QENS) experiment using a backscattering spectrometer begins by measuring the mean square displacement (MSD) of hydrogen atoms. To obtain the MSD $\langle x^2 \rangle$ of hydrogen atoms in a protein or its hydration water, one generally performs the so-called fixed window scan (an elastic scattering measurement with a fixed resolution window of FWHM of ± 0.8 meV) (Bee, 1988) in the temperature range from 40 K to 290 K, covering completely the supposed crossover temperature T_L. Since the system is in a stationary metastable state at temperature below and above T_L, one makes measurements by heating and cooling respectively at a heating/cooling rate of 0.75K min^{-1} and observe exactly the same results. One can calculate $\langle x^2 \rangle$ from the Debye–Waller factor, $S_H(Q, \omega = 0) = \exp[-Q^2 \langle x^2 \rangle]$, by a linear fitting to the low-Q part of the logarithm of $S_H(Q, \omega = 0)$ vs. Q^2 plot. $S_H(Q, \omega = 0)$ can then be easily calculated by taking the ratio of the temperature-dependent elastic scattering intensity $I_{el}(Q, T, \omega = 0)$ and its low-temperature limit, $I_{el}(Q, T = 0, \omega = 0)$. One then calculates the ratio:

$$S_H(Q, \omega = 0) = I_{el}(Q, T, \omega = 0)/I_{el}(Q, T = 0, \omega = 0).$$

Figure 7.13 shows an example of the data taken from the D_2O and H_2O hydrated lysozyme powder samples, from which one can extract both MSDs from lysozyme molecule and its hydration water. In order to show the synchronisation of the temperature dependence of the two MSDs thus extracted, we multiply the $\langle x^2_{lysozyme} \rangle$ by a factor of 4.2, so both curves superpose onto each other. This figure nicely illustrates that the crossover temperatures for both protein and its hydration water defined by a sudden change of slope of MSD from a low-temperature behaviour to a high-temperature behaviour is coincident within the experimental errors.

7.3.3 The pressure dependence of the MSD for protein and its hydration water

It is well known that some bacteria can survive under extremely high-pressure and low-temperature conditions in the deep ocean. The microorganisms living in the deepest ocean yet isolated and characterised were sampled at 11 000 m depth or 1100 bar in the deep-sea sediments of the Marianas trench, where the Pacific oceanic lithosphere subducts into the Earth's mantle (Daniel *et al.*, 2006). How

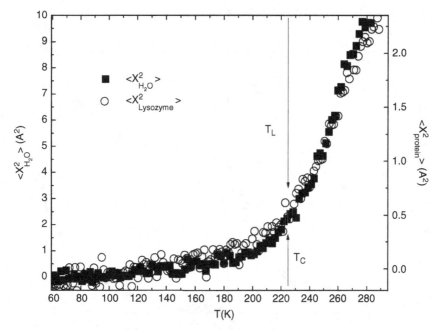

Figure 7.13 Comparison of MSDs measured for the protein and its hydration water. Note that the MSD for hydration water is plotted using the scale on the left-hand side and MSD for the protein is using the scale on the right-hand side (the multiplication factor of the left and right scales is 4.2). MSD for the protein is taken from the elastic scan of D_2O hydrated sample. Note the crossover temperature of the hydration water (T_L) and the crossover temperature of the protein (T_c) agrees with each other. Reprinted from Chen *et al.* (2010a), Copyright 2010, with permission from IOS Press.

can proteins in the microorganisms still function under these extreme conditions? Besides the fact that high pressures denature most of the dissolved proteins above 3000 bar, the behaviours of proteins under pressure below the denaturation limit (2000 bar) both for structure and dynamics are relevant to the biological functions of proteins and are of great interest (Mozhaev *et al.*, 1996; Heremans and Smeller, 1998; Boonyaratanakornkit *et al.*, 2002). One can show by measured MSD that the temperature dependence of the protein dynamics closely follows that of the hydration water under different pressures. Figure 7.14 shows the temperature dependence of MSD in lysozyme and its hydration water at different pressures up to 1600 bar.

7.3.4 Incoherent quasi-elastic scattering study of fragile-to-strong crossover (FSC) in the hydration water of biopolymers

In this section, we discuss both qualitative and quantitative analyses of the quasi-elastic spectra of different hydrated biopolymers, protein (lysozyme) (Chen

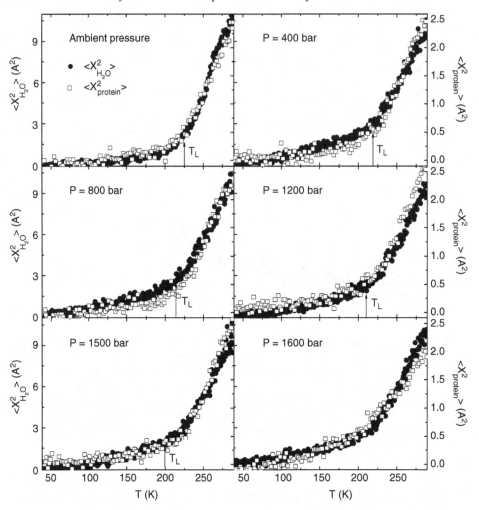

Figure 7.14 A reduced plot of the pressure dependence of the MSD of protein and its hydration water. Note in this figure that the crossover temperature of the protein and its hydration water is closely synchronised at a range of pressures below 2000 bar. The crossover temperature is seen to decrease as the pressure becomes higher. Reprinted from Chen *et al.* (2010b), Copyright 2010, with permission from IOS Press.

et al., 2006c), DNA (Chen *et al.*, 2006b) and RNA (Chu *et al.*, 2008). Our objective is first to show that the peak height and the peak width of the incoherent quasi-elastic spectrum are necessarily related to each other because the area under $S_H(Q, E)$ is in principle normalised to unity at each Q value. From this property we can already show by plotting the peak height of $S_H(Q, E)$, which is $S_H(Q, E = 0)$, as a function of temperature that there is a dynamic crossover phenomenon without detailed analysis of the spectrum.

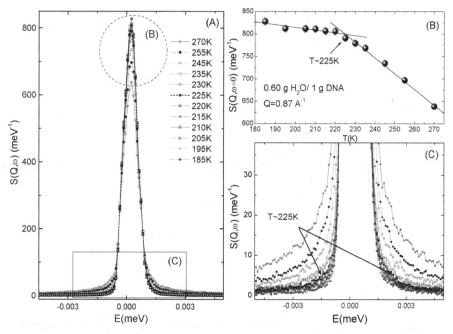

Figure 7.15 The difference neutron spectra between the H_2O hydrated and the D_2O hydrated DNA samples. Panel (A) displays the area-normalised QENS spectra at $Q = 0.87\,\text{Å}^{-1}$ at a series of temperatures. Panels (B) and (C) display the heights of the peak as a function of temperature and the wing spectral region, respectively, at those temperatures. One notes from panel (B) a cusp-like transition signalling the rate of change of peak height from a steep high-temperature region to a slower low-temperature region at a crossover temperature of about 225 K. The error bars are of the size of the data points. In panel (C), we may notice, from the wings of these spectral lines, that two groups of curves, 270–230 K and 220–185 K, are separated by the curve at temperature 220 K. In this panel, the scatter of the experimental points gives an idea of the error bars. Reprinted with permission from Chen *et al.* (2006b). Copyright 2006, AIP Publishing LLC.

In Figure 7.15 we use hydrated DNA as an example to illustrate the above-mentioned fact that by plotting the peak height as a function of temperature, we can already detect the presence of the dynamic crossover temperature at $T = 225$ K, in a qualitative way.

We now illustrate the detailed analysis of the quasi-elastic peak using relaxing cage model (RCM). Figure 7.16 shows a series of spectra of lysozyme hydration water taken at different temperatures at a pressure of 400 bar and $Q = 0.56$ and $1.11\,\text{Å}^{-1}$. One notices immediately that the peak height increases as temperature decreases, indicating the narrowing of the peak width, which is a qualitative measure of the α-relaxation time.

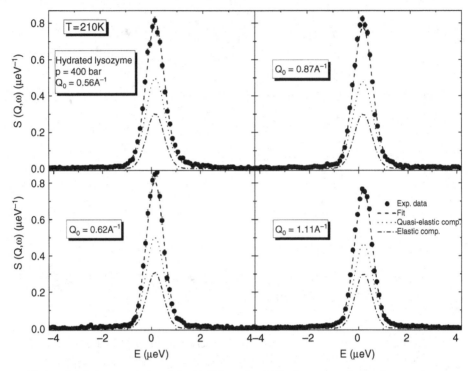

Figure 7.16 An example of the RCM analysis for lysozyme hydration water at 400 bar at two different temperatures $T = 210$ K (below T_L) and 240 K (above T_L) respectively. Notice that for the high-temperature case ($T = 240$ K), the quasi-elastic components are much broader. And we can also see clearly that for higher T, the peak height is much lower than the low-temperature case ($T = 210$ K).

In Figure 7.17, $\log(\tau_T)$ vs. $1/T$ in the same scale for six different pressures 1, 400, 800, 1200, 1500 and 1600 bar was plotted. Since the protein dynamics is strongly coupled to that of its hydration water, a short structural relaxation time of the hydration water enables the protein to maintain its flexibility and thus it is able to sample more conformational substates. For lower temperatures, τ_T obeys an Arrhenius behaviour, which can be fitted by a straight line in the $\log(\tau_T)$ vs. $1/T$ plot; while for high temperatures, the behaviour of $\log(\tau_T)$ switches over to obey a Vogel–Fulcher–Tammann (VFT) law

$$\log \langle \tau_T \rangle = \log \tau_0 + \frac{DT_0}{T - T_0},$$

which are shown with dashed curves in Figure 7.17.

A very distinct phenomenon in these experimental results is that τ_T shows a completely different behaviour at pressure above 1600 bar. The crossover phenomenon disappears above this pressure and τ_T appears to transform into

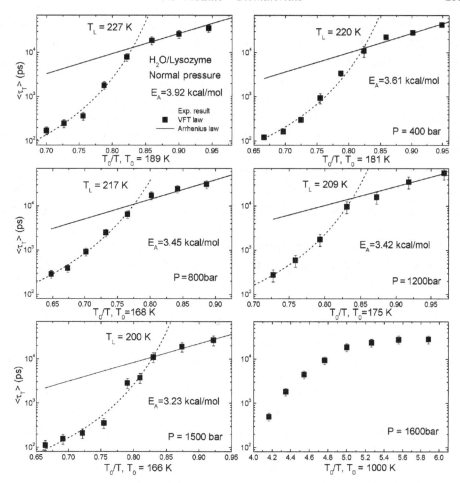

Figure 7.17 Extracted relaxation time plotted in a log scale against T_0/T (where T_0 is the VFT temperature) under six different pressures $P =$ ambient pressure, 400, 800, 1200, 1500 and 1600 bar. Note for the pressures up to 1500 bar, there is a well-defined crossover temperature; but at the pressure of 1600 bar, $\langle t_T \rangle$ appears to be a smooth curve, neither having the super-Arrhenius behaviour at high temperatures nor Arrhenius behaviour at low temperatures.

a smoothly bending over curve (concave downwards). Moreover, for the same temperature, τ_T is no longer decreasing as pressure increases to 1600 bar. This suggests that the hydration water must have crossed a *liquid–liquid critical point* around 200 K and 1600 bar (Poole *et al.*, 1992; Chu *et al.*, 2009).

Previous experiments on confined water in MCM-41-S porous silica material (Faraone *et al.*, 2004; Liu *et al.*, 2005a) have shown that an increased applied pressure will shift the FSC temperature to a lower value. Figure 7.5 shows that this is also true for the interfacial water on the surfaces of protein, and that while a well-defined FSC phenomenon is observed for the applied pressure up to 1500 bar, when exceeding this pressure, the FSC phenomenon disappears. Xu *et al.* (2005) thus identify a Widom line in the *T*–*P* plane with an end point for the case of protein hydration water which is nearly identical to that of the confined water in MCM-41-S. This implies the existence of liquid–liquid critical point in both the 1-D (MCM-41-S case) and 2-D (protein hydration water case) confined water.

Xu *et al.* (2005) have previously shown by a molecular dynamics (MD) simulation that this super-Arrhenius to Arrhenius crossover is due to crossing of the Widom line in the single-phase region. Upon the crossing of the Widom line from the high-*T* side to low-*T* side, the local structure of water evolves from a predominately high-density form (HDL, fragile liquid) to a predominately low-density form (LDL, strong liquid) as the temperature crosses this characteristic temperature T_L (Mallamace *et al.*, 2007b). At the pressure of 1600 bar, the τ_T vs. $1/T$ plot appears to be a smooth curve, having neither the super-Arrhenius behaviour of high temperatures nor the Arrhenius behaviour of low temperatures. Liu *et al.* (2005) attribute it to the phase-separated mixture of the HDL and LDL due to the crossing of the hypothetical first-order liquid–liquid transition line (Poole *et al.*, 1992). If these arguments are valid, then the disappearance of the FSC phenomena signals the crossing of the state point from the Widom line to the first-order liquid–liquid transition line. These two lines are separated by the liquid–liquid critical point if it exists.

Figure 7.18 thus plots the trajectory of the crossover temperature T_L as a function of *P* (full circles). It is remarkable to see that this Widom line of the protein hydration water seems to coincide with the Widom line of the confined water in MCM-41-S found in previous experiment of Liu *et al.* (2005). We also show the homogeneous nucleation temperature line (T_H), crystallisation temperatures of amorphous solid water (T_X), and the temperature of maximum density line (TMD), taken from the known phase diagram of bulk water.

7.3.5 *Extension of the study of the crossover phenomenon in hydration water on DNA and RNA*

At ambient pressure, Chen *et al.* measured the average translational α-relaxation time τ_T of the hydration water by QENS, and found a dynamic crossover in hydration water at a universal temperature $T_L = 225 \pm 5$ K in all three biomolecules – lysozyme (Chen *et al.*, 2006c), B-DNA (Chen *et al.*, 2006b) and RNA (Chu *et al.*, 2008). Thus they have shown that $T_D \approx T_L$ at ambient pressure. As discussed in Section 7.3.3, traditionally the dynamic transition temperature of a protein is discussed in terms of the turning point of MSD vs temperature plot (Rasmussen *et al.*, 1992). One can also use a similar plot to discuss the dynamic crossover temperature in protein hydration water, or more generally for

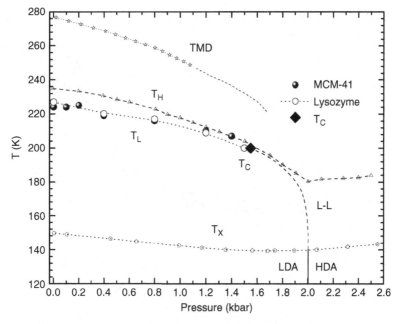

Figure 7.18 The pressure dependence of the measured FSC temperature T_L, plotted in the *T–P* plane, comparing with previous results from water in MCM-41-S (Liu *et al.*, 2005). Copyright 2005 by The American Physical Society.

hydration water in biopolymers (Caliskan *et al.*, 2002, 2006). The question naturally arises whether the dynamic crossover temperatures measured by these two different methods are identical or not. Figure 7.19 presents a plot of MSD and the average relaxation time of hydration water in DNA and RNA together in the same figure.

It can be seen clearly from the graph that the crossover temperatures as determined from the elastic scan and the quasi-elastic scattering methods are identical within the experimental error bars.

Figure 7.20 shows the MSD of the three biopolymers and their hydration water in the form of scaled plots. From these plots one can see nicely the synchronisation of T_D (the glass transition temperature of the three biopolymers) and T_L (the dynamic crossover temperatures of their hydration water).

7.3.6 Molecular dynamics simulations of hydrated protein powder

To better understand above experimental results on hydrated protein powder, Lagi *et al.* (2008) subsequently perform MD simulations on the random powder model developed by Tarek and Tobias (2000). This realistic model can reproduce experimental data within the statistical error bars, including the measured mean square displacements of the protein and its hydration water and the translational

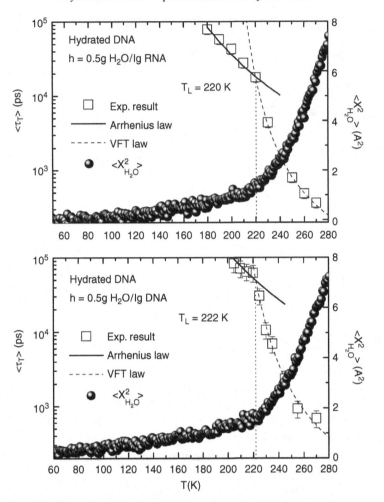

Figure 7.19 An example taken from RNA and DNA hydration water which illustrates that the values of the dynamic transition temperature (MSD vs. T) and the dynamic crossover temperature (log τ_T vs. $1/T$) are closely synchronised (Chu *et al.*, 2008). Copyright 2008 by The American Physical Society.

α-relaxation time of the hydration water, τ_T, calculated from the self-intermediate scattering functions (SISF). The dynamic crossover that Chen *et al.* observed in experiments can thus be attributed solely to the long-time decay (in the range of 100 ps to 50 ns) of the SISF of the hydrogen atoms attached rigidly to a typical water molecule (Chen *et al.*, 2006d; Swenson *et al.*, 2006) not to the long-range proton diffusion coupled to the motion of the so-called Bjerrum-type defects (Swenson *et al.*, 2006).

Lagi *et al.* (2008) put in a box two OPLS-AA (Jorgensen and Tiradorives, 1988) lysozyme molecules randomly oriented and 484 TIP4P-Ew water molecules

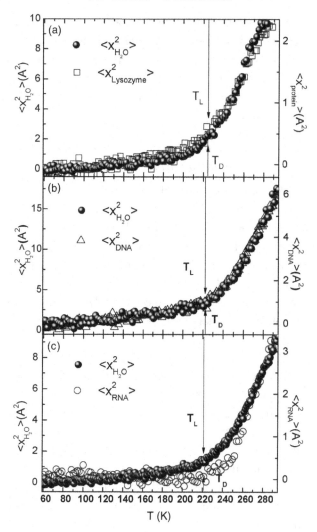

Figure 7.20 Each panel shows the temperature dependence of the MSD of hydrogen atoms in both the biopolymer and its hydration water, respectively. It shows evidence that the crossover temperatures of the two systems, the biopolymer and its hydration water, are closely synchronised. (a) MSD of hydrated lysozyme; (b) MSD of the hydrated B-DNA; (c) MSD of the hydrated RNA. The arrows indicate the approximate positions of the crossover temperature in both the biopolymer (T_D) and its hydration water (T_L). Note that the scale on the left-hand side is for MSD of the hydration water and that on the right-hand side is for the biopolymer. Original drawing by X.-Q. Chu.

($h = 0.34$ for each protein). After an energy minimisation of 5000 steps with the steepest descent algorithm, they equilibrated the system in a NPT ensemble (isobaric-isothermal). They then performed 11 simulations at different temperatures (from 180 K to 4280 K, in 10 K steps) with a parallel-compiled version of

GROMACS 3.0 (Lindahl *et al.*, 2001), starting each simulation from the final configuration of the closest temperature. Each MD simulation length was 50 ns after the equilibration time. While there are only a few water molecules sandwiched between the two proteins, there are more water molecules around other parts of protein surface. But on average, $h = 0.3$ is supposed to be only one monolayer of water covering each protein.

Demonstration of the dynamic crossover phenomenon in protein hydration water

Figure 7.21(a) shows the calculated water hydrogen self-intermediate scattering functions (SISF) as a function of time at fixed Q-value ($0.6\,\text{Å}^{-1}$), while the inset shows the ISF at $T = 220$ K for different Q-values. The solid lines are the best fits to the ISF according to the RCM described above. The RCM fits of the ISF allow them to extract τ_T as a function of temperature as shown in Figure 7.21(b). The crossover feature is clearly visible looking at the decay of the ISF below and above T_L. The crossover temperature is determined to be $T_L = 221$ K, very close to the experimental value of 220 K (Chen *et al.*, 2006b)

Hydration dependence of the crossover phenomenon

But how does the relative amount of water that hydrates the protein powder affect the dynamics crossover of that protein? To address this question, one needs first to understand that proteins are dynamic systems that are strongly coupled with their environment (Frauenfelder *et al.*, 2009). The interaction between protein and its hydration water is of great and fundamental importance to countless processes as already reported in numerous experimental studies in the literature. These include the mean square displacement (msd) of the protein constituent atoms (Paciaroni *et al.*, 2002; Roh *et al.*, 2006; Schiró *et al.*, 2012) the sub-picosecond intra-protein collective vibrations (Orecchini *et al.*, 2001, 2009; Wang *et al.*, 2013, 2014a,b) the intra-protein α- and β-fluctuations (Fenimore *et al.*, 2002, 2004) the protein enzymatic activity (Careri *et al.*, 1980), etc. Clearly there are many puzzles of protein dynamics, and the results from diverse types of measurement is crucial for solving any of these puzzles and collectively offering us a more unified picture of the hydration process.

   To contribute to a better understanding of the dynamics of protein hydration water, Chen *et al.* have reported their various results on dynamic crossover, which is referred to as a transition in the characteristic relaxation time of the hydration water molecule from an Arrhenius behaviour at low temperatures to a super-Arrhenius behaviour at high temperatures (Chen *et al.*, 2006b,c,d, 2009a, 2010b). Specifically, Chen *et al.* (2006d) state that at ambient pressure, the dynamic crossover takes place at $T_X = 220$ K for hydrated lysozyme (with hydration level, $h = 0.3$, i.e.

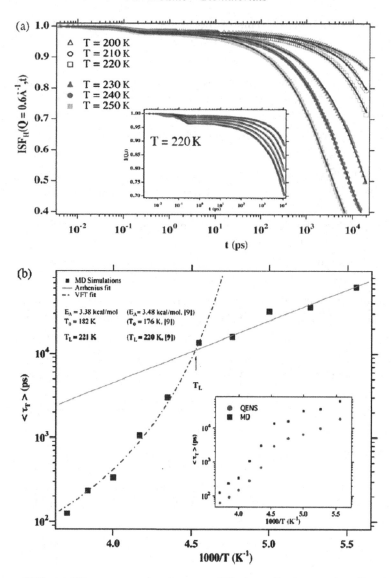

Figure 7.21 (a) Water proton incoherent self-intermediate scattering functions calculated at six different temperatures. (Inset) ISF at five different Q-values (from top to bottom, $0.4, 0.5, 0.6, 0.7$ and $0.8\,\text{Å}^{-1}$). The choice of the Q range was dictated by the low-limit value of $Q = 0.2\,\text{Å}^{-1}$ imposed by the simulation box dimensions and the high-limit value of $Q = 1\,\text{Å}^{-1}$, below which rotational motions can be neglected. The solid curves are fits to the relaxing cage model in a wide time range of seven orders of magnitude, between 2 fs and 20 ns. (b) Temperature dependence of the average translational relaxation time, $< t_T >$, calculated from MD simulation; T_0 is the ideal glass transition temperature. Numerical data are fitted with a Vogel–Fulcher–Tammann (VFT) law at high temperatures and with an Arrhenius law at low temperatures (solid lines) but with the same prefactor. Reprinted with permission from Lagi *et al.* (2008). Copyright 2008, American Chemical Society.

0.3 g of water/g of protein). This phenomenon is of particular interest, partially because:

(i) the crossover temperature T_X of the hydration water is close to the transition temperature of the so-called protein dynamic transition (PDT) temperature T_D (Doster *et al.*, 2010), thus Chen *et al.* conjecture that the PDT is induced by the dynamic crossover of the hydration water; and

(ii) this crossover is interpreted as a characteristic feature of the structural relaxation of the hydration water and is attributed to the existences of the high-density liquid and low-density liquid phases in the supercooled water.

As expected, this phenomenon and its physical implications have generated continuing debates and interchanges. They vary from criticisms to counter- or new proposals. For example, Doster *et al.* (2010) state that the crossover observed in Chen *et al.* (2006d) is due to numerical errors in experimental data analysis and would disappear when an improved analysis method is used; yet Magazù *et al.* (2011) and Schiró *et al.* (2012) disagree with Doster and draw their own conclusions on the role of dynamic crossover with their neutron scattering experimental data. Swenson *et al.* (2006), Pawlus *et al.* (2008) and Fenimore *et al.* (2013) propose that the appearance of the dynamic crossover in the hydration water is due to the existences of two different relaxation processes – the structural relaxation and a secondary relaxation – rather than a qualitative change of the alpha-relaxation time from an Arrhenius behaviour to a super-Arrhenius behaviour of the structural relaxation time.

With these diversified views, in order to address some of the existing concerns in relation to the dynamic-crossover in the protein hydration water, Chen's group has performed many more experiments (Bertrand *et al.*, 2013b; Wang *et al.*, 2013, 2014b,c). Specifically Wang *et al.* (2014b) have further investigated the characteristic relaxation time of the hydration water of lysozyme at three different hydration levels, $h = 0.18, 0.30$ and 0.45, using a quasi-elastic neutron scattering (QENS) technique. The experiment was perform on the backscattering spectrometer, BASIS, at the Spallation Neutron Source, Oak Ridge National Laboratory. In this latest study, for each hydration level, the spectra at three Q-values, $0.5, 0.7$ and $0.9\,\text{Å}^{-1}$, were measured. The energy window used for data analysis is from -20 to $20\,\mu\text{eV}$, the energy resolution of the spectrometer is $3.4\,\mu\text{eV}$. For lysozyme, the hydration level of 0.2–0.25 (denoted as h_c) is critical because the hydrogen-bonding sites on the protein surface are completely saturated with water at this hydration level (Careri *et al.*, 1980). Wang *et al.* (2014b) stress that $h = 0.30$ and 0.45, the two higher hydration levels studied, are well above h_c, while the lowest hydration level, $h = 0.18$, is slightly lower than h_c. The latter situation is expected to be similar to other cases of 'monolayer coverage' of water on hydrophilic surface (Mamontov and Herwig, 2011).

The data of this latest experiment of Wang *et al.* (2014b) are analysed by assuming the self-intermediate scattering function (SISF) of the hydration water as modelled with the following equation:

$$F_s(Q, t) \approx A(Q) \exp\left[-\left(\frac{t}{\tau(Q, T)}\right)^\beta\right]. \tag{7.15}$$

The Q-dependent relaxation time in the stretch exponential function extracted from the higher-temperature experimental data that exhibit super-Arrhenius behaviour, can be fitted by the Vogel–Fulcher–Tammann (VFT) relation:

$$\tau(Q) = \tau_0(Q) \exp\left[\frac{DT_0}{T - T_0}\right], \tag{7.16}$$

while the lower-temperature experimental data exhibit Arrhenius behaviour, can be fitted by the Arrhenius relation,

$$\tau(Q) = \tau_0(Q) \exp\left[\frac{E_0}{K_B T}\right]. \tag{7.17}$$

The Q-dependent average relaxation time is then calculated by Eq. (7.18):

$$< \tau >= \frac{1}{\beta}\Gamma\left(\frac{1}{\beta}\right)\tau_0(Q). \tag{7.18}$$

Thus, the significant results of Wang *et al.* (2014a) can be shown clearly in the three panels of Figure 7.22:

- Figure 7.22(a) shows the τ for the sample with $h = 0.30$, the crossing temperature T_X of the two fits for the measured Q values of $0.5\,\text{Å}^{-1}$ and $0.9\,\text{Å}^{-1}$ is 220 K. For comparative purpose, Wang *et al.* (2014a) also plot in this panel (a) the neutron scattering result of Doster *et al.* (2010) for the relaxation time of hydration water of deuterated C-phycocyanin (CPC) with $h = 0.3$ and at $Q = 1\,\text{Å}^{-1}$. Not surprisingly, due to the similar h, Q-value and analysis method used in these two cases, the data presented in panel (a) seem to agree perfectly despite the claim of Doster *et al.* (2010) that no dynamic crossover happens. In fact, the latest analysis of Wang *et al.* confirms that the dynamic crossover indeed takes place when the hydration water is sufficient, and it becomes more visible as h increases, due to the strong h dependence of the higher-temperature data. Similar phenomenon is also observed in the water confined in nanoporous silica material MCM-41. Faraone *et al.* (2004) show that as the diameter of the confining pore increases from 14 to 18 Å (thus the water content also increases),

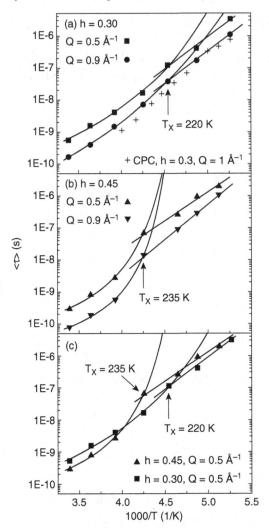

Figure 7.22 Arrhenius plot of the mean characteristic relaxation time τ at $Q = 0.5$ and $0.9\,\text{Å}^{-1}$ for the samples with $h = 0.30$ and 0.45. Panels (a) and (b) correspond to the samples with $h = 0.30$ and 0.45 respectively. The dynamic crossover takes place at about $220\,\text{K}$ in the former case and at about $235\,\text{K}$ in the latter case. The result of Doster *et al.* (2010) for the CPC with $h = 0.3$ and at $Q = 1\,\text{Å}^{-1}$ is also plotted in panel (a). Panel (c) shows the graph of τ at $Q = 0.5\,\text{Å}^{-1}$ for $h = 0.30$ and 0.45. It is seen that the hydration water at $h = 0.45$ exhibits a more visible dynamic crossover than that at $h = 0.30$. For clarity, we do not draw the data at $Q = 0.7\,\text{Å}^{-1}$ for both of the samples (Wang *et al.*, 2014b). Copyright 2014, AIP Publishing LLC.

Table 7.1 *Parameters in the VFT relation and the Arrhenius relation for*
$h = 0.30$ *and* 0.45.

h	$Q[\text{Å}^{-1}]$	D	$T_0[\text{K}]$	$E_0[\text{kcal/mol}]$
0.30	0.5	4.96	148	9.20
0.30	0.9	9.98	123	9.26
0.45	0.5	1.41	201	8.93
0.45	0.9	1.03	207	9.85

D changes from 4.62 to 1.47, and T_0 changes from 170 to 200 K. Such h dependences of D and T_0 exhibit good agreement to the results of the protein hydration
water of Wang *et al.* (2014b).

- Figure 7.22(b) shows that as h increases to 0.45, the dynamic crossover shifts
  to 235 K. The parameters D, T_0 and E_0 for these two samples are listed in
  Table 7.1. It is shown that the parameters for the higher-temperature data, D
  and T_0, display strong h dependence.
- Figure 7.22(c) shows that a significant difference in the curvature of the $< \tau >$
  in the Arrhenius plot at higher temperatures as h increases from 0.30 to 0.45.
  That the T dependence of $< \tau >$ exhibits stronger super-Arrhenius behaviour
  with $h = 0.45$ than with $h = 0.30$ is also reflected in the measured spectra.

Because of the similarity of T_X and T_D for the protein with $h = 0.3$, Chen *et al.*
(2006d) tentatively attribute the onset of the PDT to the dynamic crossover of the
protein hydration water. This assumption has also generated different views, such
as by Magazù *et al.* (2011), Schiró *et al.* (2012), and Mamontov and Chu (2012).
It is well known that most of proteins only show biological function with sufficient level of hydration. Therefore, to understand how protein works, it is essential
to know the relation between the PDT and the dynamic crossover temperature
in hydration water. Roh *et al.* (2006) investigated the PDT of lysozyme also at
$h = 0.18$, 0.30 and 0.45 (as shown in Figure 2 of Roh *et al.* (2006)). Their results
enable Wang *et al.* (2014b) to make a direct comparison between the PDT and the
dynamic crossover in protein hydration water. The common features of the PDT
and the dynamic crossover are found to include:

(i) both of these two crossover phenomena appear at $h = 0.30$ and 0.45, and are
    strongly suppressed at $h = 0.18$;
(ii) both of them are enhanced as h increases.

However, the h dependences of T_D and T_X are not similar to each other. According to Roh *et al.* (2006), the values of T_D at $h = 0.30$ and 0.45 are both around
200 K, which is substantially different from the results of T_X shown in the Wang

et al. (2014b) study. In fact, Paciaroni *et al.* (2002) report that T_D even decreases as h increases. Taking into account of these results, Wang *et al.* (2014b) conclude that the dynamic crossover in the protein hydration water does not directly induce the PDT.

Wang *et al.* (2014b) thus summarise that they have investigated the single-particle dynamics of protein hydration water at three different hydration levels. The dynamic crossover phenomenon appears when h is higher than the 'monolayer' hydration level, $h_c = 0.2-0.25$, and becomes more visible as h increases in the measured range. It disappears when h is slightly lower than h_c, and in this case τ exhibits Arrhenius behaviour in the whole range of measured temperature (see Figure 7.23). The higher-temperature data for the samples with $h = 0.30$ and 0.45, which exhibit super-Arrhenius behaviour, are sensitive to h. On the contrary, the lower-temperature data, which exhibit Arrhenius behaviour, are relatively insensitive to h and display local-like characteristics. These results highlight the importance of the tetrahedral hydrogen-bond structure in the dynamics of the hydration water. In addition, the crossover temperature exhibits different h dependence from that of the protein dynamic transition (PDT) temperature. This difference shows that the dynamic crossover in the hydration water does not directly induce the PDT.

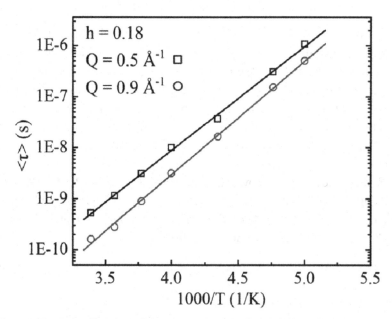

Figure 7.23 Arrhenius plot of the mean characteristic relaxation time $< \tau >$ at $Q = 0.5$ and $0.9\ \text{Å}^{-1}$ for the sample with $h = 0.18$. In this case, the T dependence of $< \tau >$ can be described with Arrhenius relation in the whole range of measured temperature (Wang *et al.*, 2014b). Copyright 2014, AIP Publishing LLO.

7.3.7 Comparing these results with those of other related studies

Several authors have questioned the existence of the reported FSC in the dynamics of protein hydration water. The evidence for a FSC in confined water comes primarily from QENS measurements (Chen *et al.*, 2006b,c; Chu *et al.*, 2008) of the (Q-independent) single-particle relaxation time of hydration water molecules and NMR measurements (Mallamace *et al.*, 2007c) of the self-diffusion constant of the hydration-water molecules. The same type of crossover has not been observed in the dynamics measured by dielectric relaxation spectroscopy (Swenson, 2006; Pawlus *et al.*, 2008).

Based on the assumption that dielectric relaxation spectroscopy probes the *main* structural relaxation, it was suggested that QENS might be probing a secondary relaxation process that breaks away from the main structural relaxation at the crossover temperature (Pawlus *et al.*, 2008). That QENS does not probe the *main* structural relaxation was also suggested based on studies of rotational dynamics of protein hydration water measured in ^{2}H NMR experiments (Vogel, 2008). Doster *et al.* (1989) have reported on QENS measurements of the hydration water dynamics for hydrated phycocyanin, a light harvesting blue copper protein. By fitting their data with the model used by Chen *et al.* (2006c), they were able to reproduce the FSC observed. However, by altering the fitting procedure, it was shown that the FSC can be significantly modified and even eliminated. Based on a careful analysis of resolution effects in neutron scattering experiments, Magazù *et al.* (2011) have recently concluded that the FSC reported (Chen *et al.*, 2006c) is a confirmed result in the framework of neutron scattering techniques, and therefore, it is due neither to the energy resolution effects nor to numerical errors in data analysis protocol, differently to what W. Doster *et al.* proposed.

In contrast to essentially immobile MCM-41, the dynamics of the hydrated protein, which serves as the confining substrate for the hydration water, are also of principal interest. At low temperatures, proteins exist in a glassy state characterised by the absence of conformational flexibility and biological function. Upon warming, hydrated protein shows a so-called *dynamic transition* at $T \approx 220$ K, above which the conformational flexibility is restored and enzymatic function is activated (Rasmussen *et al.*, 1992). The transition is marked by an upturn in the average mean square hydrogen displacement (Doster *et al.*, 1989) as shown in Figure 7.13. The nature and origin of the dynamic transition in hydrated proteins and it relationship to the dynamics of the protein hydration water have been hotly debated topics in recent years.

A series of recent experiments have established that biological function is not required for the protein dynamical transition to occur. He *et al.* (2008) have shown, using terahertz time domain spectroscopy, that the dynamic transition does not depend on the tertiary or secondary structure of the protein by demonstrating that the dynamic transition is still observed in hydrated lysozyme powder that has been chemically denatured with guanidinium hydrochloride. Similar results were found by Mamontov *et al.* (2010) in QENS experiments on hydrated lysozyme that was denatured with sodium hydroxide. Recently, Schiró *et al.* (2011) have demonstrated that the dynamic transition does not even require the protein polypeptide chain. These authors investigated hydrated myoglobin and a hydrated mixture of amino acids with the same chemical composition as myoglobin. The protein dynamic transition was observed in both samples and was confirmed to be absent when both samples were dehydrated. It was consequently concluded that the dynamical transition primarily involves the motion of the amino acid side chains.

Recent experiments have also revealed that the dynamic transition, as quantified by the mean square displacement, does not represent the appearance of a new relaxation mode, or a sudden change in the dynamics, but instead appears when a relaxation process, which also exists at low temperatures, moves into the experimentally accessible frequency window. Sokolov *et al.* (2008) have made QENS measurements of dry and hydrated lysozyme and RNA samples. The temperature dependence observed in the dry samples is largely attributable to methyl-group rotations. By subtracting the dry sample signal from that obtained for the hydrated sample, the relaxation process associated with the dynamic transition can be isolated. These authors concluded the hydration of biological macromolecules activates a new relaxation process that is present at all temperatures. These findings were confirmed and further extended with the inclusion of dielectric relaxation spectroscopy measurements as reported by Khodadadi *et al.* (2008). The recent work of Magazù *et al.* (2011), which confirmed the FSC in the single-particle translational relaxation times of the hydration water, also confirms that the protein dynamic transition does not correspond to the appearance of a new relaxation mode in the protein dynamics. One awkward feature of the dynamic transition is that it depends on the energy resolution width of the neutron scattering instrument. This issue has recently been addressed by Vural and Glyde (2012), who propose a method for extracting the intrinsic mean square displacement of H atoms in hydrated proteins. The results reported in the preceding paragraph have important implications for the relationship between the protein dynamics and the dynamics of the hydration water. If the apparent dynamic transition does not represent a fundamental change in the dynamics, then it is unlikely that the protein dynamic transition induces the FSC observed in the hydration water. This conclusion is supported by the recent QENS measurements of Chu *et al.* (2012). The protein dynamics were studied in a $LiCL–H_2O$ solution with the finding that a dynamic crossover was still observed in the aqueous solvent, but no dynamic transition was observed in the motion of the protein. Thus, the two are decoupled in this system.

7.4 Module – Dynamic crossover in hydration water of cement pastes

Hydration water in cement

The hydration water in cementitious materials is a type of strongly confined water. The major components of Ordinary Portland Cement (OPC) paste are tricalcium silicate Ca_3SiO_5 (C_3S) and dicalcium silicate Ca_2SiO_4 (C_2S), which account for 50–70% and 20–30% of the composition respectively (Ridi *et al.*, 2011). The setting and hardening of cement pastes is the result of exothermic hydration reactions. These reactions produce a binary mixture of a calcium-silicate-hydrate (C-S-H) gel and calcium hydroxide micro-crystalline particles when cement and water are mixed together. As the cement paste ages an increasing fraction of the water becomes immobilised as it is sequestered into the $Ca(OH)_2$ particles. The C-S-H gel is the major binding component of cement pastes. Consequently, many experiments have been performed on pure C-S-H gel. The exact microstructure of C-S-H gel is still an active topic of research, the quantitative structural parameters of which have recently been determined by Chiang *et al.* (2012) using small angle neutron scattering technique. However, schematic models, such as the Jennings Colloidal Model CM-II (Jennings, 2000, 2008), seem to be able to rationalise much of the available data. In this Jennings Model, C-S-H is composed of colloidal particles, or *globules*, with dimensions on the order of tens of nanometers. According to CM-II, the microstructure of cement paste can be schematically described through a hierarchy of pore sizes (see Figure 7.24 (Ridi *et al.*, 2011)). The basic structural unit is a disc-like globule with a layered internal structure (as shown in the enlargement on the right-hand side of the figure). The water inside the globule is located in both inter-lamellar spaces and in very small cavities (intra-globular pores, IGP), of dimensions ≤ 1 nm. The packing of these globules produces a porous structure with two characteristic pore types: small gel pores (SGP), of dimensions 1–3 nm,

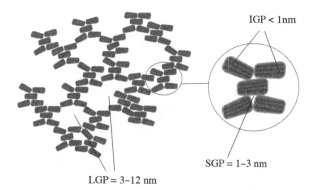

Figure 7.24 The microstructure and porosity in C–S–H gel as described in CM-II. Reprinted from Ridi *et al.* (2011). Copyright 2011, with permission from Elsevier.

and large gel pores (LGP), of dimensions 3–12 nm. In sufficiently hydrated gels, the fractal packing structure of these globules contains large and small pores that are filled with water. As can be expected from confinement, the water that occupies the inter-globular space in the C-S-H gel shows significantly slower dynamics than bulk hydration water.

The cement hydration reaction is comprised of multiple distinct phases and proceeds over the course of months. Consequently, the dynamics of water confined in the inter-globular spaces of C-S-H also evolve as a function of aging time. Fratini *et al.* (2001) have made a QENS study of the aging-dependent dynamics of water confined in hydrated C_3S. This system can be modelled as containing both immobile and glassy water populations. In this regard, the dynamic picture is similar to that of water confined in MCM-41. The results were analysed in terms of the relaxing cage model (RCM) discussed in Section 7.2. From this analysis, the authors extracted the fraction of immobile water, the average relaxation time, and the stretch exponent as they evolved over 30 h at two different temperatures, 30 °C and 15 °C. Over the course of the ageing, the fraction of immobile water and average relaxation time were found to increase, and the stretch exponent was found to decrease. It was concluded that the process of ageing continuously converts glassy water into immobile water, possibly through the shifting of pore water into the layered structure of the globules and into crystalline $Ca(OH)_2$ colloidal particles. In order to probe both the translational and rotational dynamics of water confined in cement paste, Faraone *et al.* (2002) have studied the ageing-dependent single-particle dynamics over a wide range of Q-values in hydrated C_2S at $T = 303$ K. After monitoring the evolution of the dynamics for 100 days at different intervals, it was concluded that the rotational dynamics are not affected by the ageing of the cement paste. The rotational relaxation in the paste was found to be roughly five times slower than observed in bulk water. A similar study was also made by them with C_3S (Chen *et al.*, 2010a).

7.4.1 Observation of FSC in hydration water of cement paste

Zhang *et al.* (2008) have observed a FSC in the hydration water of aged white Portland cement (WPC) paste. Water confined in white cement at 40% hydration level by mass, was studied after curing for 8 days. A dynamic crossover was observed at $T_L \approx 231 \pm 5$ K in the translational relaxation time of the confined water. Their analysis was supplemented by differential scanning calorimetry (DSC) and near infrared spectroscopy (NIR) measurements. These confirmed that the confined water did not freeze and is strongly confined after 8 days of curing and supported the interpretation that water in the large gel pores is entirely depleted after 28 days as it percolates into colloidal particles and small gel pores. The self-intermediate scatting function, $F_H(Q, t)$, was approximated by a simplified RCM model (cf. Eq. (7.2)). The crossover temperature coincides with a peak in the DSC curve. These authors also studied this phenomenon using the dynamic response function $\chi_T(Q, t)$ (Berthier *et al.*, 2005). Details of this analysis are given in Section 9 of Chen *et al.* (2010a).

The dynamic response function $\chi_T(Q, t)$ (Berthier *et al.*, 2005) also known as the dynamic susceptibility, is defined by

$$\chi_T(Q, t) = -\left(\frac{\partial F_H(Q, t)}{\partial T}\right)_P, \tag{7.19}$$

where $F_H(Q, t)$ is the self-intermediate scattering function of the hydrogen atoms. The application of the dynamic response function to the study of water is discussed further in Chen *et al.* (2010a). Here we present a brief discussion of this paper. If one could experimentally measure the four-point correlation function $\chi_4(Q, t)$, which is easily relatable to the concept of dynamic heterogeneity, then it would not be necessary to study the properties of $\chi_T(Q, t)$. However, the dynamic response function is much easier to measure than the four-point correlation function. As a function of time, $\chi_4(Q, t)$ is known to reach a maximum at the characteristic relaxation time of the system. The height of this maximum is related to the volume in which correlated fluctuations take place. The dynamic response function $\chi_T(Q, t)$ serves as a bound on $\chi_4(Q, t)$ through the equation (Berthier *et al.*, 2005)

$$\chi_4(t) \geq \frac{k_B}{c_P}T^2\chi_T^2, \tag{7.20}$$

where k_B is Boltzmann's constant and c_P is the isobaric specific heat.

This relation implies that the features of $\chi_T(Q, t)$ might be roughly interpretable as those of $\chi_4(Q, t)$. In particular, $\chi_T(Q, t)$ also exhibits a peak around the Q-dependent relaxation time, with the height of this peak being related to the strength of the dynamic heterogeneities. An advantage of using $\chi_T(Q, t)$ is that it provides a model-independent means of analysing relaxation phenomena in glassy systems. If the RCM is assumed to provide an accurate description of the physical process, then it can be shown that $\chi_T(Q, t)$ is essentially probing the temperature dependence of the Q-dependent relaxation time (Zhang *et al.*, 2009b). Zhang *et al.* (2009b) also use MD simulations to show that the peak height of $\chi_T(Q, t)$ increases as the crossover temperature is approached and decreases for decreasing temperatures below the crossover temperature. However, some care is required in assuming that there is a direct proportionality between $\chi_T(Q, t)$ and $\chi_4(Q, t)$. Based on MD simulations of the TIP4P-Ew potential model for water, both $\chi_T(Q, t)$ and χ_4 can be calculated directly. While χ_T was found to have a peak at the dynamic crossover temperature and the height of the peak reached its

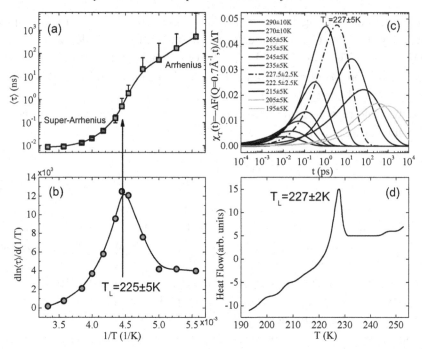

Figure 7.25 (a) Arrhenius plot of the average translational relaxation of water under confinement in aged cement paste. The relaxation times undergo a FSC at $T_L \approx 225$ K. (b) Derivative of the Arrhenius plot shown in (a). The derivative exhibits a maximum at the crossover temperature. (c) Dynamic response function calculated from the SISF using finite differences. (d) DSC curve of water confined in hydrated cement paste (aged 8 days). The dynamic response function and the DSC curve also locate the FSC in the vicinity of $T_L \approx 225$ K (Zhang *et al.*, 2009b). Copyright 2009 by The American Physical Society.

maximum value at the dynamic crossover temperature, the height of the peak in χ_4 continued to grow below T_L. Zhang *et al.* (2009b) have used $\chi_T(Q, t)$ to obtain the crossover temperature for water in hydrated white Portland cement paste with the result $T_L = 227 \pm 5$ K (Figure 7.25(c)). The observed crossover temperature coincides with that of the peak found in differential scanning calorimetry measurements as shown in Figure 7.25(d).

7.5 Module – Dynamic crossover in confined water and its relation to the second critical point of water

The liquid–liquid critical point hypothesis

A comprehensive explanation of the anomalous properties of bulk-supercooled water has not yet been realised. The same is necessarily true of deeply cooled water under strong confinement. It is well established that as liquid water is cooled the average degree of hydrogen-bonding increases and that the preferred geometry and strength of these hydrogen bonds have a profound impact on water's properties. Formalising this microscopic picture into a consistent macroscopic theory is an active and ongoing pursuit. Theoretical models of water date back over 100 years (Röntgen, 1892) and are so numerous and varied as to warrant their own review. A handful of *scenarios*, such as the *singularity-free scenario* (Sastry *et al.*, 1996) and the *stability limit conjecture* (Speedy, 1982) have been proposed in the last 40 years with the aim of accounting for the low temperature properties of water. Of these, the one that has stimulated the greatest research effort is the liquid–liquid critical point (LLCP) hypothesis. This hypothesis and its application to confined deeply cooled water are the foci of this section.

Bulk water

Before addressing the connection between the LLCP hypothesis and the properties of deeply cooled water under strong confinement, it is useful to review some general features of the critical phenomena found in the vicinity of a known critical point. Thermodynamic and dynamic critical phenomena have been extensively categorised and characterised both experimentally and theoretically in an astounding variety of condensed matter systems (Domb, 1996). Within this seemingly diverse array of phase transitions, these phenomena exhibit a great degree of universality. Hence, there is relatively little flexibility in the criterion required to establish the existence of a critical point in a particular system. The constraints provided by the rigorous definition of a critical point are essential for establishing whether the LLCP hypothesis is truly applicable to supercooled water or whether the observed behaviour is indicative of other phenomena.

Many of the thermodynamic anomalies observed in bulk-supercooled water are reminiscent of those found in the vicinity of a critical point. By definition, a critical point terminates a line of first-order transition between two distinct phases. For example, the high-temperature critical point of water (located at $T_c = 647$ K and $P_c = 22$ MPa) terminates the liquid–vapour coexistence curve. Thermodynamic response functions, such as the isobaric heat capacity, diverge at a critical point, and show anomalous enhancement in the region surrounding the point.

On the basis of molecular dynamics simulations, Poole *et al.* (1992) suggested that the existence of a liquid–liquid critical point in the deeply supercooled region of liquid waters phase diagram could account for water's observed behaviour. In this instance, the proposed critical point terminates a line of first-order transitions between a high-density liquid (HDL) phase and a low-density liquid (LDL) phase. These phases are proposed to be the high-temperature liquid-state continuations of the experimentally accessible low-density amorphous (LDA) and high-density amorphous (HDA) ice phases (Mishima and Stanley, 1998a). As regards the LLCP hypothesis, it is tantalising that these amorphous phases are separated by a first-order transition (Mishima, 1994).

This *second critical point* in supercooled water, if it exists, is generally believed to be located below the homogeneous nucleation temperature and at relatively high pressures with the first-order line of liquid–liquid phase transitions extending upwards in pressure. By one estimation the critical temperature and pressure are $T_c \approx 220$ K and $P_c \approx 100$ MPa (Mishima and Stanley, 1998a). Many of the experiments performed on supercooled water have been made at atmospheric pressure, corresponding to the one-phase region of the proposed phase diagram. Experiments that are not conducted along the critical isotherm ($T = T_c$) or the critical isobar ($P = P_c$) do not exhibit the singular behaviour that is typically associated with critical phenomena. Instead the thermodynamic response functions exhibit maxima along various loci emanating from the critical point. The line of isothermal compressibility maxima is a *natural* extrapolation of the first-order transition line into the one-phase region. Along this line, the long-range fluctuations responsible for critical phenomena have their greatest extent. In the supercooled water literature, the line of isobaric heat capacity maxima has sometimes been given the honorific title of the *Widom line* (Xu *et al.*, 2005).

That the proposed second critical point is located in a region of the phase diagram that is experimentally inaccessible, at least in the liquid state, certainly presents some conceptual challenges. In equilibrium thermodynamics, the concept of a critical point is well defined. This definition can be extended to metastable states located in the supercooled region for which metastable equilibrium (Williams and Evans, 2007) can be achieved. For instance, protein solutions and colloids with short-range attractive potentials are known to exhibit a metastable fluid–fluid critical point within the fluid–solid coexistence curve (ten Wolde and Frenkel, 1997). A rigorous definition of a metastable critical point requires consideration of the stable equilibrium state. The same is necessarily true of a critical point with a proposed location below the homogeneous nucleation limit.

A phenomenological means of explaining the behaviour of supercooled water is to invoke a so-called *virtual* critical point. While this type of critical point is a theoretical construct, in that phase coexistence can never be observed, it does produce well-defined and testable predictions. This approach has recently been investigated by Anisimov and collaborators (Fuentevilla and Anisimov, 2006; Bertrand and Anisimov, 2011; Holten *et al.*, 2012). These authors posit the existence of a critical point in the unstable region of liquid water's phase diagram. The equation of state induced by this critical point, as appropriate for the universality class of the 3D Ising model, is then extrapolated into the experimentally accessible region of liquid water's phase diagram. By fitting the equation of state to the available bulk thermodynamic data, it was concluded that the order parameter of the phase transition is dominated by the entropy (as opposed to the density) and that the experimental data are consistent with $P_c = 27.5$ MPa and $T_c = 224$ K. This critical pressure is notably much lower than other estimates found in the literature. The fitted value of P_c is heavily influenced by the isothermal compressibility measurements at elevated pressures. Since the apparent strength of the compressibility anomaly decreases with increasing pressure, the critical point is constrained to lie between the experimental pressures (10 MPa and 50 MPa) exhibiting the most anomalous behaviour. Additionally, this approach is limited by the fact that it ignores the effects of the stable crystalline states to which the metastable liquid state decays.

While some of the anomalies found in low-temperature water, such as the increase in the isobaric heat capacity, are potentially attributable to a LLCP, others, such as the density maximum, are not. The density maximum is more indicative of the crossover from the dominance of *density ordering* to the dominance of *bond-orientational ordering* as the temperature is decreased at constant pressure. Since the hypothetical critical point is conjectured to be well below the temperature of maximum density, the order parameter for the liquid–liquid transition might closely reflect the associating degrees of freedoms. A comprehensive theory, which is consistent with the LLCP hypothesis, might be expected to illuminate how a critical point emerges in a fluid dominated by hydrogen bonding.

In addition to thermodynamic anomalies, critical phenomena are also characterised by anomalies in transport coefficients and relaxation rates (Hohenberg and Halperin, 1977). The long-time relaxation rate of a system is generally determined by the ratio of a transport coefficient and a thermodynamic response function. For example, the relaxation of density fluctuations in a single-component fluid, as characterised by the full-intermediate scattering function $F(Q, t)$, is governed by the decay rate $\Gamma_T = D_T Q_2$, where D_T is the thermal diffusivity. The thermal diffusivity is in turn given by the ratio $D_T = \lambda/\rho C_P$, where λ is the thermal conductivity and C_P is the isobaric heat capacity. Both λ and C_P diverge at the liquid–vapour critical point. However, C_P diverges more strongly than λ and the relaxation rate Γ_T goes to zero, producing the so-called *critical slowing down*. It is notable that the viscosity of a single-component liquid also diverges at the critical point. An enhancement of the relaxation time is also found when crossing the line of heat capacity maxima. The search for critical slowing down in supercooled water is complicated by the fact that the dynamics of supercooled water become increasingly glassy as the temperature is lowered. Since a comprehensive understanding of the glass transition is lacking, it is unclear how the interplay between glassy behaviour and critical dynamics would manifest itself. Just as with the isothermal compressibility, the decrease of the shear viscosity of supercooled water with increasing pressure also apparently constrains the critical pressure of the proposed LLCP.

Confined water

A central question in the study of condensed matter under confinement is the extent to which the observed properties can be related to those seen in bulk. Developing an understanding of the two main differences found between confined systems and bulk systems, i.e. geometric constraints on the size of fluctuations (global) and interactions between the substance and substrate (local), is important for establishing this connection. The first of these issues has been well studied in the context of critical phenomena and produces what are referred to as finite-size effects (Fisher and Barber, 1972).

The anomalous behaviour of the thermodynamic response functions observed near a critical point stem from the growth of the correlation length ξ of spontaneous fluctuations. When ξ approaches the characteristic size of the system, the observed critical phenomena begin deviate from those observed in bulk. Singular critical phenomena are not observed, instead the singularities are replaced by maxima, the location of these maxima are shifted relative to the bulk critical point, and the anomalous behaviour is smeared out over a larger range of temperatures and pressures. Strictly speaking, a critical point cannot exist in a finite-size system. However, phase coexistence can still occur.

Liquid–liquid equilibria and phase separation in confinement have previously been reviewed by Gelb *et al.* (1999). In addition to finite-size effects, these authors discuss two additional features of liquid–liquid equilibria in confinement. The first is the prevalence and strength of metastability within the two phase-regions, which leads to hysteresis. The second is the slowness of macroscopic phase separation, which does not occur over typical laboratory time scales. The role of dimensionality is also addressed. In a slit-geometry, a crossover from three-dimensional behaviour to two-dimensional behaviour is expected as the slit width is narrowed. As evidenced by the Ising model, a finite temperature critical point can still exist in two dimensions. A crossover between three-dimensional and quasi-one-dimensional behaviour is expected as the radius is reduced in a cylindrical pore geometry. A finite temperature critical point cannot exist in a truly one-dimensional system due to the disruptive influence of fluctuations.

All of the unique phenomena observed in deeply cooled water under strong confinement have been interpreted in terms of the LLCP hypothesis. Via molecular dynamics simulations of pure (non-confined) water, Xu *et al.* (2005) have found that the line of heat capacity maxima is coincident with a dynamic crossover in the diffusion coefficient for various models of water. Since the line of heat capacity maxima is anticipated from the LLCP hypothesis, this work connects the fragile-to-strong crossover (FSC) to the crossing of this line. Both a maximum in the heat capacity and a FSC have been observed for water confined in MCM-41 and so it is certainly plausible that the same connection exists between the two, although heat capacity measurement have not yet been made above atmospheric pressure. We remark that a crossover from fragile to strong behaviour is not typically associated with dynamic critical phenomena. In addition to a line of heat capacity maxima, a line of thermal expansivity minima is expected to extend from a LLCP into the one-phase region. The density minimum, which implies a minimum in the thermal expansivity has also been used as supporting evidence for the LLCP hypothesis (Liu *et al.*, 2007).

The density hysteresis observed in strongly confined water at higher pressures has been interpreted as resulting from the crossing of a first-order liquid–liquid transition line (Zhang *et al.*, 2011). In general, metastable states exist on either side of a line of first-order phase transitions. If the relaxation from the metastable state into the stable state is sufficiently slow relative to the rate at which the transition line is crossed, then a hysteresis will appear in the measured order parameter profile. The measured profiles will be different depending on the direction in which the line was crossed. As discussed by Gelb *et al.* (1999) strong metastability and hysteresis are characteristic of phase separation in confinement.

Implicit in the application of the LLCP hypothesis to water under strong confinement is the assumption that the behaviour of water confined in MCM-41 pores is simply an extrapolation of bulk behaviour to lower temperatures. In light of the previously discussed confinement effects, this assumption requires closer scrutiny. Clearly confinement alters the free energy surface of the confined water, in that the sector associated with freezing in the bulk no longer corresponds to energetically favourable states. The extent to which this modification leaves the liquid sector unchanged is uncertain. A rigorous explanation of the statement that a portion of the water confined in MCM-41 pores is *bulk-like* should address this issue quantitatively. Binder *et al.* (2011) have recently reviewed the case for differentiating between bulk and confined systems using illustrative simulations of near-critical systems in long cylindrical pores. They found that the physical picture arising from very long, and therefore more physical as regards MCM-41, pores and short pores are distinctly different. While the axial density profile is homogeneous in short pores, the long pores are characterised by multiple regions of axial inhomogeneity.

An alternative explanation of the behaviour of supercooled water under strong confinement has recently emerged from the simulations work of Limmer and Chandler on the behaviour of the mW water model which was studied in narrow cylindrical *silica* pores as a function of temperature, pressure and pore diameter (Limmer and Chandler, 2011, 2012). They claim that their mW model of water did not exhibit a LLCP even though it did qualitatively reproduce many of water's anomalies. Consequently, any experimentally observed phenomena that are reproduced in these simulations must have an alternate explanation that does not rely on an existence of LLCP. A fragile-to-strong crossover was indeed observed in the simulations and was shown to be coincident with the freezing transition in the larger diameter pores and to represent a continuous transition to *crystal-like* states in smaller diameter pores. Notably, these authors argue that the transition to these *crystal-like* states is not expected to produce a freezing peak in the heat capacity. The low temperature Arrhenius relaxation observed in the frozen states is associated with the motion of the disordered surface layer. It was also argued that the transition to these *crystal-like* states could account for the observed density hysteresis at elevated module pressures. As discussed in Erko *et al.* (2012), they have found a temperature-dependent density minimum for water confined in larger diameter MCM-41 pores where crystallisation definitely occurs.

However, in response to these claims (Limmer and Chandler, 2011, 2012; Sciortino *et al.*, 2011; Liu *et al.*, 2012) conducted additional thorough simulation work on the ST2 model of water. Their findings support the liquid–liquid transition scenario of water in deeply supercooled conditions instead of the liquid–crystal transition of water as suggested in (Limmer and Chandler, 2011, 2012).

7.6 Module – Density measurement of confined water in MCM-41S

7.6.1 Introduction

In many biological or geological systems, water resides in pores of nanoscopic dimensions close to either hydrophilic or hydrophobic surfaces, comprising a layer of water near the surface and a region of 'free (core) water', as already described in Section 7.1, Figure 7.2. Such 'confined' water has attracted considerable attention, due to its fundamental importance in many processes, such as protein folding, concrete curing, corrosion, molecular and ionic transport, etc. (Chandler, 2005; Brovchenko and Oleinikova, 2008; Fayer and Levinger, 2010). Yet, our understanding of numerous physicochemical anomalies of confined water particularly of its relation to bulk water, is still sketchy and basic gaps persist. We have addressed much unusual behaviour of such confined water in the supercooled region where water remains in the liquid state below the melting point in previous sections of Chapter 7. This last section aims only at explaining anomalies such as the density maximum and minimum (Liu *et al.*, 2007; Mallamace *et al.*, 2007a; Zhang *et al.*, 2009a), density hysteresis associated with crossing the conjectured first-order liquid–liquid phase transition line (LLPT), the apparent peaking of the thermodynamic response function, and in particular, the isobaric thermal expansion coefficient, on crossing the Widom line associated with the conjectured second liquid–liquid critical point (LLCP) (Zhang *et al.*, 2011; Liu *et al.*, 2013; Wang *et al.*, 2014a).

7.6.2 Data analysis method

Liu *et al.* (2007) of the Chen's Group at MIT first developed their methods for detecting density minimum in supercooled D_2O in MCM-41-S at ambient pressure. Then Zhang *et al.* (2011) used the same methods in their experiment to detect the density hysteresis of heavy water. They used long-wavelength neutrons ($\lambda = 4.7$ Å), and focused attention on the small-angle region (Q from 0.15 Å^{-1} to 0.35 Å^{-1}). In such a configuration, neutrons view the water and the silica matrix as continuous media and only the long-range (> 18 Å) order is probed. The short-range water–water, silica–silica and water–silica correlation peaks are located at Q values larger than 1.5 Å^{-1}, which are beyond the Q-range they studied and thus do not concern their measurements. In this small-angle diffraction experiment, the neutron scattering intensity distribution $I(Q)$ is given by

$$I(Q) = nV_p^2(\Delta\rho_{sld})^2 \bar{P}(Q)S(Q), \tag{7.21}$$

where n is the number density of the scattering unit (number density of the cylindrical pores in the MCM-41S), V_p is the volume of the scattering unit, $\Delta\rho_{sld} = \rho_{sld}^{D_2O} - \rho_{sld}^{MCM}$ is the difference of scattering length density (SLD) between the scattering unit $\rho_{sld}^{D_2O}$ and the environment ρ_{sld}^{MCM}, $\bar{P}(Q)$ is the normalised particle structure factor (or form factor) of the scattering unit (i.e. the form factor of the cylindrical pore), and $S(Q)$ is the inter-cylinder structure factor of a 2-D hexagonal lattice (Chen, 1986). The SLD of the scattering unit $\rho_{sld}^{D_2O}$ is proportional to its mass density $\rho_m^{D_2O}$ as $\rho_{sld}^{D_2O} = \alpha\rho_m^{D_2O}$, where $\alpha = N_A \sum b_i/M$, and N_A is Avogadro's number, M is the molecular weight of D_2O, and b_i is the coherent scattering length of the i-th atom in the scattering unit. The SLD of the silica material has been determined by a separate contrast matching experiment by hydrating the sample with different ratio of D_2O and H_2O. When the molar ratio is $[D_2O] : [H_2O] = 0.66 : 0.34$, the Bragg peak is matched out. For silica, it is a rather rigid material when compared to water. Its thermal expansion coefficient is in the order of 10^{-6}K^{-1} as compared to 10^{-3}K^{-1} of water. In the experiments of Zhang *et al.* (2011), the position and the width of the Bragg peaks do not change with temperature, indicating that the structure change of the confining matrix is negligible in the measured temperature range. Therefore, based on the above relations, Zhang *et al.* (2011) found that all the variables in the expression for $I(Q)$ are independent of temperature except for $\rho_m^{D_2O}$. Hence Zhang *et al.* (2011) were able to determine the density of confined D_2O by measuring the temperature-dependent neutron scattering intensity $I(Q)$ at the Bragg peak.

The form factor $\bar{P}(Q)$ of a long and thin ($QL > 2\pi$) cylinder is given by $\bar{P}(Q) = \pi/QL(2J_1(QR)/QR)^2$, where L and R represent the length and the radius of the cylinder respectively, and $J_1(x)$ is the first-order Bessel function of the first kind. The structure factor $S(Q)$ can be well approximated by a Lorentzian function. Therefore, the measured neutron scattering intensity is expressed as

$$I(Q) = nV_p^2 \left(\alpha \rho_m^{D_2O} - \rho_{sld}^{MCM}\right) \frac{\pi}{QL} \left(\frac{2J_1(QR)}{QR}\right)^2 \left(\frac{\frac{1}{2}\Gamma}{(Q - \frac{2\pi}{d})^2 + (\frac{1}{2}\Gamma)^2}\right)$$
$$+ BQ^{-\beta} + C, \tag{7.22}$$

and at the Bragg peak $Q_0 = 2\pi/d$,

$$I(Q_0) = A\left(\alpha \rho_m^{D_2O} - \rho_{sld}^{MCM}\right) + BQ_0^{-\beta} + C, \tag{7.23}$$

where $A = nV_p^2$, Γ is the FWHM of the Lorentzian peak, and C is the Q-independent incoherent background (Liu *et al.*, 2007). The approximation of the Bragg peak by a Lorentzian function is purely empirical. There are many factors that can affect the broadening of a diffraction peak. These include the imperfection of the lattice, the instrument resolution, etc. But the choice of the peak formula form will not affect the extraction of the density of water, which merely depends on the peak height.

By fitting the neutron scattering intensity with the above model at the highest and lowest temperature at each pressure, Zhang *et al.* (2011) were able to obtain the parameters B, β and C, and thus able to confidently subtract the 'background' (the second and third terms in the above equations). They determined the last unknown temperature-independent constant A by normalising the density of the highest temperature at each pressure to that of the bulk D_2O taken from NIST Scientific and Technical Database (NIST Chemistry WebBook, http://webbook.nist.gov/chemistry/fluid/).

Even earlier than Liu *et al.* (2007) and Morineau *et al.* (2002) used a similar method to measure the density of confined toluene and Xia *et al.* (2006) to measure that of benzene. Recently, the reliability of this method to determine the density of water confined in MCM-41S has been criticised because of a possible layering effect of water in the pores (Mancinelli *et al.*, 2010). However, the scenario hypothesized in Mancinelli *et al.* (2010) assumes the existence of voids in the hydrated pores; this possibility is not consistent with the measurement of Zhang *et al.* (2011) of a contrast matched sample ($[D_2O] : [H_2O] = 0.36 : 0.34$) in which the diffraction peak is almost completely masked and no evidence of the scattering from the voids can be recognised. The hypothesis of the existence of voids in the hydrated pores originates from the layering density profile suggested by Mancinelli *et al.* (2010) that implies that water can penetrate into the wall surrounding the cylindrical pore of MCM-41S. The problem of whether there is void (micropore) on the wall of MCM-41 has been investigated by gas absorption techniques many times since 1993. The great majority investigations concluded that MCM-41 materials do not have any microporosity. Recent experiments suggest that MCM-41 is exclusively mesoporous with no water penetration into the wall (Silvestre-Albero *et al.*, 2008). On the basis of these results Zhang *et al.* (2011) believed that the layering density profiles suggested in Mancinelli *et al.* (2010) was unrealistic and inconsistent with the scattering pattern of their measurements.

7.6.3 *Results and discussion*

The existence of the density minimum in supercooled water was not much described prior to 2007 before Liu *et al.* (2007) first reported an experimental result using small-angle neutron scattering (SANS) to measure the density of heavy water contained in 1D cylindrical pores of mesoporous silica materials MCM-41S with pores of nominal diameter of 15 ± 1 Å. Figure 7.26 shows the results of Liu *et al.* (2007) in which the density maximum and minimum at ambient pressure are clearly indicated. The results of Liu *et al.* (2007) are consistent with the predictions of molecular dynamics simulations of supercooled bulk water (Paschek, 2005), see Figure 7.27.

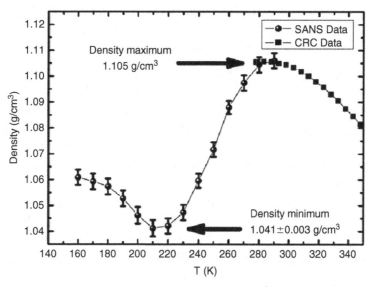

Figure 7.26 Average D_2O density inside the 15 ± 1 Å pore measured by the SANS method as a function of temperature. A smooth transition of D_2O density from the maximum value at 284 ± 5 K to the minimum value at 210 ± 5 K is clearly shown. The filled squares are the density data for bulk D_2O taken from the CRC Handbook (Lide, 2007), Liu *et al.* (2007). Copyright 2007, National Academy of Sciences, USA.

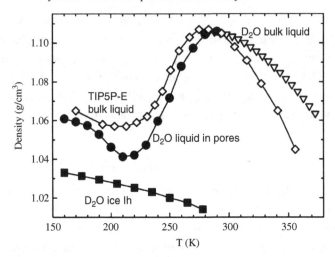

Figure 7.27 Comparison of density vs. temperature curves at ambient pressure for bulk liquid D_2O (open triangles) (Lide, 2007), confined liquid D_2O (filled circles) from this work, D_2O ice Ih (filled squares) (Angell and Kanno, 1976), and MD-simulations of liquid TIP5P-E water (open diamonds) (Paschek, 2005). The density values for the TIP5P-E model (which is parameterised as model of H_2O) have been multiplied by 1.1 to facilitate comparison with the behaviour of D_2O (Liu *et al.*, 2007). Copyright 2007, National Academy of Sciences, USA.

Evidence of the first-order liquid–liquid phase transition in supercooled heavy water

The work of Liu *et al.* (2007, 2008a) was greatly expanded by Zhang *et al.* culminated with their report in 2011 on observation of and the density hysteresis phenomenon (Zhang *et al.*, 2011). They used a simpler but more efficient neutron elastic diffraction technique employing the cold neutron triple-axis spectrometer SPINS at NIST Center for Neutron Research (NCNR) to measure the density of heavy water confined in a nanoporous silica matrix in a temperature–pressure range, from 300 K to 130 K and from ambient 1 bar to 2900 bar.

They observed a prominent hysteresis phenomenon in the measured density profiles between warming and cooling scans above 1000 bar. They interpreted this hysteresis phenomenon as support for the hypothetical existence of a LLPT of water in the macroscopic system if crystallisation could be avoided in the relevant phase region.

Density plays a central role in many classical phase transitions. In particular, it is the order parameter in the gas–liquid and liquid–solid transitions. Therefore, its experimental determination assumes primary importance regarding the hypothesised liquid–liquid phase transition. In making such measurements, Zhang *et al.* (2011) sought evidence of a remnant of a first-order liquid–liquid phase transition of water that would exist in the macroscopic system if it were possible to avoid crystallisation. Their results together with the updates of Wang *et al.* (2014a) are further described in the following.

Zhang *et al.* (2011) used Figure 7.28 to explain the purpose of their experimental procedures. It shows the hypothesised phase diagram of low-temperature water in the presence of a first-order HDL–LDL transition. Normally, a discontinuous change of the state functions, such as density, associated with a first-order transition is difficult to detect directly. When the equilibrium phase boundary is crossed, due to the metastability or the kinetics of the phase transition, the phase separation may take very long time to happen, especially in confinement (Poole *et al.*, 1993; Gelb *et al.*, 1999; Evans, 1990). However, a first-order line should still manifest itself with a significant hysteresis when it is crossed from opposite directions of the transition line as shown in Figure 7.28(b). Although a hysteresis phenomenon, if it exists, may not prove the existence of a first-order transition, the absence of hysteresis can serve to rule out a first-order phase change. Accordingly, Zhang *et al.* (2011) performed a series of warming and cooling scans over a range of pressures. For each pressure, the sample was cooled from 300 K to 130 K at ambient pressure, and then pressurised to the desired value (light dashed line in Figure 7.28(a)). They then waited about 2 hours for the system to equilibrate. The warming scan with 0.2 K min^{-1} was first performed from 130 K to 300 K (up-arrow line in Figure 7.28(a)). When the warming scan was finished, they waited another 2 hours at room temperature for system equilibration. After that, the cooling scan with 0.2 K min^{-1} was performed from 300 K to 130 K (down-arrow line in Figure 7.28(a)). When the full cycle was finished, the sample was brought back to ambient temperature and pressure before measuring another pressure. The system Zhang *et al.* (2011) chose to study is water confined in long cylindrical pores of silica with a diameter of 15 Å. In such confined water, two of its three dimensions are finite, leaving one along the pore axis that can be considered macroscopic. Thus, it is intrinsically difficult to observe a phase transition in such a restricted geometry. However, they may still find a remnant of a phase transition in a confined system, which in principle could be demonstrated by its size scaling.

The contrast between the neutron coherent scattering length density (SLD) of heavy water and that of the silica matrix gives rise to a strong signal in this experiment of Zhang *et al.* (2011). Specifically, they observe a well-defined first Bragg diffraction peak arising from the plane of a 2D hexagonal lattice of water cylinders in the grains of MCM-41S silica matrix (Figure 7.29(a)). Figure 7.29(b–d) illustrates the elastic neutron diffraction intensities measured at the highest and lowest temperatures at three representative pressures. One immediately notices that the width and the position of the Bragg peak do not change with temperature. Hence, for their purposes, the structure of the confining matrix

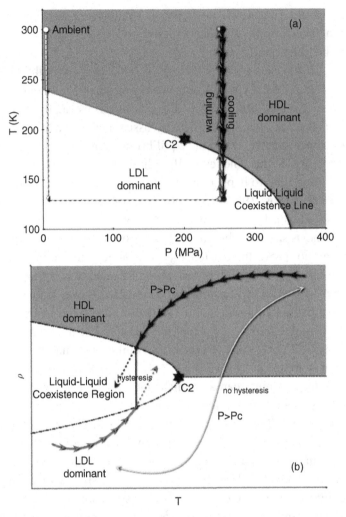

Figure 7.28 The hypothesised (a) *T–P* and (b) *ρ–T* phase diagrams of water in the presence of a first-order liquid–liquid phase transition are used to explain the motivation of the experimental procedures of Zhang *et al.* (2011). In order to detect the hypothesised transition, both warming (up-arrow) and cooling (down-arrow) scans were performed. If a first-order HDL–LDL transition indeed exists at high pressures, a hysteresis phenomenon should be observed because of the long time required for the phase separation, shown as the difference between the two arrow curves in panel (b). However, at relatively low pressures, when no first-order line is crossed, there should be no hysteresis, shown with the curve with no arrow sign in panel (b). Further studies are needed to identify the hypothesised liquid–liquid critical point C2 (Zhang *et al.*, 2011). Copyright 2011, National Academy of Sciences, USA.

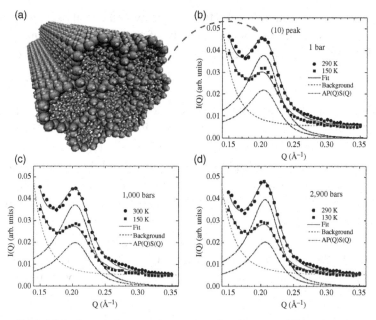

Figure 7.29 This figure demonstrates that the neutron diffraction intensities can be fitted with the model described in the text. (a) Schematic representation of a D_2O hydrated MCM-41S nanoporous silica crystallite (pore diameter $2R15$ Å ± 2 Å). (b–d) The elastic neutron diffraction intensity $I(Q)$ at three pressures measured by SPINS at NCNR. The structure factor peak at around 0.21 Å^{-1} comes from the plane of the 2-D hexagonal arrangement of the water cylinders in the crystallite (Zhang *et al.*, 2009a). The peak height is proportional to the square of the difference of neutron scattering length density (SLD) between the confined D_2O and the silica matrix, and therefore is a sensitive indicator of the average mass density of D_2O in the pores. By fitting with Eq. (7.22), the temperature-independent background (light dashed line) and the temperature-dependent elastic diffraction intensities (darker dash-dotted line) can be separated accordingly (Zhang *et al.*, 2011). Copyright 2011, National Academy of Sciences, USA.

can be regarded as unaffected by temperature. Once the data are corrected for the temperature-independent background arising from the fractal packing of the MCM-41S crystallites (grains) and the incoherent scattering, the only temperature-dependent quantity is the height of the Bragg peak, which is proportional to the square of the difference of SLD between the heavy water and the silica matrix, and therefore a sensitive indicator of the average mass density of the confined water.

Zhang *et al.* (2009a) can therefore sit at the Bragg peak position and monitor the peak intensity as a function of temperature, rather than performing a scan in Q at each temperature. While their measurements are highly precise and sensitive with regard to relative changes, there is an overall uncertainty that they estimate (from the results of repeated measurements on the same and different sample batches) to be about 0.02 g cm^{-3} (standard deviation) in the overall density scale, arising from uncertainties in the scattering length density of the silica matrix, and the model they have used to analyse the data. Even after careful considerations of all these sources of uncertainties, the estimation of the uncertainty of the absolute density is still a challenge because of the possible systematic errors arising from the model used in the analysis. It should be pointed out, however, that this uncertainty can be considered as a scaling factor, and that the relative shape of the density curves is almost directly related to the measured scattering intensity.

Figure 7.30 (Zhang *et al.*, 2011) shows the measured density of confined D_2O with both cooling and warming scans at a series of pressures. The fact that the warming and cooling curves join at both the high- and low-temperature ends implies that the expansion–contraction processes are reversible. Up to 1000 bar, the density difference between the cooling and warming scans is small, which could be attributed to the temperature lag when ramping the temperature continuously. The density difference due to this reason is small and relatively independent of

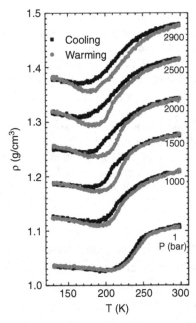

Figure 7.30 The density profiles of confined D_2O in a hydrophilic substrate MCM-41S are measured in both warming and cooling scans. The data are shifted by 0.05 g cm^{-3} between adjacent pressures for clarity. A hysteresis phenomenon becomes prominent above 1000 bar. Error bars, due to counting statistics, in the density are smaller than the point size. The two horizontal arrows indicate the locations of the sudden change of slope in the density profiles ('kink') at 1 bar and 1000 bar (Zhang *et al.*, 2011). Copyright 2011, National Academy of Sciences, USA.

pressure. However above 1000 bar, the density difference (hysteresis) opens up progressively as the pressure is increased.

Wang *et al.* (2014a) extended this to expect that the magnitude of the density difference depends on the temperature ramping rate. Considering the feasibility of neutron scattering experiments, they chose a ramping rate of 0.2 K min^{-1}. With such a slow rate, a rather uniform temperature distribution over the sample is assured, but the system may require much longer times to reach physical equilibrium after crossing a phase boundary. There seems to be a small regression of the opening up of the density difference at 2900 bar between 190 K and 210 K. Further investigations are needed to find out whether this regression is real or an experimental artefact. Nevertheless, the biggest density difference (confined D$_2$O) is found to be on the order of a few percentage above 1000 bar. In comparison to the density difference between high- and low-density amorphous H$_2$O ice (about 25%, measured at much lower temperatures), the observed difference is small. This might be from the combined effects of confinement, isotopic difference and temperatures. Note that the accuracy of the absolute density depends on the background subtraction and the scaling. However, the relative shape of the density profiles is independent of the analysis.

Another remarkable feature of the density profiles is that a clear minimum is observed at each measured pressure. The minimum temperature T_{min} decreases from 210 K to 170 K as the pressure is increased from ambient to 2900 bar. Poole *et al.* (2005) have proposed that the occurrence of such a density minimum is an indication of the full development of a defect-free random tetrahedral network (RTN) of hydrogen bonds. Below T_{min} the completed RTN shows normal thermal contraction as the temperature is further lowered. The results therefore imply that at higher pressures, the RTN can only be reached at lower temperatures. This is a consequence of the fact that the enthalpically favourable hydrogen bonded RTN has a lower density compared to its less developed counterpart.

As shown in Figure 7.30, Zhang *et al.* (2011) observed prominent density hysteresis between heating and cooling scans in the pressure range from 1500 bar to 2900 bar for the confined D$_2$O system. The appearance of the density hysteresis indicates the crossing of a first-order phase transition line because of the thermodynamic discontinuity at the phase boundary and the metastability of the first-order phase transition. By finding the temperature of the maximum density difference for each pressure, Zhang *et al.* determined the approximate position of the LLPT line of the confined D$_2$O in the pressure range from 1500 bar to 2900 bar and the temperature range from 190 K to 210 K.

Wang *et al.* (2014a) applied a similar method to higher pressures. For each pressure, the sample was first pressurised to the desired value. After 2 hours waiting, the cooling scan was performed with a cooling rate of 0.2 K min^{-1} from 300 K

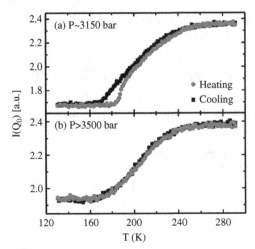

Figure 7.31 The profiles of $I(Q_0)$, which directly reflect the profiles of according to (7.23), in the heating scan and cooling scan when (a) p is between 3650 bar and 2990 bar and (b) p is above 3500 bar. A prominent hysteresis appears in the case shown in (a), which indicates a crossing of the phase boundary. The maximum density difference appears at $T = 184$ K and $p = 3150$ bar. The case shown in (b) exhibits no hysteresis, which indicates there is no phase separation (Wang *et al.*, 2014a). Copyright 2014, AIP Publishing LLC.

to 130 K. After the cooling scan, they waited for another 2 hours and then performed the heating scan with again a rate of 0.2 K min^{-1} from 130 K to 300 K. The results are shown in Figure 7.31. In the case shown in Figure 7.31(a), as the temperature decreases from 300 K to 130 K, the pressure decreases from 3650 bar to 2990 bar due to the thermodynamic effect of the pressurisation gas helium (the pressure of the gas will decrease as the temperature decreases). However, for a specific temperature, the corresponding pressures in the cooling scan and in the heating scan are effectively the same. So that such thermodynamic effect of the helium even becomes significant when the pressure is higher than ≈ 3000 bar, will not cause an additional density difference.

Figure 7.30(a) shows the profiles of $I(Q_0)$, which directly reflect the profiles of $\rho_m^{D_2O}$ according to Eq. (7.22), in the cooling scan and the heating scan in the temperature range from 130 K to 295 K and the pressure range from 2990 bar to 3650 bar. The maximum density difference appears at $T = 184$ K and $P = 3150$ bar (the pressure was recorded for different temperatures when performing the temperature scan so that the value of pressure at $T = 184$ K in this case can be estimated). This point is denoted by a triangle in the (P, T) plane in Figure 7.32. In the case shown in Figure 7.31(b), as the temperature changes between 130 K and 300 K, the pressure changes between 3492 bar and 4416 bar. Figure 7.31(b) shows that the density profiles in the cooling scan and heating scan are effectively identical in

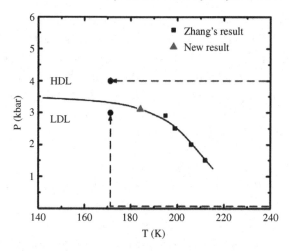

Figure 7.32 The phase diagram of deeply cooled confined D_2O. The squares denote the positions of the maximum density difference according to Zhang *et al.* (2011) and the triangle denotes the position of the maximum density difference according to the result in this work. The black solid curve is the estimated LLPT line. The dashed lines show the routes for the density measurement (Wang *et al.*, 2014a). Copyright 2014, AIP Publishing LLC.

this case. Notice that the slope of the hypothetical LLPT line in the (P, T) plane is negative, thus the disappearance of the density hysteresis in Figure 7.30(b) suggests that there is no phase transition above ≈ 3500 bar and between 130 K and 300 K. In this region, the system is in the hypothetical HDL state. Combining the result of Zhang *et al.* (2011) and the result shown in Figure 7.31, Wang *et al.* (2014a) draw the estimated locus of the hypothetical LLPT line in Figure 7.32.

In order to verify the phase diagram shown in Figure 7.32, Wang *et al.* (2014a) measured the average density of the confined D_2O by performing the Q scan and fitting the Q scan intensity distribution with Eq. (7.22) in both sides of the proposed LLPT line. In the (P, T) plane, for the measured point below the LLPT line (in the pure LDL phase), they approach it by first cooling the system to the desired temperature and then pressurising the system to the desired pressure. For the measured point above the LLPT line (in the pure HDL phase), then they approach it by first letting the system stay under the desired pressure at room temperature for two hours, and finally cooling the system to the desired temperature. These two measurement routes are denoted by the dashed lines in Figure 7.32. With such routes, they try not to cross the LLPT line so that the complicated phase separation can be avoided. Before each measurement, so they waited for 40 minutes to let the system equilibrate. One measurement took about 80 minutes. In this work, they measured two temperatures, 172 K and 200 K, in the hypothetical two-phase region of the phase diagram. For each temperature, four pressures were measured. Two of them are below the LLPT line (in the LDL phase) while the other two are above the LLPT line (in the HDL phase). Figure 7.33(a-1) shows the four Q scan intensity distributions at under $P = 2000$ bar, 3000 bar, 4000 bar and 5000 bar, $T = 172$ K. One can find that the Q scan intensity exhibits a substantial increase when the system leaves the LDL phase (2000 bar and 3000 bar) and enters the HDL phase (4000 bar and 5000 bar). This phenomenon reflects a large increase of the average density of the confined D_2O. The average densities of the confined D_2O, obtained by fitting the Q scan data with Equation (7.22) under $P = 2000$ bar, 3000 bar, 4000 bar and 5000 bar at $T = 172$ K, are shown in Figure 7.33(a-2). One can find that the value of $\rho_m^{D_2O}$ abruptly increases from 1.091 g cm^{-3} to 1.270 g cm^{-3} as the pressure changes from 3000 bar to 4000 bar. For the case at $T = 200$ K (as shown in Figure 7.33(a-2), the situation is similar. The average density of the confined D_2O abruptly increases from 1.100 g cm^{-3} to 1.283 g cm^{-3} as the pressure changes from 2000 bar (in LDL phase) to 3000 bar (in HDL phase). In both cases, the average density of the confined D_2O increases by $\approx 16\%$ as the system transforms from the LDL to the HDL. In addition, in the measured pressure range and temperatures, the LDL exhibits smaller isothermal compressibility (χ_T) than the HDL ($\chi_T = (1/\rho)(\partial\rho/\partial P)_T$, which can be estimated from the (ρ, P) plots in the right-hand panels of Figure 7.33). This phenomenon can be attributed to the difference between the local structures of the LDL and HDL. According to Soper and Ricci (2000), the LDL has an open, hydrogen-bonded tetrahedral structure extending to the second coordination shell. This local structure is similar to that of ice Ih and makes the LDL relatively rigid. On the contrary, for the HDL, the second coordination shell collapses due to the breaking of the hydrogen bonds between the first and second coordination shells. This feature makes the HDL 'softer' than the LDL.

In order to further verify the phase diagram shown in Figure 7.32, Wang *et al.* (2014a) measured the average density of the confined D_2O $\rho_m^{D_2O}$ at $T = 225$ K from 1 bar to 4000 bar. This region is supposed to be in the one-phase region of the phase diagram. For each measured point, they approach it by first cooling the system to 225 K and then pressurising the system to the desired pressure. The result is shown in Figures 7.33(c-1) and (c-2). At this temperature, the value of $\rho_m^{D_2O}$ smoothly increases as pressure increases. This indicates that there is no phase transition from a low-density phase to a high-density phase in this region. Moreover, in this case the system exhibits higher isothermal compressibility than both the LDL and HDL in the pressure range from 1 bar to 3000 bar. This feature suggests that at $T = 225$ K, as pressure increases, the local structure of the confined D_2O transforms from the predominantly LDL to the predominantly HDL. At pressures higher than 3000 bar, the isothermal compressibility of the system is similar to that of the HDL, which suggests that under such high pressures the system is dominated by the HDL.

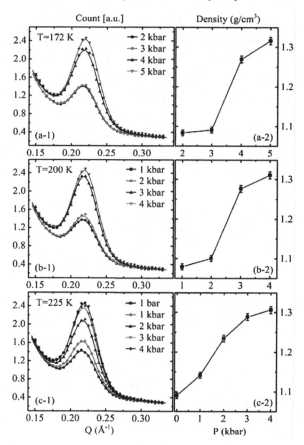

Figure 7.33 Q scan intensity distribution and the extracted average density of the confined D_2O at different temperatures and pressures. For each row, the left panel shows the Q scan intensity distributions and corresponding fitted curves at a specific temperature; the right panel shows the corresponding average densities of the confined D_2O at this temperature (Wang *et al.*, 2014a). Copyright 2014, AIP Publishing LLC.

Density extrema and peaking of thermal expansion coefficient of H_2O by SAXS method

Up to now, we have discussed in this module the use of small-angle neutron scattering and elastic neutron diffraction methods to measure the density of confined D_2O in MCM-41-S-15. But, it is also possible to measure the density of confined H_2O as well using small-angle X-ray scattering (SAXS) technique. This is mainly because X-ray is preferentially scattered by the oxygen atom of both H_2O and D_2O and the contribution of scattering intensity from the hydrogen or deuterium ti very small. Liu *et al.* (2013) therefore used SAXS to study the density of H_2O confined in MCM-41-S-15. Their results are shown in Figures 7.34 and 7.35 below.

Figure 7.34 shows the average H_2O density confined in MCM-41-S-24 as compared with the density of water and ice Ih in their bulk form. Figure 7.35 shows the isobaric thermal expansion coefficients calculated from the measured $\rho(T)$ by $\alpha_p = -\frac{1}{\rho}\frac{d\rho}{dT}$. The points at which thermal expansion coefficient crosses zero represent the density extrema. The deduced isobaric thermal expansion coefficient is quite close to the results reported in the work of Mallamace *et al.* (2007a). The thermal expansion coefficient is one of the thermodynamic response functions of water and Figure 7.35 shows the peaking of the coefficient at the ambient pressure, which means that at this pressure the confined water is in the one-phase region of

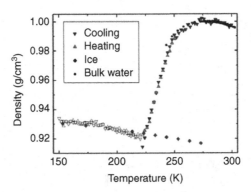

Figure 7.34 The average H_2O density confined in MCM-41-S-24 compared with water and ice Ih density in their bulk form. In both the heating and cooling scans, filled symbols represent liquid water, and empty symbols represent solid ice (Liu *et al.*, 2013). Copyright 2013, AIP Publishing LLC.

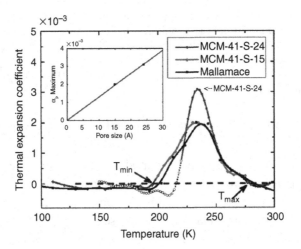

Figure 7.35 Thermal expansion coefficient of H_2O confined in MCM-41-S-15 and MCM-41-S-24 compared with Mallamace's calculation. In the case of MCM-41-S-24, filled symbols and solid line represent liquid water, while empty symbols and dotted line represent ice. The horizonal dashed line is zero of thermal expansion coefficient. The two solid arrows point out the temperatures of the density maximum and minimum, which are denoted as T_{max} and T_{min}, respectively. The inset shows the maximum of ∞_p are proportional to the pore sizes with zero intercept (Liu *et al.*, 2013). Copyright 2013, AIP Publishing LLC.

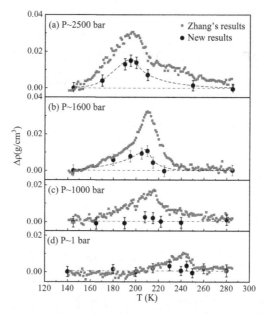

Figure 7.36 Latest result of the density hysteresis measurement of D_2O. The grey points are the density differences between cooling scan and heating scan measured by Zhang *et al.* (2011). The black points denote the new result using a better temperature change protocol (Wang *et al.*, 2015).

the low temperature confined water and the peaking of this quantity signals the crossing of the Widom line. Furthermore, the inset of Figure 7.35 shows that this peak height seems to go to zero as the diameter of the pore goes to zero. This suggests empirically that the second critical point, if it exists, disappears when the pore size go to zero because the correlation length associated with the critical phenomenon has no where to grow, an interesting observation indeed.

Latest results of the density hysteresis measurement of D_2O

Figure 7.30 shows the density profiles of D_2O confined in MCM-41-S with hysteresis phenomenon at all different pressures from ambient up (Zhang *et al.*, 2011). Yet, Figure 7.6 shows clearly that the dynamic crossover occurs at ambient pressure, possibly due to crossing of the Widom line that exists only in the one-phase region of water. So no hysteresis should be seen in the cooling–heating cycle. Thus these dynamic experimental results seem to be in contradiction with the density measurement. To investigate further, Wang *et al.* (2014e) repeated the measurement using a better temperature change protocol. As seen in Figure 7.36, they showed that there is no hysteresis when the pressure is less than 1.5 kbar, while the hysteresis appears and increases when the pressure goes above 1.6 kbar. These results now agree with the existence of a Widom line on the lower-pressure side having a common end point with an LLPT line on the higher-pressure side. LLCP is thus situated at $P_c = 1.3 \pm 0.3$ kbar, and $T_c = 214 \pm 3$K.

8

Dynamic crossover phenomena in other glass-forming liquids

8.1 Introduction

There are several theories and models of glassy behaviour. For example, Binder and Kob discussed these in great details in their book (Binder and Kob, 2011), with extensive references. Theories often cited are, for example, Adam–Gibbs theory (Adam and Gibbs, 1965), free-volume theory (Cohen and Turnbull, 1959; Turnbull and Cohen, 1961, 1970), mode-coupling theory (MCT), random first-order theory (RFOT), etc. While all these theories and models can certainly enhance our understanding of glass-forming systems, none seems to be flawless. Some are too sophisticated and complex, while others are too macro in their approach and make simple assumptions. Thus there is plenty of room for much more in-depth research and experiments. In this chapter, we shall place more emphasis on recent experimental results, which can help us to understand better the complex landscapes and systems of the glass-forming liquids. Whenever possible, we will relate the results to some of the available theories.

In Chapter 7, we have already discussed the dynamic crossover (often called fragile-to-strong dynamic crossover, FSC) phenomenon in supercooled water found by Chen and coworkers (Faraone *et al.*, 2004). Therefore, in this chapter, we are exploring such phenomena in other glass-forming liquids. It has been found that there is a similar dynamic crossover behaviour in other glass-forming liquids, substantiated not only by mode-coupling theory and MD simulations, but also collaborated with many experimental results (Chen *et al.*, 2009b). We shall expand on this in the following.

8.1.1 Experimental results on transport coefficients

The existing data on transport coefficients of common bulk glass-forming liquids exhibit the same FSC phenomena as water does. For example, in Figure 8.1, for

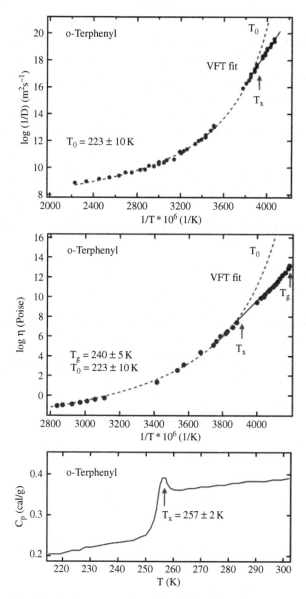

Figure 8.1 o-Terphenyl dynamic crossover phenomenon. (Top) Arrhenius plot of the inverse of the self-diffusion constant; (middle) Arrhenius plot of the viscosity; (bottom) specific heat at constant pressure. The dotted lines represent the fit of the self-diffusion and viscosity data with a VFT law at high temperatures; the continuous lines are the fit with an Arrhenius law at low temperatures. In each case, T_x is defined by the peak of the specific heat (Chen *et al.*, 2009b). ©IOP Publishing. Reproduced with permission. All rights reserved.

the well-known case of o-terphenyl, the Arrhenius plots of the inverse of the self-diffusion constant (top panel) (Mapes *et al.*, 2006) together with the viscosity (middle panel) and followed by the specific heat (lower panel) are shown (Laughlin and Uhlmann, 1972). One immediately notices that the crossover temperatures as indicated by $1/D$ and the viscosity are quite close to each other and agree with the peak position of the specific heat. This is another proof that the Adam–Gibbs theory is valid for the case of organic glass-forming liquids.

One consequence of the Adam–Gibbs theory (Adam and Gibbs, 1965) is that if the specific heat shows a peak at certain temperature T_x then the Arrhenius plot of the viscosity or the relaxation time will show a change of slope at the temperature of the peak.

To prove this theorem, recall that for the Adam–Gibbs theory the shear viscosity is given by an equation as follows:

$$\ln \frac{\eta(T)}{\eta(T_0)} = \frac{A}{T \Delta S_{conf}(T - T_0)}, \tag{8.1}$$

where A is a constant and the configurational entropy can be calculated from the excess specific heat at constant pressure (defined approximately as the difference of the specific heats in the supercooled liquid state and the crystal state at the same temperature) as

$$\Delta S_{conf}(T - T_0) = \int_{T_0}^{T} \frac{\Delta c_p(T')}{T'} dT'. \tag{8.2}$$

Since the Arrhenius plot of a quantity X is $\ln(X)$ vs. $1/T$, its slope is given by $d\ln(X)/d(1/T)$. Then from equation Eq. (8.1) one has

$$\frac{d \ln(\eta/\eta_0)}{d(1/T)} = \frac{A}{\Delta S_{conf}(T - T_0)}, \tag{8.3}$$

which implies

$$\frac{d \ln(\eta/\eta_0)}{d(1/T)}\bigg|_{T>T_x} > \frac{d \ln(\eta/\eta_0)}{d(1/T)}\bigg|_{T<T_x}. \tag{8.4}$$

Therefore, assuming that the specific heat has a sharp peak at T_x, the slope is larger above T_x and smaller below, as displayed in Figure 8.1. As an example, calculated values of the configurational entropy and the specific heat peak at 240 K are shown in Figure 8.2.

Thus, we have given a plausible argument that the Adam–Gibbs theory can predict a dynamic crossover phenomenon (change of slope) in an Arrhenius plot of viscosity whenever C_p shows a peak.

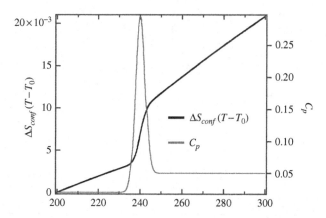

Figure 8.2 An example of the calculated values of the configurational entropy and the specific heat peak. At the peak position of the specific heat the configurational entropy shows a sudden jump. Original drawing provided by M. Lagi.

Figure 8.2 shows the Arrhenius plot of the viscosities of four well-known glass-forming liquids – o-terphenyl, salol, α-phenyl-o-cresol and tri-α-naphtylbenzene – as measured by (Laughlin and Uhlmann, 1972). The one related to α-phenyl-o-cresol is already shown in Chapter 7 as Figure 7.11 when other glass-forming liquids besides water were discussed there. All four plots in Figure 8.3 show a FSC phenomenon because at higher temperature all four viscosities show a super-Arrhenius behaviour (which can be fitted by a Vogel–Fulcher–Tamman law, as indicated by the dashed lines). At low enough temperatures, one can see that they all switch to an Arrhenius law all the way down to the glass transition temperature T_g. Other than the case of tri-α-naphtylbenzene, which does not have measured specific heat at constant pressure, the insets of the other three panels the measured specific heats at constant pressure show, as also given by (Laughlin and Uhlmann, 1972).

As compared with the specific heat in supercooled confined water in Figure 8.4 and Figure 7.25, these peaks in organic liquids are rather small (of the order of 0.4 Cal/g). If T_x is defined as the temperature of the peak of the specific heat, then the arrow signs in the main panels confirm within the error bars the coincidence with the FSC temperature T_L of the viscosity, namely $T_x = T_L$. Here the FSC temperature T_L may be defined as the temperature where the Arrhenius plot of the viscosity has a maximum slope.

8.1.2 Computer simulations on bulk water

To ensure that these phenomena described above are inherent properties of water and not due to the confinement, Lagi ran a simulation of a model of bulk water

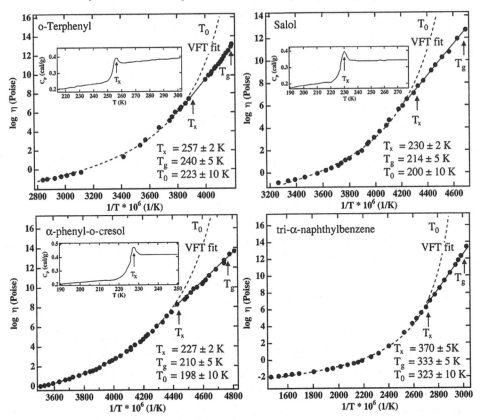

Figure 8.3 Evidence of the dynamic crossover phenomenon in the Arrhenius plot of the viscosity for four different well-known glass-forming liquids. (Top left) o-Terphenyl; (top right) Salol; (bottom left) α-phenyl-o-cresol; (bottom right) tri-α-naphtylbenzene. Insets show their specific heat peaks at T_x (Laughlin and Uhlmann, 1972). These temperatures in the corresponding viscosity profiles are indicated with arrow signs at T_x. Note that the arrows coincide with the turning point from a super-Arrhenius to an Arrhenius behaviour in viscosity within the error bars. Specific heat data were not available for the case of tri-alpha-naphthylbenzene. The dotted lines represent the fit of the viscosity data with a VFT law at high temperatures; the continuous lines are the fit with an Arrhenius law at low temperatures (Chen *et al.*, 2009b). ©IOP Publishing. Reproduced with permission. All rights reserved.

based on the TIP4P-Ew model by Horn *et al.* (2004) in 2009, as shown in Chen *et al.* (2009b). This specific water model was chosen because of the excellent agreement of its calculated diffusion constant with the experimental values for bulk water, over a wide range of temperatures (between 240 K and 320 K). The dynamic crossover in the Arrhenius plot of the self-diffusion constant has been previously observed by Xu *et al.* (2005) and Kumar *et al.* (2006) with simulations of bulk water using other water models.

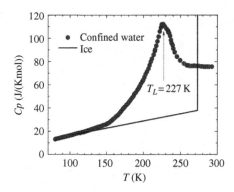

Figure 8.4 The substantial heat capacity peak of supercooled water confined within pores of silica gel (Maruyama *et al.*, 2004). The pore size is around 3 nm. Note that the peak position is at $T_L = 227$ K, identical to the one in the DSC cooling scan in Figure 7.25. Notice that this specific heat peak for confined water is four times higher than that of the typical organic glass forming liquid shown in Figures 8.1 and 8.3 (Chen *et al.*, 2009b). ©IOP Publishing. Reproduced with permission. All rights reserved.

Chen *et al.* (2009b) ran long MD trajectories for a box of 512 water molecules of up to 1 μs in the NVT ensemble. The systems were considered equilibrated when the mean square displacement of the water molecules was larger than 0.1 nm² (Sciortino *et al.*, 1996) (see inset of Figure 8.5, bottom right panel). The results of the simulation are shown in Figure 8.5.

Chen *et al.* (2009b) also calculated the self-intermediate scattering functions (ISF) for the oxygen atoms for five Q-values (0.4, 0.5, 0.6, 0.7, 0.8 Å^{-1}) and fit the data according to the RCM (top-left panel). The top right and bottom-right panels show the Arrhenius plots of the transport properties obtained from the trajectories: the translational relaxation time $< \tau >$ and the inverse of the self-diffusion constant $1/D$, respectively. Both the plots show a dynamic crossover at $T_L = 215 \pm 5$ K, analogous to the one in Figure 7.25. Furthermore, $< \tau_{TL} >$ is in the 1 to 10 ns range for both experiments and simulations, thus the results confirm the general behaviour of many glass formers (Novikov and Sokolov, 2003). The bottom-left panel shows instead the dynamic response function $\chi_T(Q, t) = -dF_s(Q, t)/dT$ as extracted from the trajectories. This agrees with who found Zhang *et al.* (2008), the experimental observations of that the maximum of $\chi_T(Q, t)$, namely $\chi_T^*(Q)$, decreases after the dynamic crossover temperature $T_L = 215$ K. With these results, we can conclude that bulk water simulations are able to reproduce qualitatively our experimental findings of the 3D confined water in cement paste. The maximum of $\chi_T^*(Q)$ happens at the dynamic crossover temperature T_L, and is not caused by the confinement.

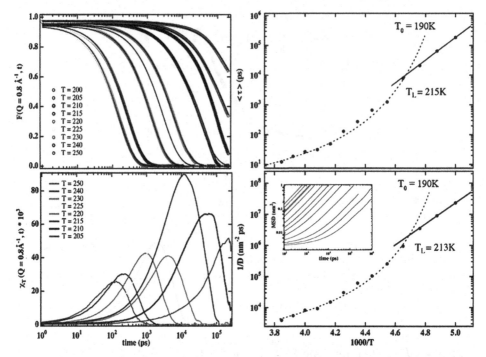

Figure 8.5 Numerical results from molecular dynamics simulations of bulk TIP4P-Ew water. (Top left) Oxygen self-intermediate scattering functions at $Q = 0.8\,\text{Å}^{-1}$ for several temperatures. (Bottom left) Dynamic response function, $\chi_T(Q, t)$, calculated with finite differences of the ISFs displayed in the top left panel. (Top right) Arrhenius plot of the alpha-relaxation time, extracted by fitting the ISFs with the relaxing cage model. (Bottom right) Arrhenius plot of the inverse of the self-diffusion constant, extracted by fitting the mean square deviation (inset) with the Einstein relation (Chen *et al.*, 2009b). Copyright IOP Publishing. Reproduced with permission. All rights reserved.

8.1.3 *Extended mode-coupling theory (eMCT) applied to Lennard-Jones system*[1]

Having found some supporting experimental results together with those from computer simulations, one can raise a number of important questions, such as: can the dynamic crossover phenomenon that we detected experimentally in all sorts of confined water be described as a fragile-to-strong dynamic crossover (FSC)? Do the single particle and collective α-relaxation times behave the same way as the shear viscosity as a function of temperature? If so, can the crossover temperature be identified as the T_c defined in idealised mode-coupling theory? To answer yes to

---

[1] Chong (2008); Chong *et al.* (2009).

all these questions, Chen *et al.* (2009b) demonstrated in Figure 8.6 the agreement with eMCT at least for the case of the Lennard-Jones system.

Figure 8.6 shows an Arrhenius plot of the three scaled quantities of the coherent, incoherent α-relaxation times and the viscosity, as a function of T_c/T, as predicted by the eMCT. It is clearly visible in both panels that the behaviours of all three quantities are very close to each other, at least in the vicinity of the MCT T_c. Furthermore, the crossover temperature is best identified to be at T_c and for practical purpose the maximum slope in the Arrhenius plot of the viscosity (bottom panel) can be used to detect the position of the crossover temperature, as was already demonstrated in Figure 7.24.

An extended mode-coupling theory of Lennard–Jones system predicts that the FSC temperature T_L of viscosity and α-relaxation times coincides with the mode-coupling T_c. Thus the dynamic crossover phenomenon in common glass-forming liquids can be attributed to change of dynamic behaviour of the supercooled liquid from a regime dominated by caging effect in a dense liquid to a quasi-arrested regime where the relaxation is only possible with the help of collective hopping processes (dynamic heterogeneities).

It should be pointed out in this occasion that, from the physical point of view, a genuine change of thermodynamic and dynamic behaviour is taking place mainly at T_L or T_x, not at the traditionally emphasised T_g. In the supercooled glass-forming liquids, the thermodynamics and dynamics are mutually coupled through the Adam–Gibbs relation, Eq. (8.3).

8.2 Absence of structural arrest at T_g

Along the line of our discussion thus far on the dynamic crossover phenomena in glass-forming liquids other than water, the latest experimental evidence of the absence of structural arrest at T_g will be discussed next. Mallamace *et al.* (2014) explored the system's ergodicity by measuring $|g_1(q, t)|^2$ of ortho-terphenyl (OTP) for five temperatures, $T = 255, 250, 248, 246$ and 243 K, above and below the calorimetric glass transition temperature T_g (= 247 K), as plotted in Figure 8.7. In the inset of this figure they also show the measured power spectral density (PSD), $S(q, \omega)$ at the same temperatures. All the $|g_1(q, t)|^2$ curves are characterised by a bimodal relaxation and all show the decay to zero at finite times, even below T_g. Thus, the system's ergodicity is fully preserved, for $T < T_c$ (FSC temperature 285 K) and also below T_g as proposed by the theory (Eckmann and Procaccia, 2008; Boue *et al.*, 2011). The OTP evolution towards a very slow dynamics is also observable, in the same T region, by the marked $S(q, \omega)$ increases in the lowest frequency region on decreasing T (inset, Figure 8.7).

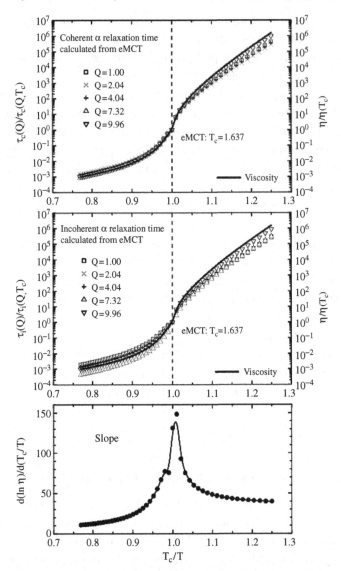

Figure 8.6 (Top and middle panels) Scaling plot of $\tau(Q)/\tau(Q, T_c)$ and $\eta/\eta(T_c)$ vs. T_c/T of the normalized α-relaxation time $\tau_{coh}(Q)$, $\tau_{inc}(Q)$ and viscosity calculated from the eMCT. When their Q-dependence is scaled out in this way, the coherent and incoherent α-relaxation times and the viscosity behave essentially the same way, showing a similar crossover phenomenon. (Bottom panel) The slope in the Arrhenius plot of the viscosity shows a sharp peak at T_c. Therefore it is reasonable to locate the T_c by the peak position of the maximum slope in the Arrhenius plot of the viscosity (Chen *et al.*, 2009b). Copyright IOP Publishing. Reproduced with permission. All rights reserved.

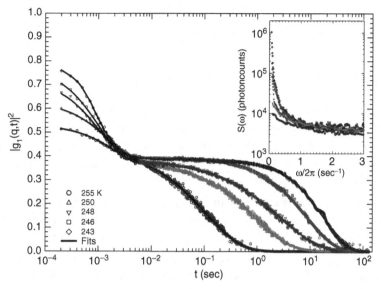

Figure 8.7 This figure shows the result of a photon correlation spectroscopy (PCS) measurement (Leone *et al.*, 2012) of the absolute square of the scattered electric field amplitude correlation function $|g_1(q, t)|^2$ for $T = 255, 250, 248, 246$ and 243 K. The inset reports the power spectral density (PSD), $S(q, \omega)$, measured at the same temperatures. All the $|g_1(q, t)|^2$ curves are characterised by a bimodal relaxation and also by the decay to zero at finite times and show that the system relaxation times span about six decades. It should be noted that the ergodicity is preserved even below glass transition temperature T_g as proposed by the theory (Eckmann and Procaccia, 2008; Boue *et al.*, 2011). The general layout of the experimental setup can be found in (Leone *et al.*, 2012). Reprinted with permission from Leone *et al.* (2012). Copyright 2012, AIP Publishing LLC.

In short, Mallamace *et al.* (2014) have presented a set of relaxation functions measured in the supercooled OTP, above and below the calorimetric glass transition temperature T_g, which firmly establishes the absence of the dynamical arrest phenomenon in any of these temperatures. The comparison of these data with the main findings of the study on grafted colloids by Brambilla *et al.* (2009) shows that molecular and colloidal systems behave, in this respect, the same way. This provides another support to the idea that the crossover temperature T_c, rather than the traditionally emphasised T_g, plays an essential role for the understanding of the dynamical crossover phenomenon in general glass-forming systems.

8.3 Fractional Stokes–Einstein relation[2]

It is well known that in the case of a simple fluid like water, at ambient conditions, the test-particle diffusion coefficient of a spherical particle of radius R immersed in

---

[2] Chen *et al.* (2006c); Mallamace *et al.* (2010).

water is related to the shear viscosity of water by the well-known Stokes–Einstein relation (SER),

$$D = \frac{k_B T}{6 \pi \eta R},$$

(8.5)

where k_B is the Boltzmann constant. When we consider water by itself then the SER can be generalised to give the self-diffusion coefficient of a water molecule (D_S) in terms of the inverse of its own shear viscosity by an equation similar to that of Eq. (8.5), provided that the radius of the test particle R is replaced by an equivalent radius of a water molecule a.

This phenomenon of the self-diffusion coefficient being inversely proportional to the shear viscosity at a constant temperature T for the case of water has already been discussed in Chapter 7. It was shown there that the validity of this SER, namely, $D_S \eta / T = \text{const}$ is only true for temperatures above the dynamic crossover temperature $T_L = 224.5\,\text{K}$. However the value of the constant starts to increase rapidly as the temperature of water goes below T_L. This was shown in Figure 7.9, taken from Chen *et al.* (2006c).

In this section, we will discuss the validity of SER of all other glass forming liquids studied. In particular, we shall show that below the dynamic crossover temperature of the glass-forming liquid, T_x the familiar Stokes–Einstein relation $D_S / T \sim \eta^{-1}$ breaks down and is replaced by a fractional form $D_S / T \sim \eta^{-\zeta}$, with $\zeta \approx 0.85 \pm 0.02$.

Figure 8.8 analyses the data in terms of the scaling law approach for understanding the violation of the Stokes–Einstein relation (SER) and the Debye–Stokes–Einstein (DSE) relation. More precisely D_s vs. η (lower data) and $1/\tau$ vs. η (upper data) are plotted in a log–log scale to detect the change of behaviour to the *fractional* SER and DSE respectively (Garrahan and Chandler, 2003; Jung *et al.*, 2004) below T_x. The data clearly show two different scaling behaviours above (dashed lines) and below (solid lines) the fragile-to-strong dynamic crossover temperature T_x. From both figures in Figure 8.8, it is evident that, for all liquids studied, the onset of the breakdown takes place at approximately the same viscosity Poise and that the curves are universal in the sense that the data for different liquids superpose.

Mallamace *et al.* (2010) further evaluated the SER ratio and the DSE ratio $R_{SER} \equiv D_S \eta / T$ and plotted as a function of the temperature ratio T / T_X in the upper and lower insets of Figure 8.8. It was found that deviation from the constancy occurs very close to T_X in both these quantities. A recent study of the extended MCT (Chong *et al.*, 2009) demonstrated that, in agreement with other experimental observations in some glass-forming liquids (Chang and Sillescu, 1997; Douglas and Leporini, 1998; Swallen *et al.*, 2003; Mapes *et al.*, 2006; Fernandez-Alonso

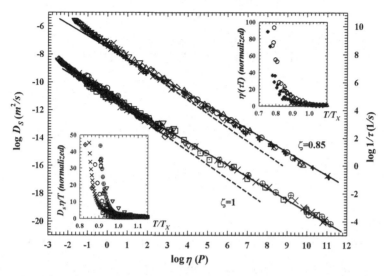

Figure 8.8 The main plot shows a scaled representation of the fractional SER and DSE, depicted by the lower and upper lines of data, respectively. In both cases, for all the liquids studied, the scaling exponent ζ takes almost the same value, $\zeta = 0.85 \pm 0.02$. We note that the onset of the fractional SER and DSE takes place at the same value of viscosity, $\eta_X \approx 10^3$ Poise. These data demonstrate a remarkable degree of universality in the temperature behaviour of the transport properties of supercooled liquids. The deviation of the SER ratio from constancy for nine liquids analysed is shown in the lower inset while that of the DSE ratio for six liquids is shown in the upper inset. In both cases the deviation occur just below the corresponding crossover temperature T_X (Mallamace *et al.*, 2010). Copyright 2010, National Academy of Sciences, USA.

et al., 2007), these SER and DSE violations occur very close to T_c. They are an example of the decoupling of transport coefficients (Xu *et al.*, 2009), the microscopic origins of which can be due to the onset of typical length scales that increase rapidly as T decreases below T_c (Garrahan and Chandler, 2003; Eckmann and Procaccia, 2008).

9

Inelastic neutron scattering in water

9.1 Introduction to Q-dependent density of states $G_S(Q, \omega)$ of water

The incoherent inelastic neutron spectrum, $S_S(Q, \omega)$, is directly related to the so-called Q-dependent density of vibrational states of hydrogen atoms $G_S(Q, \omega)$, in a molecular system (Chen and Yip, 1976). Since in water a large fraction of hydrogen atoms participate in hydrogen bonding at a given time, and furthermore since this fraction is a strong function of temperature, it is expected that measurements of the hydrogen atom density of states by inelastic neutron scattering (INS) would provide a sensitive means of studying the temperature dependence of hydrogen-bond dynamics.

The basic principle of the method is to extract the Q-dependent density of states through the relation (Chen $et\ al.$, 1984; Toukan $et\ al.$, 1988)

$$G_S(Q, \omega) = \frac{\omega^2}{Q^2} S_S(Q, \omega). \tag{9.1}$$

$G_S(Q, \omega)$ can then be shown to be given by a Fourier transform of the self part of the longitudinal current correlation function $J_S(Q, t)$, defined as:

$$J_S(\vec{Q}, t) = \frac{1}{N_H} \left\langle \sum_{j=1}^{N_H} [\hat{Q} \cdot \vec{v}_j(0) e^{-i\vec{Q} \cdot \vec{r}_j(0)} \hat{Q} \cdot \vec{v}_j(t) e^{i\vec{Q} \cdot \vec{r}_j(t)}] \right\rangle. \tag{9.2}$$

This can be seen by using the following relation:

$$J_s(\vec{Q}, t) = -\frac{1}{Q^2} \frac{d^2}{dt^2} \frac{1}{N} \left\langle \sum_{j=1}^{N} [e^{-i\vec{Q} \cdot \vec{r}_j(0)} \cdot e^{i\vec{Q} \cdot \vec{r}_j(t)}] \right\rangle$$

$$= -\frac{1}{Q^2} \frac{d^2}{dt^2} F_s(\vec{Q}, t), \tag{9.3}$$

286

where $F_s(\vec{Q}, t)$ is the intermediate scattering function of the hydrogen atom. Taking the Fourier transform of both sides of the above equation, one finds that

$$
\begin{aligned}
FT\left(J_s(\vec{Q}, t)\right) &= -\frac{1}{Q^2} FT\left(\frac{d^2}{dt^2} F_s(\vec{Q}, t)\right) \\
&= -\frac{1}{Q^2}(i\omega)^2 FT\left(F_s(\vec{Q}, t)\right) \\
&= \frac{\omega^2}{Q^2} S_s(\vec{Q}, \omega) = G_s(\vec{Q}, \omega).
\end{aligned}
\tag{9.4}
$$

Hence we have proven that the Q-dependent density of states is the Fourier transform of the self part of the longitudinal current correlation function $J_S(Q, t)$, namely

$$
G_S(Q, \omega) = \frac{1}{\pi} \int_0^\infty J_S(Q, t) \cos \omega t \, dt.
\tag{9.5}
$$

Since by definition, the $Q \rightarrow 0$ limit of the $J_S(Q, t)$ is the velocity autocorrelation function of the hydrogen atom, namely,

$$
\lim_{Q \to 0} J_S(Q, t) = \langle v_{HH}^z(0) \cdot v_{HH}^z(t) \rangle.
\tag{9.6}
$$

In this last equation, the Q-vector is taken to be pointing along the z direction. We have thus a very useful relation

$$
\lim_{Q \to 0} G_S(Q, \omega) = f_H(\omega),
\tag{9.7}
$$

where $f_H(\omega)$ is the true density of states of the hydrogen atom, which is defined as

$$
f_H(\omega) = \frac{1}{\pi} \int_0^\infty dt \, \cos \omega t \langle v_{HH}^z(0) \cdot v_{HH}^z(t) \rangle,
\tag{9.8}
$$

with a normalisation condition of

$$
\int_0^\infty f_H(\omega) d\omega = \frac{k_B T}{2 M_H}.
\tag{9.9}
$$

The vibrational density of states of a given species of atom in liquid state is defined as the spectral density of the velocity autocorrelation function of that particle. In practice, the $Q \rightarrow 0$ limit is satisfied when the product $Qb < 1$ (Chen and Yip, 1976), where b is the bond length between O and H atoms in a water molecule. Study of $f_H(\omega)$ in liquids is particularly important in their supercooled states, where they may approach amorphous or glassy states. For water, $b = 1$ Å and measurement of $f_H(\omega)$ is possible only for $E = \hbar\omega$ less than about 50 meV. However, computer molecular dynamics calculations of both $f_H(\omega)$ and

$G_S(Q, \omega)$ are straightforward (Toukan and Rahman, 1985; Toukan *et al.*, 1988). Thus the strength of INS is that it gives directly the proton density of states, or its Q-dependent extension, both of which are directly calculable in CMD once the inter- and intra-molecular potentials of water molecules are specified. This feature of INS is in sharp contrast to the traditional optical spectroscopic techniques of infrared and Raman spectroscopies for which CMD calculations are possible only in some approximations (Bansil *et al.*, 1986). We shall begin by describing the general feature of $G_S(Q, \omega)$ in water. With increasing frequency, after the prominent quasi-elastic peak, the inelastic spectrum shows three inter-molecular bands in water: a sharp and strong band at about 6 meV (48.4 cm^{-1}) called A, a broad and weak band spreading between 20 and 35 meV (161–282 cm^{-1}) called B, and a strong and broad band from 50 to 130 meV (403–1049 cm^{-1}) called L. The A and B bands are translational in character and have their counterparts in Raman spectrum (Walrafen, 1964; Krishnamurthy *et al.*, 1983). The Raman band at 60 cm^{-1} corresponding to A is assigned to be the transverse vibration of water molecule as a whole with respect to the hydrogen bond direction. This transverse motion induces a large-amplitude oscillation of the hydrogen atom perpendicular to the hydrogen bond and is reflected in the INS spectrum as a strong peak. The Raman bands at 190 cm^{-1} and 260 cm^{-1}, corresponding to B, are due to O–O stretching modes that induce only small amplitude motions of the H-atom. Thus these bands have weak intensities in the INS spectrum. On the other hand, the L band is librational in character. The water molecule rotates around the centre of mass (nearly coincident with the oxygen atom) resulting in large-amplitude motion of the H-atoms. This gives rise to an increase in the INS spectrum known as the hinder rotation band. In the Raman spectrum, it corresponds to a broad librational band extending from 300 cm^{-1} to 900 cm^{-1} (Walrafen, 1972). The reason for such a broad band is that a water molecule has three different moments of inertia and librations around each axis have different frequencies. Figure 9.1 shows a proton density of states $f_H(\omega)$ extracted from an INS spectrum for bulk water at 25 °C and water contained in 52% hydrated Vycor at the same temperature (Bellissent-Funel *et al.*, 1995). One readily notices the sharpness of the A band at 6 meV, a very weak B band at around 20–35 meV and an intense L band in the range 50–130 meV.

We now turn our attention to the intra-molecular vibrations in water. We show in Figure 9.2 a Q-dependent proton density of states in a two-dimensional plot, as functions of Q and E, obtained for water at -15 °C using a high-energy medium resolution chopper spectrometer (HEMRCS) operating at the Intense Pulse Neutron Source (IPNS) of Argonne National Lab. Starting from the left-hand corner of the energy scale, the first peak about 80 meV is the L band, the second peak at 220 meV is the OH bending mode (BEN) and the broad peak at around 420 meV is the OH stretch vibration (STR). For bulk water in condensed

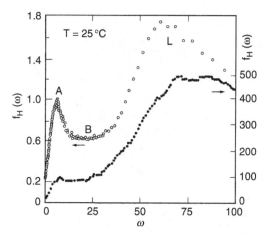

Figure 9.1 Proton density of states for bulk water at 25 °C (open circles), and for water contained in 52% hydrated Vycor at the same temperature (solid circles) (Bellissent-Funel *et al.*, 1995). Copyright 1995 by The American Physical Society.

Table 9.1 *Experimental vibrational energies in ice and bulk water in condensed phase.*

	T (K)	E_L (meV)	E_{BEN} (meV)	E_{STR} (meV)
Ice	253	82	207	407 (49)
H_2O (supercooled)	258	74	207	418 (49)
H_2O (bulk)	313	–	207	441 (66)
H_2O (bulk)	353	–	207	443 (70)
Q (Å^{-1})	–	2.6	3.5	6.6

phase, INS spectra show that there is only one single broadened peak. Table 9.1 lists all the observable peaks in INS spectra. Values in parenthesis are bandwidths. We summarise the main features of the experimental spectra as follows:

(i) Significant temperature dependence is observed for the L band and STR band while the BEN band is nearly temperature independent. The frequency of the L band decreases as the structure changes from ice to water. The opposite behaviour is observed for the STR band.

(ii) The width of the STR band, which can be more reliably estimated, increases with temperature. The energy resolution at the stretch band is 16 meV.

A computer molecular dynamics simulation was also performed in addition to the experimental measurements to interpret the experimental results. The simulated spectra at constant Q values predict quite correctly the main features of

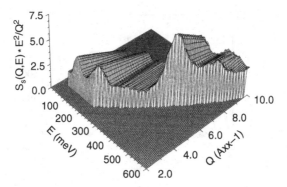

Figure 9.2 Density of states $G_S(Q, E)$ for water at $-15\,°C$ measured by INS technique using HRMECS at IPNS. The incident neutron energy E_0 is 800 meV. From the low-energy side, there are three bands: L (librational band), BEN (bending vibrational band) and STR (stretching vibrational band) as described in the text (Chen *et al.*, 1984). Copyright 1984 by The American Physical Society.

the measured experimental spectra for light water. Doppler broadening of the librational band at the higher Q values is clear evident. This feature would not have been observed without carrying out a simulation at finite Q such as the present one (Toukan and Rahman, 1985). The detailed study of the Q-dependent effect in the librational band should be a matter of further interesting of experimental investigation. The simulation also shows a decrease in intensity due to Debye–Waller factor at higher Q values. However, direct comparison of the intensity obtained by classical CMD simulation with the corresponding experimental spectra in the stretch region shown in Figure 9.3 indicates that the simulated intensity is lower by a factor of 2 as compared to the experimental spectrum.

Furthermore, the CMD spectrum shows only a weak combination band at 525 meV for any of the Q values considered. This discrepancy can only be accounted for by considering a quantum-mechanical excitation in liquid water at these temperature and confirms earlier calculations that attributed the combination band to this feature (Ricci and Chen, 1986; Ricci *et al.*, 1986). The CMD results of Toukan and Rahman (1985) using the SPC flexible water molecule potential are available for both gas and liquid phases.

Careful examination of these results reveals the following two important points:

(i) The intra-molecular O–H bond length in liquid is 2% longer than that in the gas phase. This elongation leads to a weakening of the O–H bonds in the liquid state and hence to mode softening of the stretch vibration. This is observed in the experiment.

(ii) The formation of hydrogen bonds in the liquid phase leads to further softening of the O–H stretch vibrational mode. As the temperature of the system is decreased, more intact hydrogen bonds are formed, leading to a decrease in

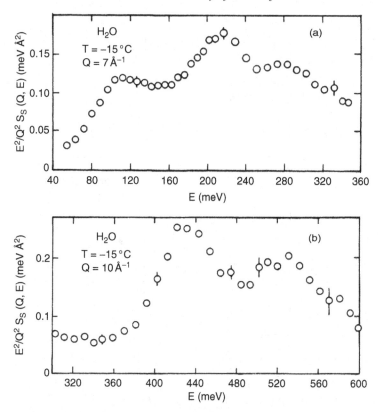

Figure 9.3 (a) The measured constant Q density of states in supercooled water showing the librational and bending bands at a Q value of $7\,\text{Å}^{-1}$. (b) The measured constant Q density of states in supercooled water showing the stretch and combination bands at a Q value of $10\,\text{Å}^{-1}$ (Ricci and Chen, 1986). Copyright 1986 by The American Physical Society.

the STR-band energy. The experimental results listed in Table 9.1 also support this result. This trend is correlated with a corresponding increase in L band energy (Table 9.1).

It is of interest to compare the absolute intensity of an INS spectrum with that generated by CMD. Using the SPC flexible water molecule potential, Toukan *et al.* (1988) have generated $G_s(Q, \omega)$ using the definition of $J_s(Q, t)$ given in Eq. (9.2) combined with the Fourier transform relation of Eq. (9.5). The result for H_2O at $T = 250\,\text{K}$ is shown in Figure 9.4. Two features are immediately noticeable from inspection of these graphs:

(i) the L band is considerably broadened by increased Q value;
(ii) the BEN band is moderately affected by Q but the STR band is hardly affected by it. This latter fact implies that the STR band represents a well-localised

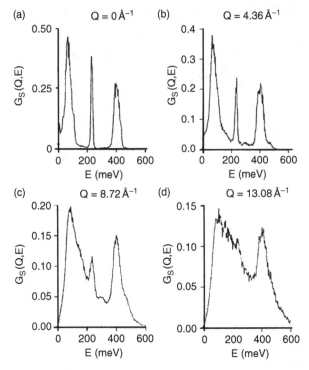

Figure 9.4 Q-dependent proton density of states $G_s(Q, E)$ calculated by CMD simulation for a temperature of 250 K for Q values of 0, 4.36, 8.72, and 13.08 Å^{-1}. The absolute cross-section calculated by the classical CMD simulation is lower by a factor of 2 as compared to that measured experimentally. The spectrum also does not show the combination band at 525 meV indicating the quantum-mechanical origin of the excitations in liquid water (Toukan and Rahman, 1985). Copyright 1985 by The American Physical Society.

mode of vibration. It is for this reason that experimentally one can still observe a sharp STR band at an energy as high as 500 meV and a Q value as large as 10 Å^{-1}. Since it is a localised mode, the mode can be well approximated by a harmonic oscillation.

For a harmonic oscillator (or harmonic lattice vibrations) the quantum mechanical dynamic structure factor can be written as follows:

$$S_s(Q, \omega) = \exp[\frac{\hbar\omega}{2k_B T}] \exp[-\gamma Q^2] S_S^{cl}(Q, \omega), \tag{9.10}$$

$$\gamma = \int_0^\infty d\omega \frac{f_H(\omega)}{\omega^2} [\cosh\frac{\hbar\omega}{2k_B T} - 1], \tag{9.11}$$

where $S_S^{cl}(Q, \omega)$ is the classical dynamic structure factor calculated by CMD. To generate the CMD calculation with the experiment, we must multiply the $S_S^{cl}(Q, \omega)$ with the detailed balance factor and the Q-dependent recoil factor given in Eqs. (9.10) and (9.11). The detailed calculation shows that (Ricci *et al.*, 1986)

$$G_S(Q, \omega)|_{EXP} = 2.1 G_S^{cl}(Q, \omega)|_{CMD}. \qquad (9.12)$$

Comparing the STR band heights given by experimental spectra and simulated spectra, the relation Eq. (9.12) is seen to be approximately satisfied.

Another point to notice is that while the STR band in classical CMD calculation consists of only one single peak, at about 420 meV, the measured spectra (quantum mechanical in nature) shows an additional satellite peak occurring at above 530 meV. Ricci *et al.* (1986) have interpreted the presence of the satellite peak as arising from the OH vibration of the free water molecule when an incident neutron (800 meV) happens to break the hydrogen bond on the other side of the OH covalent bond of the water molecule (Chen, 1991).

9.2 Density of state measurements of water: a plausible evidence for the existence of a liquid–liquid phase transition in low-temperature water

9.2.1 Introduction

It has been conjectured that a first-order liquid-to-liquid (L–L) phase transition (LLPT) between a high-density liquid (HDL) and a low-density liquid (LDL) in supercooled water may exist, as a thermodynamic extension to the liquid phase of the first-order transition established between the two bulk solid phases of amorphous ices, the high-density amorphous ice (HDA) and the low-density amorphous ice (LDA). In a preliminary experiment described in Chen *et al.* (2013), incoherent inelastic neutron scattering (INS) was used to measure the density of state (DOS) of hydrogen atoms in H_2O molecules in confined water as functions of temperature and pressure, by which they were able to follow the emergence of the LDL and HDL phases from the corresponding LDA and HDA at supercooled temperatures and high pressures. They reported the differences of librational and translational DOSs between the hypothetical HDL and LDL phases, which were similar, but much smaller, to the corresponding differences between the well established HDA and LDA ices. This constitutes possible evidence that the HDL and LDL phases are the thermodynamic extensions of the corresponding amorphous solid water HDA and LDA ices.

At present the anomalous and counter-intuitive behaviours of the thermodynamic response functions, such as isothermal compressibility, isobaric specific heat and isobaric thermal expansion coefficient, of low-temperature water can be understood if one accepts the idea that a first-order low-density liquid to high-density liquid phase transition (LLPT) line exists in deeply supercooled water. Originally, this idea arises from a molecular dynamics simulation of the ST2 model of water (Poole *et al.*, 1992). In this ST2 model of water, it was predicted that the end point of this LLPT line, which is supposed to be a liquid–liquid critical point (LLCP), might exist in bulk water if it can be supercooled down below the homogeneous nucleation line. In 1996 a strong relationship between the dynamic and thermodynamic behaviours at ambient pressure of SPC/E model water was reported (Gallo *et al.*, 1996), indicating that the well-known singular temperature T_s (Speedy and Angell, 1976) at which extrapolations of various thermodynamic and dynamic anomalies diverge, can be identified with the crossover temperature T_c of mode-coupling theory (Götze, 2009). Furthermore, unlike the crossover phenomenon predicted in MCT, this crossover temperature marks the fragile-to-strong crossover that occurs in supercooled liquids approaching the calorimetric glass transition temperature (Liu *et al.*, 2005). More 2005 studies have completed the picture for water by connecting this transition to the crossing of the Widom line (Xu *et al.*, 2005). In 2011, a constant Q elastic neutron diffraction study of isobaric temperature T scan of the D_2O density showed a pronounced density hysteresis when the pressure P is more than 1500 bar (Zhang *et al.*, 2011). This phenomenon gave an indication of crossing of the first-order LLPT line above this pressure. In a more recent paper, Chen *et al.* (2013) supplement the two previous neutron scattering results with a third experimental measurement of the hydrogen density of state of the confined water. This new inelastic neutron scattering experimental result shows a possibility of the emergence of LDL and HDL phases from the corresponding LDA and HAD ices when they are transformed into liquid state.

9.2.2 Incoherent inelastic neutron scattering measurements

The incoherent neutron scattering technique is quite suitable for the study of the hydrogen atom dynamics in confined water. The main reason is that the scattering cross-section of a hydrogen atom is about 80 barn, which is much larger than that of other atoms in the MCM-41-hydration-water system, which is composed of silica, oxygen and hydrogen atoms. Furthermore, the neutron scattering cross-section of a hydrogen atom is mostly incoherent. Therefore the inelastic neutron scattering spectra of the system essentially reflects the self-dynamics of hydrogen atoms in the sample alone.

The incoherent inelastic neutron scattering (INS) experiment was performed at the fine energy resolution chopper spectrometer (SEQUOIA) (Granroth *et al.*, 2010) at the Spallation Neutron Source (SNS), Oak Ridge National Laboratory.

The SEQUOIA spectrometer can supply incident neutron energies E_i between 10 and 250 meV with the energy-transfer resolution as fine as 1–1.5 % E_i and also with a fairly good energy transfer resolution of 2.5–5 % E_i at higher incident energies of 250 meV to 2 eV. Approximately 0.5 g water in MCM-41S was placed in an annular cylindrical gap (10 mm outer and 6 mm inner diameters, 50 mm height) inside an aluminium alloy high-pressure cell with helium gas as a pressure transmission medium. The experiments were done with three incident neutron energies $E_i = 30, 160$ and 800 meV selected by the Fermi chopper, rotating at 180, 420 and 600 Hz, respectively. The data were recorded over a wide range of scattering angles, from $-30°$ to $+60°$ the horizontally and $\pm18°$ vertically. The background spectra for the dry MCM-41S sample in the high-pressure cell were measured at the same temperature and subtracted from the original data. The collected neutron scattering data were transformed from the time-of-flight data to the self-dynamical structure factor $S_s(Q, \omega)$ and then to the Q-dependent density of state (DOS), $G_s(Q, \omega)$.

9.2.3 Discussions of the INS experiment

The INS measurements were performed at three points in the P–T phase diagram with different pressures, $P = 1600, 1800$ and 2345 bar, and all at the same temperature of 180 K. From Figure 9.5 we can see that the point $P = 2345$ bar is definitely in the hypothetical HDL phase, while the other two points are located in the hypothetical LDL phase. In this experiment, points 1 and 2 were measured by first cooling the sample to 180 K from room temperature and then applying the pressure to 1600 bar and then to 1800 bar (grey line in Figure 9.5). After the measurements of points 1 and 2, the pressure was released to ambient pressure and the temperature was recovered to room temperature. Then the pressure was increased to 2345 bar and sample was cooled down to 180 K to measure the point 3 (white line in Figure 9.5). With such measurement routes, one tried not to cross the hypothetical first-order L–L phase transition line so that the complicated kinetic process of the phase separation phenomenon can be avoided. In this way, the authors made sure that the states 1 and 2 are in the pure LDL phase and the state 3 is in the pure HDL phase according to the hypothetical phase diagram of bulk water (Mishima and Stanley, 1998b).

The INS spectra of solid phases and accessible bulk liquid phases of H_2O and D_2O have been investigated extensively in previous studies (Toukan *et al.*, 1988; Kolesnikov *et al.*, 1994, 1997, 1999; Li, 1996; Li and Kolesnikov, 2002). The DOSs of LDA, HDA and Ih ices (Kolesnikov *et al.*, 1994, 1997, 1999; Li, 1996; Li and Kolesnikov, 2002) are shown in Figure 9.6 for comparison with the DOSs of the hypothetical LDL and HDL phases. A characteristic feature of the DOS of

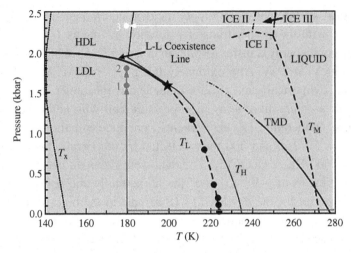

Figure 9.5 Water *P–T* phase diagram. The solid black curve above the star indicates the hypothetical first-order L–L phase transition line and the dotted line lower the star is the Widom line. Points 1, 2 and 3 correspond to the three measured points at P = 1600, 1800 and 2345 bar, respectively. The grey and white lines indicate the measurement routes (Mishima and Stanley, 1998b). Reprinted by permission from Macmillan Publishers Ltd: Nature, copyright 1988.

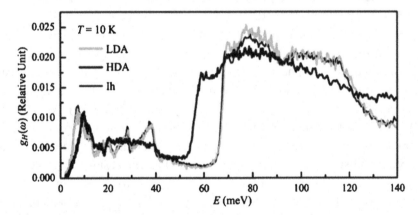

Figure 9.6 *Q*-averaged DOSs of hydrogen atoms $g_H(\omega)$ of LDA, HDA and Ih ices at T = 10 K. Notice that the DOS curves of LDA and Ih ices practically overlap within the range of $0 < E < 70$ meV. Copyright 2013, AIP Publishing LLC.

amorphous ices is that the HDA ice shows a steep cliff at $E \approx 55$ meV, while the LDA ice shows a similar cliff at $E \approx 70$ meV. As seen in Figure 9.7(a) the DOSs of the LDL and HDL phases do not have such a feature. Instead, the DOSs of the LDL and HDL phases show much smoother curves, which can be recognised as characteristic of the liquid state rather than that of the solid state. One should notice further while the DOSs of the LDA and HDA ices have significantly different structures, the DOSs of the LDA and Ih ices are quite similar. In fact, these two curves overlap in the energy range $0 < E < 70$ meV.

Figure 9.7 shows the librational bands of (a) the hypothetical HDL and LDL phases, and (b) the HDA and LDA ices. Notice that the curves for the hypothetical HDL and LDL phases are noticeably different from those of amorphous ices, which are much steeper than the former ones within the energy range 50–70 meV.

This difference indicates that the confined water sample measured in this experiment is in a liquid state rather than in a solid state. From Figure 9.7(b) one can find that, for amorphous ice, there is a substantial difference in the spectrum of the Q-averaged DOS $g_H(\omega)$ within the energy range 67–125 meV between HDA ice and LDA ice (lower darker curve in Figure 9.7(b)). What is significant is that there is also a difference within a similar energy range of 67–115 meV (the grey curve in Figure 9.7(a)) between the $P = 2345$ bar case (in the hypothetical HDL phase) and the $P = 1800$ bar case (in the hypothetical LDL phase), despite a smaller magnitude by one order compared with that of the amorphous ices. We emphasise here that only with the help of a fine-resolution, high-intensity spectrometer like SEQUOIA can such small difference can be resolved.

Furthermore, in the same energy range, the difference of the $g_H(\omega)$ spectra between the $P = 1800$ bar and $P = 1600$ bar cases (lower flat grey line in Figure 9.7(a)), which are both in the hypothetical LDL phase, is quite small. This phenomenon indicates a possible spectrum jump between the two sides of the hypothetical first-order L–L transition line. The observed similarity of the difference between the hypothetical HDL and LDL phases and the corresponding difference between the HDA and LDA ices, as well as the spectrum jump between the $P = 2345$ bar and $P = 1800$ bar cases, could be plausible evidence for the existence of pure HDL and LDL phases as the thermodynamic extensions of HDA ice and LDA ice, respectively.

Figure 9.8 shows the inter-molecular translational bands of (a) the hypothetical HDL and LDL phases, and (b) the HDA and LDA ices. In this band, the similarity of the difference between the HDA and LDA ices and the corresponding difference between the hypothetical HDL and LDL phases (but a smaller magnitude by one order than the former one, just like the situation in the librational band) is also observed around $E = 7$ meV. This supports our previous statement that the HDL and LDL phases exist as the thermodynamic extensions of HDA and LDA ices. They did not measure the DOS in this band at $P = 1600$ bar due to the limited beam time.

The intra-molecular O–H stretch vibrational bands were also measured. However, the Q-averaged DOSs at the three measured pressures are practically identical within the error bars in this band, as shown in Figure 9.9. The positions

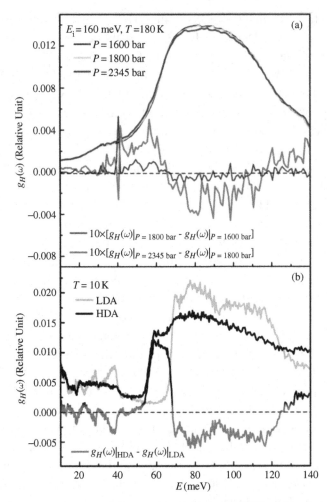

Figure 9.7 (a) Comparison between the Q-averaged DOSs of hydrogen atoms $g_H(\omega)$ at three measured pressures within the energy range $E = 10$–140 meV (the range 40–130 meV corresponds to the librational band) at $T = 180$ K. The cases of $P = 1600$ and 1800 bar (indicated by top grey curve and light grey curve, respectively) are below the hypothetical first-order L–L phase transition line, while the case of $P = 2345$ bar (black curve) is above that line. The lower grey curve indicates the ten times difference between the $P = 2345$ bar and $P = 1800$ bar cases, and the flat grey curve indicates the ten times difference between the $P = 1800$ bar and $P = 1600$ bar cases. (b) Comparison between the Q-averaged DOSs of hydrogen atoms $g_H(\omega)$ of LDA (light grey curve) and HDA (dark curve) ices at $T = 10$ K within the energy range $E = 10$–140 meV. The lower darker grey curve indicates the difference between the HDA and LDA states of ices. It is striking to observe the similarity of the difference between the hypothetical HDL phase and LDL phase (lower grey curve in (a)) and the corresponding difference between the HDA ice and LDA ice (lower darker grey curve in (b)). Copyright 2013, AIP Publishing LLC.

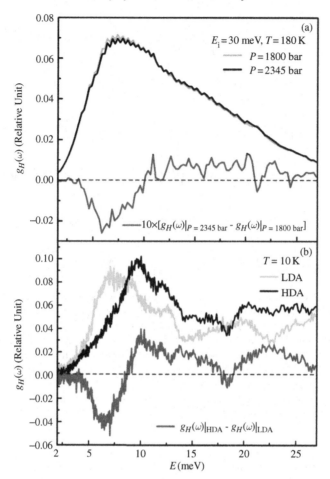

Figure 9.8 (a) Comparison between the Q-averaged DOSs of hydrogen atoms $g_H(\omega)$ at two measured pressures within the energy range $E = 2$–27 meV (which corresponds to the translational band) at $T = 180$ K. The case of $P = 1600$ bar (light grey curve) is below the hypothetical first order L–L phase transition line, while the case of $P = 2345$ bar (dark curve) is above that line. The lower darker grey curve indicates the ten times difference between $P = 2345$ bar and $P = 1800$ bar cases. (b) Comparison between the Q-averaged DOSs of hydrogen atoms $g_H(\omega)$ of LDA (light grey curve) and HDA (upper dark curve) ices at $T = 10$ K within the energy range $E = 2$–27 meV. The lower dark grey curve indicates the difference between HDA and LDA states of ices. It is striking to observe the similarity of the difference between the hypothetical HDL phase and LDL phase (lower grey curve in (a)) and the corresponding difference between the HDA ice and LDA ice (lower darker grey curve in (b)). Copyright 2013, AIP Publishing LLC.

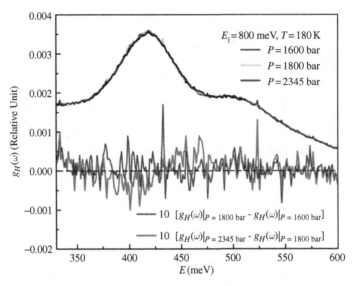

Figure 9.9 Comparison between the Q-averaged DOSs of hydrogen atoms $g_H(\omega)$ at three measured pressures within the energy range $E = 330\text{–}600$ meV (which corresponds to the O–H stretch vibrational band) at $T = 180$ K. The cases of $P = 1600$ and 1800 bar (indicated by black dots and grey curve, respectively) are below the hypothetical first-order L–L phase transition line, while the case of $P = 2345$ bar (dark curve) is above that line. The lower grey curve indicates the 10 times difference between the $P = 2345$ bar and $P = 1800$ bar cases, and the dark grey curve indicates the ten times difference between the $P = 1800$ bar and $P = 1600$ bar cases. Copyright 2013, AIP Publishing LLC.

of the O–H stretch vibrational modes at $P = 1600, 1800$ and 2345 bar are $417.25, 417.44$ and 418.15 meV, which are obtained by fitting the experimental data within $E = 350\text{–}550$ meV with two Gaussian functions. Notice that the O–H stretch vibrational modes for Ih ice (and supposedly for LDA ice) and HDA ice are at ≈ 410 meV and ≈ 421 meV, respectively (Li, 1996). The O–H stretch vibrational band arises from the motions of an H-atom which is covalently-bonded to a reference H_2O molecule on one side and is also involved in a hydrogen bonding with another one of the nearest-neighbour H_2O molecules on the other side (Ricci and Chen, 1986). Therefore, this band is mainly determined by a coupled motion of a reference H_2O molecule and its nearest neighbours, i.e. the first coordination shell. On the contrary, the other two bands studied in this paper, the translational band and the librational band, reflect inter-molecular motions that involve not only a H_2O molecule and its nearest neighbours, but also the second coordination shell of the H_2O molecule or more.

According to Soper and Ricci (2000), the primary distinction between the local structures of the hypothetical HDL and LDL phases is the distance of the second shell away from the first shell, and the breaking of the hydrogen bonds between the two shells which occurs in HDL phase. However, the first coordination shell is tetrahedral in shape for both HDL and LDL forms of water. Meanwhile, the distances from a reference H_2O molecule to its first coordination shell of the two forms of water are nearly the same. These arguments may explain why we can only see the differences in the translational band and the librational band, rather than in the O–H stretch vibrational band: because the latter is primarily determined by the structure and motions within the first coordination shell of a reference H_2O molecule, which does not change from HDL phase to LDL phase significantly; on the other hand, the other two bands reflect the motions involving the second coordination shell of a H_2O molecule of which the change from HDL phase to LDL phase is significant.

Based on the above described observation, and assuming that the phase diagram given by Mishima and Stanley (1998b) as shown in Figure 9.5 is correct, then Chen and coworkers initially interpret the result of this new INS experiment as giving a possible evidence for the emergence of the LDL phase on the lower-pressure side of the conjectured first-order liquid–liquid phase transition line, as the result of the thermodynamic continuation of the LDA ice into the liquid state. They also concluded that the HDL phase above the first-order liquid–liquid phase transition line is the continuation of the HDA ice into the liquid state as well.

However, since the above-described experiment (Chen *et al.*, 2013) was a preliminary one, Chen *et al.* have given considerable thoughts to the interpretation of these results based on the validity of the phase diagram given in Figure 9.5 for the bulk water. They have come to realise, in retrospect, that the confined water in MCM-41 should not be treated in the same way as the bulk water because it may have a quite different phase diagram due to the existence of the so-called Young–Laplace effect. This effect is to be discussed in Section 9.3 together with the new additional follow-up experiments.

9.3 Measuring boson peak as a means to explore the existence of the liquid–liquid transition in deeply cooled confined water

9.3.1 Latest experimental results

The boson peak is a broad peak observed at low frequencies ($\approx$ 2–10 meV) in the incoherent inelastic neutron scattering, and X-ray and Raman scattering spectra of many amorphous materials and supercooled liquids (Frick and Richter, 1995; Shintani and Tanaka, 2008; Chumakov *et al.*, 2011). Preliminary measurements on the boson peak in deeply cooled water confined in nanopores (Chen *et al.*,

2008) and in protein hydration water (Chen *et al.*, 2009a) under ambient pressure have been reported. The measured neutron scattering spectra showed a boson peak at about 6 meV, which only emerges below $\approx$ 230 K. A similar result was also observed in a computer simulation study (Kumar *et al.*, 2013). Since this temperature is close to the ambient-pressure Widom line temperature $T_w \approx$ 224 K (the Widom line is the locus of specific heat maxima) in a one-phase liquid (Xu *et al.*, 2005), Wang *et al.* (2014d) tentatively explained this phenomenon as due to transformation of the local structure of water from a predominantly high-density liquid (HDL) to a predominantly low-density liquid (LDL) form as T_w is crossed from above. However, this observation of the experimental results is by itself only qualitative. First, it is clear that this emergence of the boson peak in the raw spectra depends on the energy resolution of the neutron scattering spectrometer used. Second, and more importantly, the situation at high pressures has heretofore remained unknown. Therefore, without quantitative analysis of the entire spectra, including quasi-elastic and inelastic components at high pressure, definite conclusions cannot and should not be drawn.

In that study, Wang *et al.* (2014d) measured the boson peak in deeply cooled water with a series of inelastic neutron scattering (INS) experiments in the temperature range from 120 K to 230 K and the pressure range from 400 bar to 2400 bar, employing the Disk Chopper Spectrometer (DCS) (Copley and Cook, 2003) at the US National Institute of Standards and Technology (NIST) Center for Neutron Research, and the Cold Neutron Chopper Spectrometer (CNCS) (Ehlers *et al.*, 2011) at the Spallation Neutron Source of the Oak Ridge National Laboratory (ORNL). The DAVE software was used for the data reduction (Azuah *et al.*, 2009). It was found that below $\approx$ 1600 bar, the emergence of the boson peak is correlated (but not coincides) with the Widom line and with the fragile-to-strong crossover (FSC) in the deeply cooled water (Liu *et al.*, 2005). Above $\approx$ 1600 bar, the locus of the emergence of the boson peak in the (P, T) plane has a different slope as compared with its behaviour below $\approx$ 1600 bar, and the end-pressure of the Widom line is estimated by determining where the slope begin to change. Moreover, the (P, T) dependence of the shape of the boson peak is found to be related to the density minimum of water (Liu *et al.*, 2007; Zhang *et al.*, 2011; Liu *et al.*, 2013). Thus a possible way to distinguish the hypothetical low-density liquid phase and high-density liquid phase (Poole *et al.*, 1992) of deeply cooled confined water is realised.

In order to enter the deeply supercooled region of water, a nano-porous silica matrix, MCM-41S, with 15 Å pore diameter to confine the water was used. When confined, water can remain in the liquid state without crystallisation at temperatures much lower than the homogeneous nucleation temperature ($\approx$ 235 K under $P = 1$ bar) (Bertrand *et al.*, 2013a; Liu *et al.*, 2013).

The measured energy E spectrum at a fixed value of Q (the magnitude of the momentum transfer between the incident and scattered neutron), $S_m(Q, E)$, was analysed using the following equation:

$$S_m(Q, E) = I_{bg}(Q, E) + I_E(Q)R(Q, E) + R(Q, E)$$
$$\otimes \left[(I_Q(Q)S_Q(Q, E) + I_I(Q)S_I(Q, E)) \cdot D(E) \right] \qquad (9.13)$$

where $I_{bg}(Q, E)$ is the background signal from the MCM-41S matrix, $R(Q, E)$ is the energy resolution function, $I_E(Q)R(Q, E)$ represents the elastic scattering component, $S_Q(Q, E)$ is the quasi-elastic incoherent dynamic structure factor, $S_I(Q, E)$ is the inelastic incoherent dynamic structure factor, and $D(E)$ is the detailed balance factor (Squires, 1978). I_Q and I_I are the intensities of the quasi-elastic and inelastic components respectively. $S_Q(Q, E)$ represents the diffusive, non-vibrational motions of water molecules and can be modelled using the relaxing cage model (RCM) (Chen *et al.*, 1999). In this study, the parameters of the RCM are known from previous quasi-elastic neutron scattering measurements (Liu *et al.*, 2005a, 2006). $S_I(Q, E)$ describes the vibrational behaviour of water, i.e., the boson peak.

Here the damped harmonic oscillator (DHO) (Fak and Dorner, 1997) model to represent $S_I(Q, E)$ has been used to model various kinds of vibrations, from the self-motion of a single classical damped oscillator to collective excitations in different amorphous materials (Fak and Dorner, 1997; Sette *et al.*, 1998; Ruocco and Sette, 1999). Using the DHO model both the frequency and the width of the boson peak can be extracted. Detailed descriptions of the sample, instrument, models and data analysis can be found in supporting information. In the following paragraphs, we look at the results of Wang *et al.* (2014d) and consider their implications.

The left set of panels in Figure 9.10 shows inelastic neutron scattering spectra $S_m(Q, E)$ and corresponding fitted curves, under different temperatures and pressures. The right set of panels shows the theoretical incoherent dynamic structure factors extracted from the fit. From Figure 9.10 one can find that the temperatures at which the boson peak emerges (denoted as T_B) are different under different pressures. The analysis shows that under ≈ 1600 bar, the emergence of the boson peak is related to the fragile-to-strong crossover (FSC). This point can be seen in Figure 9.11: the solid curve, which represents the behaviour of $T_B(P)$, is nearly parallel to the blue dashed curve that represents the profile of $T_X(P)$ (T_X is the FSC temperature). This relation is mainly due to the drastic change of the quasi-elastic component in the spectrum induced by the FSC at T_X. Notice that, in the liquid–liquid critical point (LLCP) picture of water, the FSC coincides with the Widom line. Thus the authors can conclude that below 1600 bar, the emergence of the boson peak is correlated to the Widom line.

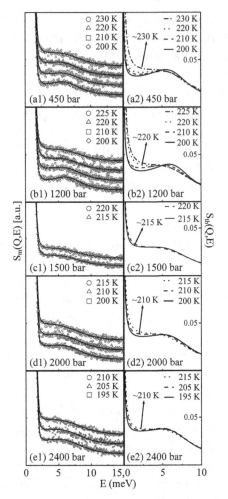

Figure 9.10 Inelastic neutron scattering spectra of the confined water system at $Q = 2\,\text{Å}^{-1}$ at different T and P. For each row, the left panel shows the measured spectrum $S_m(Q, E)$ and corresponding fitted curves under a specific pressure; for clarity the data have been shifted vertically by a fixed interval between adjacent temperatures. The right panel shows the corresponding theoretical curves $S_{th}(Q, E)$ extracted from the fit. The curve whose temperature is closest to T_B is indicated for each pressure in the right set of panels (measured at DCS) (Wang *et al.*, 2014d). Copyright 2014 by The American Physical Society.

From Figure 9.11 it is evident that above ≈ 1600 bar, the slope of the $T_B(P)$ profile is different from the slope below 1600 bar. This deviation, though not a definitive proof, is consistent with the previous result that the Widom line ends between 1500 bar and 2000 bar. Since if the Widom line continues to develop at pressures higher than 2000 bar, then due to the relation between $T_B(P)$ and the Widom line, the profile of $T_B(P)$ will smoothly develop at higher pressures in a fashion similar to that of the Widom line.

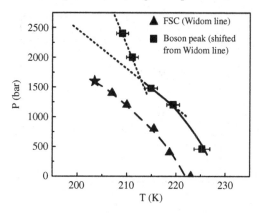

Figure 9.11 $T_B(P)$ in the (P, T) plane. The squares represent interpolated estimates of $T_B(P)$. The triangles demarcate the FSC transition. The star denotes the
estimated location of the LLCP, obtained from the FSC measurements (Liu *et al.*,
2005a, 2006). The two dotted lines denote the extrapolations of the two points at
$P = 2000$ bar and 2400 bar and of the two points at $P = 1200$ bar and 1500 bar
(Wang *et al.*, 2014d). Copyright 2014 by The American Physical Society.

From Figure 9.11 one can estimate the end-pressure of the Widom line by finding
the intersection point of the extrapolations of the two points at $P = 2000$ bar
and 2400 bar and of the two points at $P = 1200$ bar and 1500 bar. As shown
in Figure 9.11, the critical pressure estimated in this way is 1592 bar, which is
consistent with the result of 1600 ± 400 bar obtained by detecting the end-pressure
of the FSC (Liu *et al.*, 2005a).

The quantitative analysis suggests that the frequency of boson peak and the density of the deeply cooled water are positively correlated. The width of the boson
peak also exhibits such a behaviour. This result is of particular importance to water.
In the LLCP hypothesis, the order parameter to distinguish the LDL phase and the
HDL phase is just the density. With this in mind, one may distinguish the hypothetical LDL and HDL phases in deeply cooled water by looking at the shape of the
boson peak. In other words, the frequency or the width of the boson peak may
exhibit abrupt change as the water transforms between LDL and HDL, due to
significant differences in density and local structure of the different sides of the
hypothetical first-order transition line between LDL and HDL.

Recently, Wang *et al.* (2014a) found the evidence of the existence of distinct
high-density region and a low-density region in the pressure–temperature plane
of the supercooled confined D_2O with neutron scattering. The average density of
the confined D_2O in the high-density region is $\approx 16\%$ higher than that in the low-
density region. These two regions, together with their dividing line, could represent
the pure HDL and LDL phases separated by the liquid–liquid phase transition
(LLPT) line. The proposed phase diagram found in Wang *et al.* (2014d) is shown
in Figure 9.12.

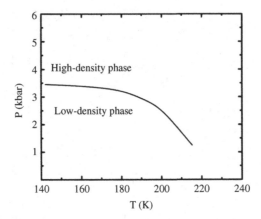

Figure 9.12 Proposed phase diagram of the liquid–liquid phase transition line of the deeply cooled confined D_2O based on its density measurement (taken from Figure 5 of Wang *et al.*, 2014a).

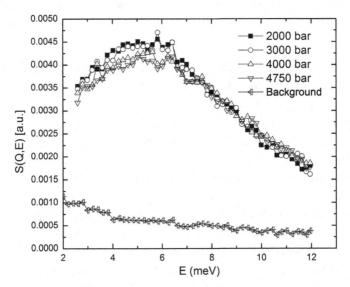

Figure 9.13 The measured spectra of the confined H_2O at $T = 165$ K and under four pressures: 2000 bar, 3000 bar, 4000 bar and 4750 bar using CNCS at SNS. The left-oriented triangles named as *background* in the Figure legend denote the spectrum of the dry MCM-41 powder (unpublished results by Wang *et al.*, 2014e).

Based on the above discussions, a series of measurements on the boson peak of the confined H_2O using CNCS was performed. The preliminary result is shown in Figure 9.13(a). In that figure, the spectra were measured at $T = 165$ K and under four different pressures: 2000 bar, 3000 bar, 4000 bar and 4750 bar. According to Figure 9.12, the former two are in the LDL phase and the latter two are in the HDL phase. From Figure 9.13 one can find that the difference between the 3000 bar and

4000 bar cases is most significant. From 3000 bar to 4000 bar, the boson peak shifts to higher frequency, whilst, the width of the boson peak also increases.

9.3.2 Young–Laplace effect

The transition pressure of the LDA and HDA for bulk D_2O, obtained by Mishima (2000), is around 2000 bar. This value is significantly lower than the transition pressure of the LDL and HDL for confined D_2O, which is between 3150 bar and 3500 bar, found in this work.

In addition, the pressure of the second critical point for bulk water conjectured by Mishima is smaller than 500 bar (Mishima, 2000, 2005, 2007), which is also largely lower than that of the confined D_2O estimated by the density hysteresis measurement that is around 1500 bar. For both cases, the pressure differences are in the order of 1000 bar. The reason might be attributed to the capillarity effect due to the confinement in pores of cylindrical geometry. For the fluid confined in a hydrophilic tube, the liquid–vapour surface forms a meniscus, which will lead to a pressure difference across this surface. An accurate calculation of this pressure difference is not trivial. Here Wang *et al.* (2014d) use the Young–Laplace equation to give a rough estimation of the pressure difference for their case: when the tube is sufficiently narrow, the pressure difference may be written as

$$\Delta P = 2\gamma \cos\theta/a, \qquad \Delta P = P_{outside} - P_{inside}, \qquad (9.14)$$

where γ is the surface tension of water in the silica capillary tube of radius a and θ is the contact angle between the two as defined in Figure 9.14.

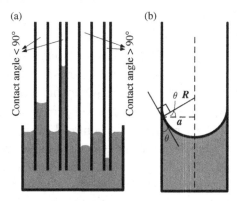

Figure 9.14 (a) Illustration of capillary rise. Left two tubes have contact angle less than 90°, right two tubes have contact angle greater than 90°. (b) Spherical meniscus with wetting angle less than 90°, which is the case for MCM-41S. The figure is modified from public domain, Wikipedia, http://en.wikipedia.org/wiki/Young-Laplace_equation.

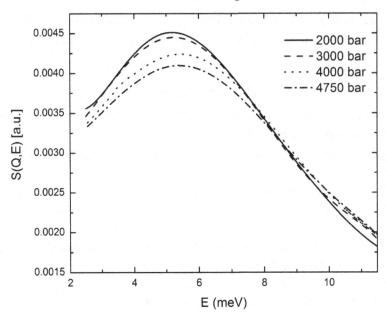

Figure 9.15 The fitted curves of the measured spectra taken at CNCS at SNS (unpublished results by Wang *et al.*, 2014e).

In the case of MCM-41S, $a = 10$ Å, $\theta = 20°$ and the surface tension of water is estimated at $T = 200$ K to be approximately $\gamma \approx 86$ dyne cm^{-2} (Hrubý *et al.*, 2014), then the pressure difference between outside pressure (the applied gas pressure) and inside pressure (the actual pressure inside the water column) is calculated to be $\Delta P = 1620$ bar. So, the Mishima value of 2000 bar in his phase diagram (Mishima, 2007) would appear in our phase diagram (see Figure 9.12 for the phase diagram of D_2O) as: 2000 bar + 1620 bar = 3620 bar, which is approximately correct.

In summary, on the basis of previous works on the density anomalies of the confined water (Liu *et al.*, 2007, 2008b, 2013; Mallamace *et al.*, 2007a,b; Zhang *et al.*, 2011); Wang *et al.* (2014a) performed a series of elastic neutron diffraction measurements on the average density of D_2O confined in the nanoporous silica matrix MCM-41S at high pressures and low temperatures. The result shows that below ≈ 210 K, the (P, T) plane of the confined heavy water can be divided into two regions (see Figure 9.12 for the associated phase diagram). The average density of the confined D_2O in the higher-pressure region is about 16% larger than that in the lower-pressure region. These two regions could represent the so-called *low-density liquid* and *high-density liquid* phases separated by the LLPT line. The incoherent inelastic neutron scattering measurements of boson peaks can indeed detect a sudden jump in peak height and peak width upon crossing the LLPT line as shown in Figure 9.15.

10

Introduction to high-resolution inelastic X-ray scattering spectroscopy

10.1 Comparison of inelastic neutron scattering and inelastic X-ray scattering techniques

10.1.1 Complementarity of inelastic neutron scattering and inelastic X-ray scattering spectroscopies

The study of atomic and molecular density fluctuations in condensed systems on the length scale of inter-particle separation is, traditionally, the domain of inelastic neutron scattering spectroscopy. The principal reason for which neutrons are particularly suitable for these studies is the very good matching between the kinematic phase space covered by thermal and cold neutron scatterings and that of phonon-like collective excitations in the condensed matter systems and bio-molecular assemblies. In fact, the energy of neutrons with wavelengths of the order of inter-particle distances of say 3 Å in liquids and solids is about 9 meV. This energy is comparable to the energies of typical phonons in condensed matter with wavelengths in the nanometre range. As a consequence, one can determine the peak position in the measured dynamic structure factor $S(Q, E)$ without requiring an excessive energy resolution of the inelastic neutron spectrometers used. Thus one is able to utilise the neutron flux from a typical thermal and cold neutron source rather effectively.

In principle, X-rays can also be used to measure the dynamic structure factor $S(Q, E)$ of the systems studied. The inelastic X-ray scattering cross-section, under the prevailing circumstances, has an expression very similar to that for neutrons, and the coupling of X-rays and neutrons to the density fluctuations in the material system is of the same order of magnitude. The X-ray scattering cross-section is derived by considering the interaction between the atomic electrons and the electromagnetic field associated with the X-rays in the Hamiltonian. In the non-relativistic limit, the interaction Hamiltonian is composed of two terms. They are the Thomson term and the photoelectric absorption term, which is the coupling of the photon field to the electron current. In the X-ray scattering case, only the charge scattering arising from the Thomson interaction term needs to be considered. This is valid for an X-ray energy sufficiently far away from the photo-absorption edges of the core electrons for the atomic system investigated. The Thomson interaction Hamiltonian, H_{Th} is given by

$$H_{Th} = \frac{1}{2} r_0 \sum_j A^2(r_j, t), \qquad (10.1)$$

where $r_0 = e^2/mc^2$ is the classical electron radius, and $A(r_j, t)$ is the vector potential of the electromagnetic field at r_j, the coordinate of the j-th electron in the atomic system. The sum extends over all electrons in the system. The double-differential cross-section is defined as a quantity proportional to the number of incident probe photons that are scattered with an energy and momentum transfers within an energy range ΔE and a solid angle $\Delta \Omega$. Considering an event where a photon of energy E_i, wave-vector $\vec{k}_i$, and polarisation $\hat{\varepsilon}_i$ is scattered into a final state of energy E_f, wave-vector $\vec{k}_f$, and polarisation $\hat{\varepsilon}_f$, while the electron system goes from the initial state $|I$ to the final state $|F\rangle$. The double-differential scattering cross-section is then given as

$$\frac{d^2\sigma(E, \Omega)}{d\Omega dE} = r_0^2 (\hat{\varepsilon}_i \cdot \hat{\varepsilon}_f)^2 \frac{k_f}{k_i} \sum_{I,F} p_I |\langle I| \sum_j e^{i Q \cdot r_j} |F\rangle|^2 \delta(E - E_i + E_f), \qquad (10.2)$$

where $\vec{Q} = \vec{k}_i - \vec{k}_f$ is the wave-vector transfer in the scattering process. The sum over the initial and final states is the statistical average, and p_I represents the population of the initial state. From this expression, which implicitly contains the dynamic structure factor of the electron density, one obtains the correlation function of the atomic density on the basis of the following considerations:

(i) We assume the validity of the adiabatic approximation. This allows us to separate a quantum state $|S\rangle$ of the system into the product of an electronic part, $|S_e\rangle$, which depends only parametrically on the nuclear coordinates, and a nuclear part $|S_N\rangle$, namely $|S\rangle = |S_e\rangle |S_N\rangle$. This approximation is particularly good for excitation energies that are small compared with the binding energies of electrons in bound core states. Considering the energy of typical phonon excitations, this is indeed the case for basically any atom.

(ii) We limit ourselves to considering the case in which the electronic part of the total wave function is not changed in the scattering process, namely, the case of Rayleigh scattering process. Then the difference between the initial state $|I\rangle = |I_e\rangle |I_N\rangle$ and the final state $|F\rangle = |I_e\rangle |F_N\rangle$ is due only to excitations associated with atomic nuclear density fluctuations.

Using these two plausible hypotheses one obtains

$$\frac{d^2\sigma(E,\Omega)}{d\Omega dE} = r_0^2(\hat{\varepsilon}_i \cdot \hat{\varepsilon}_f)^2 \frac{k_f}{k_i} \sum_{I_N,F_N} p_{I_N} |\langle I_N| \sum_n f_n(Q)e^{iQ\cdot R_n} |F_N\rangle|^2 \delta(E-E_i+E_f),$$

(10.3)

where $f_n(Q)$ is the atomic form factor of the atom n, which is the spatial Fourier transform of the atomic electronic charge density distribution. Assuming that all of the scattering units in the system are equal, this expression can be further simplified by factoring out the form factor of these scattering units. Then the double-differential cross-section for IXS from atomic density fluctuations reduces to the following simple expression:

$$\frac{d^2\sigma(E,\Omega)}{d\Omega dE} = r_0^2(\hat{\varepsilon}_i \cdot \hat{\varepsilon}_f)^2 \frac{k_f}{k_i}^2 |f(Q)|^2 S(Q,E).$$

(10.4)

In the limit $Q \to 0$, the form factor is equal to the number of electrons localised in the scattering atom, Z. On increasing the value of Q, the form factor decays rapidly with a decay constant for each electron, which is of the order of the inverse of its electronic shell dimension. At Q-values large with respect to these dimensions, therefore, the inelastic X-ray scattering from density fluctuations is strongly reduced. The cross-section derived so far is valid for a system composed of a single atomic species. Equation (10.4), however, can easily be generalised to molecular or crystalline systems by substituting for the atomic form factor with either the molecular form factor or the elementary cell form factor, respectively. The situation becomes more involved if the system is multicomponent and disordered. In this case the factorisation of the form factor is still possible, but only if one assumes some correlated distribution among the different atoms. In the limit case where the distribution is completely random, an incoherent contribution appears in the scattering cross-section, exactly as in the case of neutron scattering (Egelstaff, 1994). We remark here that the strength of the coupling of the X-rays to the atomic electrons in the derived cross-section is given by the square of the classical electron radius, $r_0 = 2.82 \times 10^{-13}$ cm multiplied by the absolute square of the atomic form factor. This means that comparing with the equivalent case of neutron scattering cross-section, the neutron scattering length of the scattering nucleus b is replaced by $r_0 f(Q)$ of the atomic electrons (Egelstaff, 1994). In spite of the strong analogies between inelastic neutron and X-ray scattering, the development of the X-ray method has so far been limited, mainly for the following reasons:

Photons with wavelength $\lambda = 0.1$ nm have energies of about 10 keV. Therefore, the study of phonon excitations in the meV region requires a relatively higher energy resolution of the order of $\Delta E/E \sim 10^{-8}$.

The total absorption cross-section of X-rays of energy 10 keV is limited in almost all cases ($Z > 4$) by the photoelectric absorption process and not by the Thomson scattering process. The photoelectric absorption, whose cross-section is roughly proportional to Z^4, determines, therefore, the actual sample thickness in the direction of the scattering path. Consequently, the Thomson scattering channel is not very efficient for a system with high Z in spite of the Z^2-dependence of its cross-section (Eq. (10.4)).

The rapid decrease (approximately exponential) of the atomic (molecular) form factor with increasing Q is responsible for a drastic reduction of the scattering cross-section, i.e. of the measured intensity, even at relatively small momentum-transfer values.

Despite these important limitations, however, there are situations where the use of X-rays has important advantages over that of neutrons. One specific case is based on the general consideration that it is not possible to study acoustic excitations propagating with a speed of sound, v_s, using a probe particle with a speed c smaller than v_s. This limitation is not particularly relevant in neutron spectroscopy in studies of crystalline samples. Here, the translational invariance of the crystalline lattice allows study of the acoustic excitations in high-order Brillouin zones, and this overcomes the difficulty of the above-mentioned kinematic limitation for measuring phonon branches with steep dispersions. In contrast, the situation is very different for topologically disordered systems: here, with only a few exceptions, and those for a limited Q, E region, it has not been possible to determine the dynamic structure factor using neutrons. For these systems, the absence of periodicity imposes the restriction that the acoustic excitations must be measured at small momentum transfers. In the case of water, this has prevented the measurement, with good energy resolution, of $S(Q, E)$ over sufficiently extended energy and momentum regions. This applies also to many other interesting liquids and glasses, where the speed of sound is too large for existing neutron spectrometers, and the interest is in the study of the collective dynamics in the small Q-region.

The above arguments explain why, in the study of disordered systems, the inelastic X-ray scattering technique can be extremely valuable. The X-ray probe, in fact, does not have the kinematic limitations of neutrons, and can access the region of small Q, providing that the required energy resolution is experimentally achievable. In fact, because the energy transfers for the X-ray case are small compared to the incident and scattered photon energies – $E_i \sim E_f$ and $|k_i| \approx |k_f|$ – a given scattering angle, θ_s, completely determines the magnitude of the momentum transfer, Q, independently of the energy transfer, E:

$$2k_i \sin\left(\frac{\theta_s}{2}\right) = \frac{2E_i}{\hbar c} \sin\left(\frac{\theta_s}{2}\right) = 1.013 E_i (\text{keV}) \sin\left(\frac{\theta_s}{2}\right). \tag{10.5}$$

From this relation one sees that, for phonon-like excitations, in inelastic X-ray scattering there is no limitation on the energy transfer at a given momentum transfer.

The other important advantages of the IXS method are:

(i) The cross-section is highly coherent, contrary to the case for neutrons, where often times it is necessary to separate the coherent, $S(Q, E)$, and incoherent, $S_s(Q, E)$, contributions from the measurements.

(ii) The multiple-scattering processes are in general strongly suppressed by the photoelectric absorption process, and, taking advantage of the small beam sizes obtained with X-rays, this allows the direct measurement of the dynamic structure factor without invoking sophisticated procedures for the reduction of the raw data.

(iii) The possibility of having very small beam sizes at the sample position allows the study of systems available in small quantities and/or their investigation in extreme thermodynamic conditions, such as under very high pressures and at high or low temperatures.

The above discussion illustrates how the inelastic X-ray scattering technique can be very useful, and complementary to the INS technique, although it can by no means be viewed as an alternative to the powerful INS spectrometry. In particular, it shows that the development of the X-ray method would give access to an extremely important region of the $E - Q$ space, and, specifically, to that of small Q values, where the acoustic excitations have energies which are not easy to access using INS. An important effort in this direction has been made at the European Synchrotron Radiation Facility (ESRF) in Grenoble, for example. There, an inelastic X-ray scattering beamline (BL21-ID16) has been constructed, and its performance will be briefly discussed in the following section.

Before we proceed to describe the basic principle of the high-resolution IXS built in ESRF, we should note here that there exists another method of achieving even higher resolution that is likely to be adopted by all the future synchrotron sources. Yuri Shvyd'ko and coworkers have recently demonstrated that a combined effect of angular dispersion (AD) and anomalous transmission (AT) of X-rays in Bragg reflection from asymmetrically cut crystals can shape spectral distributions of X-rays to profiles with high contrast and small bandwidths (Shvyd'ko *et al.*, 2011; Shvyd'ko, 2013). A prototype spectrometer, named *UHRIX*, based on this alternative principle has been build by Shvyd'ko and coworkers at Advance Photon Source (APS) in Argonne National Laboratory. The spectrometer features a spectral resolution function having Gaussian-like steep tails over two orders of magnitude in intensity, with a sub-meV (0.620 meV) bandwidth. The new capabilities are demonstrated by studies of dynamics in glass-forming liquid *glycerol*, with a momentum transfer resolution of $0.25\,\mathrm{nm}^{-1}$. The moderate working photon energy (9.13 keV) makes this spectrometer practical for all modern X-ray synchrotron and XFEL facilities.

10.1.2 Energy resolution of a perfect crystal monochromator in near backscattering geometry[1]

An X-ray beam coming from a white source such as a synchrotron can be utilised to construct a triple-axis spectrometer of an extremely high-energy resolution using Bragg reflection from perfect crystals[2] as the monochromator and the analyser in the following way.

The relation between energy E and wavenumber k for X-rays, $E = \hbar c k$, leads to a simple expression for the energy resolution

$$\frac{\delta E}{E} = \frac{\delta k}{k}. \tag{10.6}$$

If a beam of X-rays with the wavenumber k is scattered from a crystal with a Bragg angle θ_B, Bragg's law can be written as

$$2k \sin \theta_B = \tau \quad \text{or} \quad k = \frac{\tau}{2 \sin \theta_B} = \frac{\tau}{2 \cos \varepsilon}, \tag{10.7}$$

where τ is the magnitude of the reciprocal lattice vector $\vec{\tau}$ of the crystal and the angle $\varepsilon = \frac{\pi}{2} - \theta_B$ denotes the small deviation from an exact backscattering geometry. The parameters τ and ϵ for the backscattering monochromator have sources of broadening $\delta \tau$ and $\delta \epsilon$, each of which gives the corresponding width in δk associated with them. $\delta \tau$ is given by the extension of the reciprocal lattice point in the direction of the reciprocal lattice vector τ and is an intrinsic width of the Bragg peak of the perfect crystal used. $\delta \epsilon$ arises from angular divergence of the incident X-ray beam. These two broadening mechanisms produce the corresponding width of δk and, consequently, the width in the energy. The energy resolution of the monochromator given by Eq. (10.6) can thus be decomposed into contributions from two separate components:

$$\frac{\delta E}{E} = \sqrt{\left(\frac{\delta k}{k}\right)_\tau^2 + \left(\frac{\delta k}{k}\right)_\varepsilon^2}, \tag{10.8}$$

coming from the inherent widths in the τ and ϵ.

---

[1] Burkel (1999, 2000); Sinn (2001).

[2] A perfect crystal can be defined as a periodic lattice without defects and/or distortions in the reflecting volume capable of inducing relative variations of the distance between the diffraction planes, $\Delta d / d$, larger than the desired.

Crystal contribution

The contribution $(\delta k/k)_\tau$ represents the intrinsic quality of the crystal and is given by

$$\left(\frac{\delta k}{k}\right)_\tau = \frac{\delta \tau}{\tau},$$
(10.9)

derivable from Eq. (10.7). This contribution can be minimised by using a perfect crystal such as silicon. The exact description of scattering from a perfect crystal requires the dynamical theory of scattering that gives the following relation (Warren, 1969):

$$\delta \tau = 16\pi r_0 \frac{|F(Q)|}{\tau V},$$
(10.10)

where $F(Q)$ is the structure factor of the unit cell evaluated at $Q = \tau$ and V is its volume. $\delta \tau$, given in Eq. (10.10) is related to the so-called Darwin width, which is the angular range of total reflection phenomenon exhibited by the Darwin reflectivity curve. This width is characteristic quantity for the given material and the given reflection of the monochromator. $\delta \tau$ expressed by Eq. (10.10) is derived for a piece of infinitely thick crystal in the symmetric reflection position and is related to the range of total reflection for the given Bragg peak.

However, a real crystal has a finite thickness. To take this into account, a parameter Γ can be introduced:

$$\Gamma = \frac{\pi}{2d} \frac{V}{|F(Q)| r_0} = \frac{\tau}{4} \frac{V}{|F(Q)| r_0} = \frac{4\pi}{\delta \tau} = 2\pi t_E.$$
(10.11)

As the thickness of the crystal approaches Γ, a reflectivity of 1 is obtained at the centre of $\delta \tau$, however, since the Darwin reflectivity curve has strong side bands, Γ can be regarded as a lower limit for the necessary thickness of a perfect crystal used. For example, for the (11 11 11) reflection of silicon, $\Gamma = 1584$ μm. The parameter Γ is related to the extinction length t_E as given in Eq. (10.11). This extinction length is defined as the penetration depth, where the X-ray intensity is reduced to $1/e$.

From Eqs. (10.8), (10.9), (10.10) we can express the intrinsic energy resolution δE of a perfect crystal monochromator as

$$\delta E = \frac{\delta \tau}{\tau} E_i = \frac{16\pi r_0}{V} |F(Q)| \frac{E_i}{\tau^2} = \frac{4\pi r_0}{V} |F(Q)| \frac{\hbar^2 c^2}{E_i}.$$
(10.12)

Equation (10.12) shows that for X-rays the intrinsic energy resolution of a perfect crystal monochromator is inversely proportional to E_i. Therefore, the resolution will be higher if we use higher-order reflections as can be seen in Table 10.1.

Table 10.1 *Intrinsic absolute energy resolution of Si perfect crystal for different (hhh) reflections in backscattering geometry. The values for the photon wavelength λ_i, the energy E_i and the reciprocal lattice vector τ are given as well. The parameter Γ was calculated according to Eq. (10.11) and the absorption length $1/\mu$ is taken from Hildebrandt (1979).*

Reflection	(333)	(444)	(555)	(777)	(888)	(999)	(11 11 11)	(13 13 13)
λ_i [Å]	2.090	1.568	1.254	0.896	0.784	0.697	0.570	0.482
E_i [keV]	5.93	7.91	9.89	13.84	15.82	17.79	21.75	25.70
τ [Å]−1	6.01	8.02	10.02	14.03	16.03	18.04	22.04	26.05
$\delta\tau/\tau[10^{-6}]$	9.0	5.1	1.5	0.36	0.28	0.11	0.036	0.019
$(\delta E)_\tau$ [meV]	53.4	40.3	14.8	5.0	4.4	2.0	0.85	0.37
Γ[μm]	23	31	84	249	280	633	1584	2539
$1/\mu$[μm]	29	65	126	351	516	724	1311	1990

For comparison, for the case of a perfect crystal neutron monochromator, the corresponding expression is given by

$$\delta E = \frac{4\pi r_0}{V} |F(Q)| \frac{\hbar^2}{m_n}. \tag{10.13}$$

According to Eq. (10.13), for neutron, δE is independent of the incident energy, so by going to higher-order reflections the energy resolution of a monochromator does not improve.

Values of the energy widths for some reflections of the types (hhh) in the silicon crystal are shown in Table 10.1. The resolution is better for odd reflections than for even reflections, because $F(Q)$ is smaller due to $F(hhh) = e^{-W} f(Q)\sqrt{2}$ for odd h and $F(hhh) = e^{-W} f(Q) \cdot 2$ for even h; e^{-W} is the Debye–Waller factor coming from thermal vibrations of silicon atoms and $f(Q)$ is the atomic form factor of silicon.

Contribution due to angular divergence of the incident beam

From Bragg's law, Eq. (10.7), the second contribution in Eq. (10.8) is given by

$$\left(\frac{\delta k}{k}\right)_\varepsilon = \tan(\epsilon\,\delta\epsilon), \tag{10.14}$$

so by using the nearly backscattering geometry ϵ can be made as small as possible. For example, if one takes $\theta_B \approx 89.98°$, $\tan(\epsilon) \approx 10^{-4}$. In practice, one has $\Delta\epsilon \approx 10^{-5}$ rad for a typical synchrotron radiation from an undulator source. As a result, this term can be made negligibly small compared to the first term in Eq. (10.8). We shall not discuss the contribution, if any, from this term here.

10.1.3 High-energy-resolution backscattering analyser[3]

The requirements on the energy resolution of the monochromator and of the analyser are the same. However, the required angular acceptances are very different. The X-ray beam incident on the monochromator has the angular divergence of the X-ray source, and therefore one can use a perfect flat crystal. In the case of the analyser crystal, however, the optimal angular acceptance is dictated by the desired momentum resolution. Considering values of ΔQ in the region of $0.5\,\mathrm{nm}^{-1}$, well within the range for momentum transfer (1 to $10\,\mathrm{nm}^{-1}$), the corresponding angular acceptance of the analyser crystal must be $\approx 10\,\mathrm{mrad}$ or higher, which is again an angular range well above acceptable values, i.e. also larger than the deviation of the Bragg angle from 90°. The only way to obtain such large angular acceptances is to use a focusing system, which, however, has to preserve the crystal perfection, necessary to obtain the energy resolution. A solution consists in placing a large number of undistorted perfect flat crystals on a spherical surface, with the aim of using a 1:1 pseudo-Rowland-circle geometry with aberrations kept sufficiently low that the desired energy resolution is not degraded. This method has been utilised in the construction of the ESRF spectrometer on BL21-ID16: that is, approximately 10 000 perfect silicon crystals of surface size $0.7 \times 0.7\,\mathrm{mm}^2$ and thickness 3 mm have been glued on a spherical substrate of radius 6500 mm. This *perfect silicon crystal with a spherical shape* is the meV energy resolution analyser of the BL21-ID16 beamline.

Figure 10.1 shows the main optical elements of the ESRF IXS beamline. The instrument, a triple-axis spectrometer, has as its first element the monochromator crystal, whose role is to determine the energy, E_i, of the incident photons. The second element is the scattering sample, where one selects the scattering angle θ_s, and therefore the magnitude of wave vector transfer, according to Eq. (10.5).

The third element is the analyser crystal, whose role is the determination of the energy E_f of the scattered photons. To maintain the backscattering geometry for any given energy transfer, a certain energy difference between the analyser and monochromator is achieved by keeping the Bragg angle constant, and by changing the relative temperature of the two crystals. This has the effect of varying the relative lattice parameter, and therefore the values of the reflected energies. Specifically, the analyser is kept at constant temperature while the monochromator temperature, and therefore E_i, is varied. Considering that $\Delta d/d = \alpha \Delta T$, with $\alpha = 2.56 \times 10^{-6}\,\mathrm{K}^{-1}$ in silicon at room temperature, in order to obtain an energy step of about one tenth of the energy resolution, i.e. $\Delta E/E \approx 10^{-9}$, it is necessary to control the monochromator crystal temperature with a precision of about 0.5 mK. This difficult task has been achieved with a carefully designed temperature bath, controlled with an active feedback system. The X-ray source

---

[3] Ruocco and Sette (1999); Masciovecchio *et al.* (1996a,b).

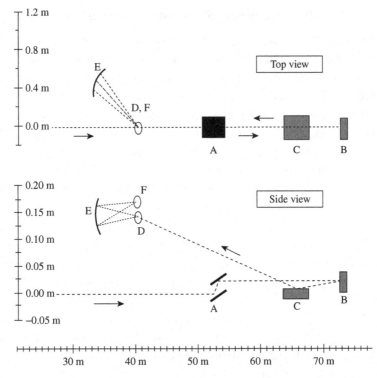

Figure 10.1 A schematic diagram of the layout of the inelastic X-ray scattering beamline ID16BL21 at ESRF. The different components and their functions are sketched in the figure: (A) pre-monochromator; (B) main monochromator; (C) toroidal mirror; (D) scattering centre (sample); (E) analyser crystal; (F) detector.

used on the ID16-BL21 beamline is made out of two undulators. The X-ray radiation utilised is that corresponding to the undulator emission of the third, fifth or seventh harmonics, chosen to optimise the photon flux at the energies defined by the reflection order of the monochromator and analyser crystals. These are the Si(hhh) reflections, with $h = 5, 7, 8, 9, 11, 12, 13$. The X-ray beam from the undulator odd harmonics has an angular divergence of approximately $15 \times 40\,\mu\text{rad}$ full width at half-maximum (FWHM), a spectral bandwidth $\Delta E/E \approx 10^{-2}$, and an integrated power within this divergence of the order of 200 W. This beam is first pre-monochromatised to $\Delta E/E \approx 2 \times 10^{-4}$ using a Si(111) double-crystal device kept in vacuum and at the cryogenic temperature of ≈ 120 K (element A in Figure 10.1. The photons from the pre-monochromator reach the high-energy-resolution backscattering monochromator (element B in Figure 10.1. This is a flat symmetrically cut silicon crystal oriented along the (111) direction, temperature controlled with a precision of 0.2 mK over the 285–295 K temperature region. The

Table 10.2 *Measured fluxes and bandwidths of the X-ray beam leaving the high-energy-resolution monochromator on ID16-B121 at the ESRF at the silicon reflection orders indicated, and with 200 mA in the storage ring.*

Reflection	Energy [keV]	Flux [photon s^{-1}]	Resolution [meV]
5 5 5	9.885	2×10^{11}	15.0 (14.8)
7 7 7	13.840	6×10^{10}	5.3 (5.0)
8 8 8	15.816	3×10^{10}	4.4 (4.4)
9 9 9	17.793	6×10^{9}	2.2 (2.0)
11 11 11	21.748	7×10^{8}	1.0 (0.8)
12 12 12	23.725	3×10^{8}	0.7
13 13 13	25.702	1×10^{8}	0.5

Bragg angle at the monochromator is $\theta_B = 89.98°$. The energy resolution of the X-ray beam leaving this monochromator depends on the reflection considered, and typical values are reported in Table 10.2. The monochromatic beam impinges on a focusing toroidal mirror (element C in Figure 10.1), which gives at the sample (element D in Figure 10.1) a beam size of 150 (vertical) $\times$ 350 (horizontal) μm^2 (full width at half maximum, FWHM). The analyser system (element E in Figure 10.1) is made up of an entrance pinhole, slits in front of the analyser crystal to set the desired momentum resolution, the analyser spherical crystal in backscattering geometry ($\theta_B = 89.98°$), an exit pinhole in front of the detector, and the detector itself (element F in Figure 10.1). There are, in fact, five independent analyser systems at fixed angular offsets in the scattering plane.

They are mounted on an arm 7 m long that can rotate around a vertical axis passing through the scattering sample. This rotation allows one to determine the scattering angle θ_s for each of the five analysers, and therefore the corresponding momentum transfer. The arm operates between 0° and 15°. The spherical analyser crystals are kept at constant temperature with a precision of 0.2 mK, and operate at the same reflection of the monochromator in the Rowland-circle geometry with 1:1 magnification. The detectors are inclined silicon diodes with an equivalent thickness of 2.5 mm. The performance of each of the five spectrometer channels corresponds to an energy resolution of 1.5 meV when one utilises the Si(11 11 11) reflection (Masciovecchio *et al.*, 1996b,a). At this order, the angular offset among the successive five analysers corresponds to a momentum transfer difference of 3 nm^{-1}. The instrumental response function of one of the five channels is reported in Figure 10.2. This has been obtained by measuring the scattering from a sample of amorphous Plexiglas at a Q-transfer of $Q = Q_m = 10$ nm^{-1}, and at $T = 20$ K, in order to maximise the elastic contribution to the scattering. The result reported

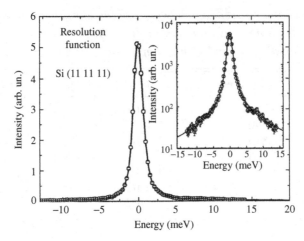

Figure 10.2 The resolution function of the whole instrument obtained using the monochromator and analyser Si(11 11 11) reflections in backscattering geometry and measuring the elastic scattering from a plastic sample. The analyser radius is 6.15 m. The energy scans are performed varying the relative temperature of the two crystals. The count rate is normalised to 100 mA current in the storage ring. Typical current values at the ESRF are between 100 and 200 mA. The data points are shown with statistical error bars (often the latter are smaller than the dot size). The solid curve is a Lorentzian fit to the data, and was used to determine the FWHM of the resolution function. The measured FWHM energy is 1.5 ± 0.2 meV, and was obtained with an analyser slit opening of $\approx 10 \times 10$ mrad^2. In the inset the same data are reported on a logarithmic scale to allow one to better appreciate the shape of the tail of the resolution function.

in Figure 10.2 summarises the best instrumental capability obtained so far from the ESRF inelastic X-ray scattering spectrometer. From this result, one can directly access the value of performing IXS experiments with meV energy resolution.

Example: analysis of inelastic X-ray scattering spectra of low-temperature water with TEE theory[4]

We shall conclude this introduction by giving an example of the IXS study of the high-frequency dynamics of liquid water measured in BL21-ID1 at ESRF.

Liao *et al.* (2000b) analyse a set of high-resolution inelastic X-ray scattering (IXS) spectra from H_2O measured at $T = 259, 273$ and 294 K using a phenomenological model. This model is a version of generalised hydrodynamic theory called the *three effective eigenmode (TEE) theory* by de Schepper *et al.* (1988) and Kamgar-Parsi *et al.* (1987). This model is appropriate for normal liquid water where the cage effect is less prominent and there is no evidence of the alpha relaxation from the MD data. They use the model to analyse IXS data at all three

---

[4] Liao *et al.* (2000b).

temperatures, from which they are able to extract the relaxation rate of the central mode and the damping of the sound mode as well as the dispersion relation for the high-frequency sound. It turns out that the dispersion relations extracted from this model for the respective temperatures give the high-frequency sound speed of 2900 ± 300 m s^{-1}. The k-dependent sound damping and central mode relaxation rate extracted from the model analyses are compared with the known values in the hydrodynamic limit.

The TEE model has already met with successes in describing the behaviour of dynamic structure factor at finite k values for a hard-sphere system (Kamgar-Parsi *et al.*, 1987; Alley and Alder, 1983; Cohen *et al.*, 1984), Lennard-Jones fluids (de Schepper *et al.*, 1988), classical fluids like Ar, Ne, Kr, ^{4}He at high temperature, and super fluid ^{4}He (Montfrooij *et al.*, 1996). For these cases it has been shown that the ISF of the density fluctuation can be well described by a sum of three exponential functions associated with three slow conserved hydrodynamic eigenmodes of the fluid, the so-called extended heat mode and two extended sound modes. Although this description is an extended hydrodynamic model, it has been shown that it provides a good approximation for the $S(k, \omega)$ in the wide-k range from 0 up to 15 $k\sigma_{LJ}$ (de Schepper *et al.*, 1988), where σ_{LJ} is the Lennard–Jones diameter. The TEE model can be derived from Zwanzig–Mori projection operator formalism (Zwanzig, 1961; Mori, 1965). In the TEE model, the dynamic structure factor is then given as

$$S(k, \omega) = \frac{S(k)}{\pi} \operatorname{Re} \left\{ \frac{\overset{\leftrightarrow}{I}}{i\omega \overset{\leftrightarrow}{I} + \overset{\leftrightarrow}{H}(k)} \right\}_{1,1}, \tag{10.15}$$

where $\overset{\leftrightarrow}{I}$ is the 3×3 identity matrix, and label 1,1 means the (1,1) element of the matrix. The matrix $\overset{\leftrightarrow}{H}(k)$ is given as

$$\overset{\leftrightarrow}{H}(k) = \begin{pmatrix} 0 & i f_{un}(k) & 0 \\ i f_{un}(k) & z_u(k) & i f_{uT}(k) \\ 0 & i f_{uT}(k) & z_T(k) \end{pmatrix} \tag{10.16}$$

where $f_{un}(k)$ is determined by the second moment of $S(k, \omega)$ to be $k v_0 / [S(k)]^{1/2}$. Three independent parameters: $z_u(k)$, $f_{uT}(k)$, $z_T(k)$ are all real numbers. For small k, Eq. (10.16) tends to the hydrodynamic matrix where the matrix elements have values given by de Schepper *et al.* (1988) and Cohen and de Schepper (1990):

$$f_{un}(k) = k c_s / \sqrt{\gamma}, \tag{10.17a}$$

$$z_u(k) = \phi k^2, \tag{10.17b}$$

$$z_T(k) = \gamma D_T k^2, \tag{10.17c}$$

$$f_{uT}(k) = k c_s \sqrt{(\gamma - 1)/\gamma}. \tag{10.17d}$$

Here, $c_s = v_0[\gamma/S(0)]^{1/2}$ is the adiabatic speed of sound; $\gamma = c_p/c_V$ is the ratio of the specific heat per unit mass at constant pressure and volume; $\phi = [(4/3)\eta + \zeta]/nm$ is the kinematic longitudinal viscosity, where η and ζ are the shear and bulk viscosity, respectively; $D_T = \lambda/nmc_p$ is the thermal diffusivity, where λ is the thermal conductivity. The three eigenvalues of the hydrodynamic matrix are therefore the three hydrodynamic modes. Here, only the three eigenvalues up to the order of $O(k_2)$ are given:

$$z_h(k) = D_T k^2 \quad \text{(heat mode)}, \tag{10.18a}$$

$$z_\pm(k) = \pm ic_s k + \Gamma_s k^2 \quad \text{(sound mode)}, \tag{10.18b}$$

where $\Gamma_s = (1/2)\phi + (1/2)(\gamma - 1)D_T$ is the sound damping.

For finite k, $z_u(k)$, $f_{uT}(k)$, $z_T(k)$ become arbitrary functions of k. However, in most cases, the eigenvalues of the matrix (Eq. (10.16)) consists of one real number z_h and a couple of conjugate complex numbers $\Gamma_s \pm i\omega_s$. Therefore the solution of Eq. 10.16 in general in the hydrodynamic like form (Bafile *et al.*, 1990; Boon and Yip, 1980) can be written as

$$S(k, \omega)/S(k)$$
$$= \frac{1}{\pi}\left\{A_0\frac{z_h}{\omega^2 + z_h^2} + A_s\frac{\Gamma_s + b(\omega + \omega_s)}{(\omega + \omega_s)^2 + \Gamma_s^2} + A_s\frac{\Gamma_s - b(\omega - \omega_s)}{(\omega - \omega_s)^2 + \Gamma_s^2}\right\}. \tag{10.19}$$

The corresponding time domain density correlation function is the intermediate scattering function $F(k, t)$ given by (Balucani and Zoppi, 1994):

$$F(k, t)/S(k)$$
$$= A_0\exp(-z_h t) + 2A_s\exp(-\Gamma_s t) \times [\cos(\omega_s t) + b\sin(\omega_s t)]. \tag{10.20}$$

Although Eq. (10.19) contains six parameters, they are all functions of the three independent parameters given in Eq. (10.16). Therefore, the normalised dynamic structure factor is the function of the three independent adjustable parameters for which the low-k limits are known exactly.

One can also cast the TEE model in the form of a continued fraction (Balucani and Zoppi, 1994),

$$S(k, z) = \left[z + \frac{f_{un}^2(k)}{z + z_u(k) + \frac{f_{uT}(k)}{z + z_T(k)}}\right]^{-1}. \tag{10.21}$$

From this, the second-order memory function of the correlation function of the density fluctuation is given as

$$K_L(k, z) = z_u(k) + \frac{f_{uT}(k)}{z + z_T(k)}, \tag{10.22}$$

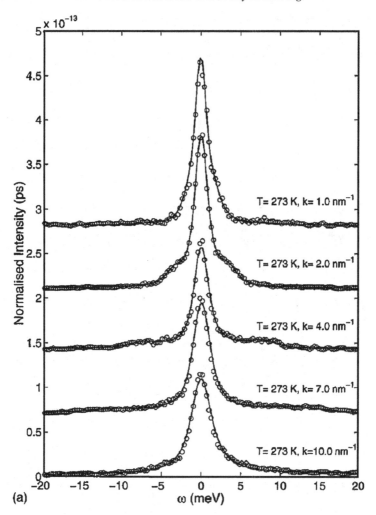

(a)

Figure 10.3 The IXS spectra of H_2O at $T = 273\,K$ taken at the indicated k values. The experimental data (open circle) are superimposed to the fit (solid line) by the TEE model as explained in the text.

with a Markovian viscosity term $z_u(k)$ and a thermal fluctuation term having the finite decay time $z_T(k)$ (Balucani and Zoppi, 1994).

Figure 10.3 shows typical Rayleigh–Brillouin spectra taken from the low temperature water (circles) and their TEE model fit (solid lines). Figure 10.4 gives the TEE model parameters as functions of the magnitude of wave vector transfer k extracted from fitting of the IXS spectra shown in Figure 10.3. Figure 10.5 shows the result of the acoustic phonon dispersion relation by plotting the Brillouin shift ω_s as a function of k obtained from the fits.

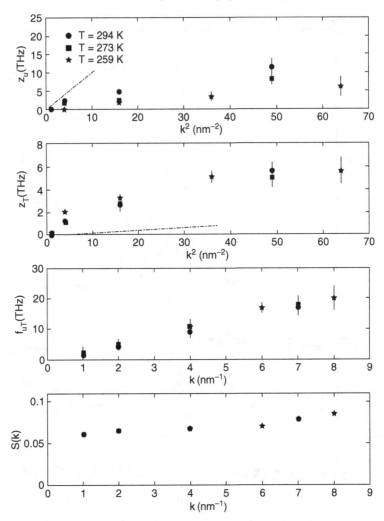

Figure 10.4 The extracted fitting parameters from IXS spectra by the three effective eigenmode model at indicated temperatures, plotted together with the input static structure factor. The dotted lines shown in the first two boxes are the expected hydrodynamic behaviours.

Table 10.3 gives the known values of thermodynamic and hydrodynamic constants.

10.2 Module – Theory of inelastic X-ray scattering from monoatomic liquids

10.2.1 Formal derivation of the dynamic structure factor of TEE theory

Here, we derive the three effective eigenmode (TEE) model based on a method described in Liao and Chen (2001); Chen *et al.* (2001) and Kamgar-Parsi *et al.* (1987). We consider the dynamic correlation functions $F_{ab}(k, z)$, defined as

Table 10.3 *Physical properties of H_2O at 1 atm, 273 K*

Density ρ (kg/m^3)	1.0
Specific heat ratio $\gamma = c_p/c_v$	≈ 1.0
Isothermal compressibility K_T (10^{-6} bar^{-1})	52.24
Adiabatic sound speed $c_s = (\gamma/\rho K_T)^{1/2}$ (m/s)	1380
Shear viscosity η (10 Kg m^{-1} s^{-1})	18.284
Ratio of bulk to shear viscosity ζ/η	1.90
Longitudinal viscosity $\phi = [(4/3)\eta + \zeta]/\rho$ (10^{-3} cm s^{-1})	59.1
Specific heat at constant pressure c_p (kJ kg^{-1} K^{-1})	4.22
Thermal conductivity λ (10^{-3} W m^{-1} K^{-1})	561.1
Thermal diffusivity $D_T = \lambda/\rho c_p$ (10^{-3} cm^2 s^{-1})	1.33

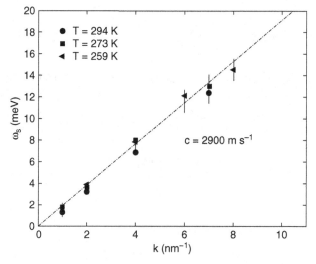

Figure 10.5 The dispersion curve of H_2O extracted from analysis of IXS spectra at the indicated temperatures. The solid symbols denote ω_s from the TEE model. The dash-dotted line represents a dispersion curve with a sound speed $c = 2900$ m s^{-1}.

$$F_{\alpha\beta}(k, z) = \left\langle b_\alpha^*(k) \frac{1}{z - L} b_\beta(k) \right\rangle \qquad (10.23a)$$

where the bracket $\langle\rangle$ indicates an equilibrium average over a canonical ensemble of N particles, the star denotes complex conjugation, L the Liouville operator of the system given by

$$L = \sum_{i=1}^N \left[\mathbf{v}_i \cdot \frac{\partial}{\partial \mathbf{r}_i} - \frac{1}{m} \sum_{i \neq j} \frac{\partial \varphi_{ij}}{\partial \mathbf{r}_{ij}} \cdot \frac{\partial}{\partial \mathbf{v}_i} \right] \qquad (10.23b)$$

and $b_a(k)$ are the microscopic fluctuations given by the first three conserved quantities defined below:

$$b_1(k) = \frac{1}{\sqrt{NS(k)}} \sum_{j=1}^{N} e^{-ik\cdot r_j}, \qquad (10.24a)$$

is the microscopic density fluctuation;

$$b_2(k) = \frac{1}{\sqrt{N}} \sum_{j=1}^{N} \frac{\mathbf{k}\cdot\mathbf{v}_j}{kv_0} e^{-ik\cdot r_j}, \qquad (10.24b)$$

is the microscopic longitudinal velocity fluctuation with $v_0 = (k_BT/m)^{1/2}$ the thermal velocity; and

$$b_3(k) = \frac{1}{\sqrt{N}} \sum_{j=1}^{N} \frac{3 - mv_j^2/(k_BT)}{\sqrt{6}} e^{-ik\cdot r_j}, \qquad (10.24c)$$

is the microscopic temperature fluctuation. One can also introduce the other microscopic dynamic quantities $b_a(k)$ with $a > 3$, so that they all together form an orthonormal and complete set including the above defined three hydrodynamic conserved quantities:

$$\langle b_\alpha^*(k)b_\beta(k) \rangle = \delta_{\alpha\beta}, \qquad (10.25a)$$

$$\sum_\alpha |b_\alpha(k)\rangle \langle b_\alpha(k)| = \mathbf{I}. \qquad (10.25b)$$

Equation (10.23a) can be rewritten as

$$F_{\alpha\beta}(k, z) = \langle b_\alpha(k)|b_\beta(k, z)\rangle \qquad (10.26a)$$

where

$$|b_\beta(k, z)\rangle = \frac{1}{z - L} |b_\beta(k)\rangle. \qquad (10.26b)$$

Rearranging Eq. (10.26b), we have

$$z|b_\beta(k, z)\rangle = L|b_\beta(k, z)\rangle + |b_\beta(k)\rangle. \qquad (10.27)$$

Then after multiplying both sides of Eq. (10.27) by $\langle b_a^*(k)|$ and using Eqs. (10.25b) and (10.26a) we get

$$zF_{\alpha\beta}(k, z) = \sum_\gamma L_{\alpha\gamma}^\infty(k)F_{\gamma\beta}(k, z) + \delta_{\alpha\beta}, \qquad (10.28a)$$

where the element of the infinite symmetric matrix $L_{\alpha\gamma}^\infty(k)$ is defined as

$$L_{\alpha\gamma}^\infty(k) = \langle b_\alpha(k)|L|b_\gamma(k)\rangle. \qquad (10.28b)$$

In order to contract the infinite matrix equation of Eq. (10.28a) to a 33 matrix equation, we can decompose an infinite matrix $\overset{\leftrightarrow}{A}$ into four matrix blocks as

$$\overset{\leftrightarrow}{A} = \lim_{M \to \infty} \begin{pmatrix} \overset{\leftrightarrow}{A}_{3\times3} & \overset{\leftrightarrow}{A}_{3\times M} \\ \overset{\leftrightarrow}{A}_{M\times3} & \overset{\leftrightarrow}{A}_{M\times M} \end{pmatrix}, \tag{10.29}$$

where $\overset{\leftrightarrow}{A}_{3\times3}$, $\overset{\leftrightarrow}{A}_{3\times M}$, $\overset{\leftrightarrow}{A}_{M\times3}$, $\overset{\leftrightarrow}{A}_{M\times M}$ represents the matrix of 3×3, $3\times M$, $M\times3$, $M\times M$, respectively. One can then write out the equation for the correlation function $\overset{\leftrightarrow}{F}_{3\times3}(k, z)$, $\overset{\leftrightarrow}{F}_{M\times3}(k, z)$, from Eq. (10.28a) as

$$z\overset{\leftrightarrow}{F}_{3\times3}(k, z)$$
$$= \overset{\leftrightarrow}{L}_{3\times3}(k, z)\overset{\leftrightarrow}{F}_{3\times3}(k, z) + \overset{\leftrightarrow}{L}_{3\times M}(k, z)\overset{\leftrightarrow}{F}_{M\times3}(k, z) + \overset{\leftrightarrow}{I}_{3\times3}(k, z) \tag{10.30a}$$
$$z\overset{\leftrightarrow}{F}_{M\times3}(k, z)$$
$$= \overset{\leftrightarrow}{L}_{M\times3}(k, z)\overset{\leftrightarrow}{F}_{3\times3}(k, z) + \overset{\leftrightarrow}{L}_{M\times M}(k, z)\overset{\leftrightarrow}{F}_{M\times3}(k, z). \tag{10.30b}$$

The correlation function $\overset{\leftrightarrow}{F}_{M\times3}(k, z)$ can be solved out from Eq. (10.30b), and inserted into Eq. (10.30a) to get

$$z\overset{\leftrightarrow}{F}_{3\times3}(k, z) = \overset{\leftrightarrow}{H}_{3\times3}(k, z)\overset{\leftrightarrow}{F}_{3\times3}(k, z) + \overset{\leftrightarrow}{I}_{3\times3}(k, z), \tag{10.31a}$$

with a z-dependent 3×3 matrix

$$\overset{\leftrightarrow}{H}_{3\times3}(k, z)$$
$$= \overset{\leftrightarrow}{L}_{3\times3}(k) + \overset{\leftrightarrow}{L}_{3\times M}(k) \left[z\overset{\leftrightarrow}{I}_{M\times M}(k) - \overset{\leftrightarrow}{L}_{M\times M}(k) \right]^{-1} \overset{\leftrightarrow}{L}_{M\times3}(k). \tag{10.31b}$$

From Eqs. (10.23b), (10.24a) and 10.24b, it is easy to see immediately that

$$Lb_1(k) = \frac{-ikv_0}{\sqrt{S(k)}}b_2(k). \tag{10.32}$$

One can then see that the elements of the first column and the first row of the matrix $\overset{\leftrightarrow}{L}^{\infty}(k)$ are all zero, except the two elements, L_{12}^{∞} and L_{21}^{∞}, which are purely imaginary (see Eq. (10.32)). Hence define a real number $f_{un}(k) = \frac{kv_0}{\sqrt{S(k)}}$, then we can write

$$L_{21}^{\infty}(k) = L_{12}^{\infty}(k) = -i f_{un}(k) = \frac{-ikv_0}{\sqrt{S(k)}}. \tag{10.33a}$$

From Eq. (10.33a) we can find that $\overset{\leftrightarrow}{L}^{\infty}(k)$ has the following form:

$$
\overset{\leftrightarrow}{L}^{\infty}(k) =
\begin{pmatrix}
0 & -i f_{un(k)} & 0 & \cdots & 0 & \cdots \\
-i f_{un(k)} & L_{22}^{\infty} & L_{23}^{\infty} & \cdots & L_{2i}^{\infty} & \cdots \\
0 & L_{32}^{\infty} & L_{33}^{\infty} & \cdots & \cdots & \cdots \\
\vdots & \vdots & \vdots & \ddots & & \\
0 & L_{12}^{\infty} & \vdots & & L_{ii}^{\infty} & \\
\vdots & \vdots & \vdots & & & \ddots
\end{pmatrix},
\tag{10.33b}
$$

so that the second term in Eq. (10.31b) can be written as

$$
\overset{\leftrightarrow}{L}_{3\times M}(k)\left[z\overset{\leftrightarrow}{I}_{M\times M}(k) - \overset{\leftrightarrow}{L}_{M\times M}(k)\right]^{-1}\overset{\leftrightarrow}{L}_{M\times 3}(k)
$$

$$
=
\begin{pmatrix}
0 & 0 & \cdots & 0 & \cdots \\
L_{24}^{\infty} & L_{25}^{\infty} & \cdots & L_{2i}^{\infty} & \cdots \\
L_{34}^{\infty} & L_{35}^{\infty} & \cdots & L_{3i}^{\infty} & \cdots
\end{pmatrix}
\left[z\overset{\leftrightarrow}{I}_{M\times M}(k) - \overset{\leftrightarrow}{L}_{M\times M}(k)\right]^{-1},
\tag{10.34a}
$$

$$
\begin{pmatrix}
0 & L_{42}^{\infty} & L_{43}^{\infty} \\
0 & L_{52}^{\infty} & L_{53}^{\infty} \\
\vdots & \vdots & \vdots \\
0 & L_{i2}^{\infty} & L_{i3}^{\infty} \\
\vdots & \vdots & \vdots
\end{pmatrix}
=
\begin{pmatrix}
0 & 0 & 0 \\
0 & L_{22}^{\cdot} & L_{23}^{\cdot} \\
0 & L_{32}^{\cdot} & L_{33}^{\cdot}
\end{pmatrix}.
\tag{10.34b}
$$

We can easily see that since the first matrix in Eq. (10.34a) is a matrix of dimension $\infty \times 3$, while the third matrix is of the dimension $3 \times \infty$, the result of the three-matrix multiplication will result in a dimension of 3×3 matrix of the form given in Eq. (10.34b), whatever the form of the second matrix in Eq. (10.34a).

Combining the second term of the Eq. (10.31b) given in Eq. (10.34b) and the first term of Eq. (10.31b), the corner 3×3 part of matrix $\overset{\leftrightarrow}{L}^{\infty}(k)$ given in Eq. (10.33b), one can find the matrix $\overset{\leftrightarrow}{L}_{3\times3}(k, z)$ in Eq. (10.31b) contains only three unknown z-dependent quantities: $\overset{\leftrightarrow}{H}_{22}(k, z)$, $\overset{\leftrightarrow}{H}_{23}(k, z) = \overset{\leftrightarrow}{H}_{32}(k, z)$, $\overset{\leftrightarrow}{H}_{33}(k, z)$, which come from the memory function and are generally z dependent. This z dependence of the elements comes from the z dependence of the second term of Eq. (10.31b), which is the z dependence of the correlation function of the three hydrodynamic fluctuation with other dynamic quantities (a > 3). When the time decay of the correlation function $F_{ab}(k, t)$ (a < 3, b > 3) is much faster than the decay of the correlation function $F_{ab}(k, t)$ (a < 3, b < 3), which means that the three hydrodynamic correlation spectra of interest $F_{ab}(k, t)$ (a < 3, b < 3) are much closer to the low-frequency region, one can introduce an approximation by setting $z = 0$ on the right-hand side of Eq. (10.31b) and hence the z-dependent matrix $\overset{\leftrightarrow}{H}_{3\times3}(k, z)$ is reduced to a z-independent matrix $-\overset{\leftrightarrow}{H}(k) = \overset{\leftrightarrow}{H}_{3\times3}(k, z = 0)$. Hence Eq. (10.31a) is reduced to

$$z\overset{\leftrightarrow}{F}_{3\times3}(k, z) = -\overset{\leftrightarrow}{H}(k)\overset{\leftrightarrow}{F}_{3\times3}(k, z) + \overset{\leftrightarrow}{I}_{3\times3}, \tag{10.35a}$$

where

$$\begin{aligned}
\overset{\leftrightarrow}{H}(k) &= -\overset{\leftrightarrow}{H}_{3\times3}(k, z = 0) \\
&= -\overset{\leftrightarrow}{L}_{3\times3}(k) + \overset{\leftrightarrow}{L}_{3\times M}(k)\left[\overset{\leftrightarrow}{L}_{M\times M}(k)\right]^{-1}\overset{\leftrightarrow}{L}_{M\times3}(k).
\end{aligned} \tag{10.35b}$$

The solution of Eq. (10.35a) gives us the correlation matrix $\overset{\leftrightarrow}{F}_{3\times3}(k, z)$. Using the symmetry properties of the Liouville matrix $\overset{\leftrightarrow}{L}^{\infty}(k)$, one finds out that the diagonal elements in matrix $\overset{\leftrightarrow}{H}(k)$ are real and the nondiagonal elements in matrix $\overset{\leftrightarrow}{H}(k)$ are purely imaginary. We finally rewrite the 3×3 matrix $\overset{\leftrightarrow}{H}(k)$ explicitly as

$$\overset{\leftrightarrow}{H}(k) = \begin{pmatrix} 0 & i\,f_{un}(k) & 0 \\ i\,f_{un}(k) & z_u(k) & i\,f_{uT}(k) \\ 0 & i\,f_{uT}(k) & z_T(k) \end{pmatrix}, \tag{10.35c}$$

where $z_u(Q)$, $f_{uT}(Q)$, and $z_T(Q)$ are all real. Solving Eq. (10.35a) for $\overset{\leftrightarrow}{F}_{3\times3}(k, z)$, we have

$$\overset{\leftrightarrow}{F}_{3\times3}(k, z) = \frac{\overset{\leftrightarrow}{I}_{3\times3}}{z\overset{\leftrightarrow}{I}_{3\times3} + \overset{\leftrightarrow}{H}(k)}. \tag{10.35d}$$

From the definition of the dynamic structure factor, we can write

$$\frac{S(k, \omega)}{S(k)} = \frac{1}{2\pi}\int_{-\infty}^{+\infty} dt\, e^{-i\omega t}\,\langle b_1^*(k)b_1(k, t)\rangle. \tag{10.36}$$

Using the definition of $F_{ab}(k, z)$ in Eq. (10.26a)

$$F_{11}(k, z) = \int_0^{+\infty} dt\, e^{-zt}\,\langle b_1^*(k)b_1(k, t)\rangle, \tag{10.37}$$

it follows that

$$\frac{S(k, \omega)}{S(k)} = \frac{1}{\pi}\text{Re}\,[F_{11}(k, z = i\omega)] = \frac{1}{\pi}\text{Re}\left(\frac{\overset{\leftrightarrow}{I}_{3\times3}}{i\omega\overset{\leftrightarrow}{I}_{3\times3} + \overset{\leftrightarrow}{H}(k)}\right)_{1,1}. \tag{10.38}$$

The subscript 1,1 means the 1,1 element of the matrix. In the following text, we will change k to Q, as the conventional way.

10.2.2 Explicit evaluation of the dynamic structure factor

In the low-Q limit, $\overset{\leftrightarrow}{H}(Q)$ tends to the hydrodynamic matrix, where the matrix elements have the following values:

$$f_{un}(Q) = c_s Q / \sqrt{\gamma}, \tag{10.39a}$$

$$z_u(Q) = \varphi Q^2, \tag{10.39b}$$

$$z_T(Q) = \gamma D_T Q^2, \tag{10.39c}$$

$$f_{uT}(Q) = c_s Q \sqrt{(\gamma - 1)/\gamma}, \tag{10.39d}$$

Where $c_s = v_0[\gamma/S(0)]^{1/2}$ is the adiabatic speed of sound; $\gamma = c_p/c_v$ is the ratio of the specific heats at constant pressure and volume; $\varphi = \left[(4/3)\eta + \varsigma\right]/\rho$ is the kinematic longitudinal viscosity (η and ς are the shear and bulk viscosity, respectively, and ρ is the mass-density); and $D_T = \lambda/\rho c_p$ is the thermal diffusivity, where λ is the thermal conductivity. At low Q, the three eigenvalues of the hydrodynamic matrix $\overset{\leftrightarrow}{H}(Q)$, z_h and $z_\pm$, correspond to decay rates of the three hydrodynamic modes of the fluid. These eigenvalues are given below, and we can verify that they are correct up to the order Q^2:

$$z_h(Q) = D_T Q^2 \quad \text{(heat mode),} \tag{10.40a}$$

$$z_\pm(Q) = \pm i c_s Q + \Gamma Q^2 \quad \text{(sound mode),} \tag{10.40b}$$

where, $\Gamma = (1/2)\left[\varphi + (\gamma - 1)D_T\right]$ is the sound-wave damping constant.

For finite values of Q, however, the parameters $z_u(Q)$, $z_T(Q)$, and $f_{uT}(Q)$ become arbitrary functions of Q. In most cases, the eigenvalues of the matrix $\overset{\leftrightarrow}{H}(Q)$ consist of one real number, z_h, and two conjugate complex numbers, $\Gamma_s \pm i\omega_s$.

Now we are in place to calculate the expression of $S(Q, \omega)$. Setting $z = i\omega$, we have

$$\frac{S(Q, \omega)}{S(Q)} = \frac{1}{\pi}\text{Re}\left(\frac{1}{z\overset{\leftrightarrow}{I}_{3\times 3} + \overset{\leftrightarrow}{H}(Q)_{1,1.}}\right). \tag{10.41}$$

Recall the following theorem:

$$\overset{\leftrightarrow}{A}^{-1} = \frac{\overset{\leftrightarrow}{A}^{\text{adjoint}}}{\det \overset{\leftrightarrow}{A}}, \tag{10.42a}$$

where $\overset{\leftrightarrow}{A}^{\text{adjoint}}$ is the adjoint matrix of $\overset{\leftrightarrow}{A}$:

$$\overset{\leftrightarrow}{A}^{\text{adjoint}} = \begin{pmatrix} A_{11} & A_{21} & \cdots & A_{n1} \\ A_{12} & A_{22} & \cdots & A_{n2} \\ \vdots & \vdots & \vdots & \vdots \\ A_{1n} & A_{2n} & \cdots & A_{nn.} \end{pmatrix}, \tag{10.42b}$$

where A_{mn} is the cofactor of element m, n. With Eq. (10.42a), we can rewrite Eq. (10.41) as

$$\frac{S(Q, \omega)}{S(Q)} = \frac{1}{\pi} \text{Re} \left(\frac{\left[z\overleftrightarrow{I} + \overleftrightarrow{H}(Q) \right]^{\text{adjoint}}_{1,1}}{\det \left[z\overleftrightarrow{I} + \overleftrightarrow{H}(Q) \right]} \right). \tag{10.43}$$

We can easily obtain that

$$\left[z\overleftrightarrow{I} + \overleftrightarrow{H}(Q) \right]^{\text{adjoint}}_{1,1} = z^2 + z(z_u + z_T) + z_u z_T + f_{uT}^2. \tag{10.44}$$

Knowing the three eigenvalues of matrix $\overleftrightarrow{H}(Q)$, z_h and $\Gamma_s \pm i\omega_s$, we can then find that

$$\det \left[z\overleftrightarrow{I} + \overleftrightarrow{H}(Q) \right] = (z + z_h)\left[z + (\Gamma_s + i\omega_s) \right]\left[z + (\Gamma_s - i\omega_s) \right]. \tag{10.45}$$

Combining Eqs. (10.29)–(10.30b), we then get

$$\frac{S(Q, \omega)}{S(Q)} = \frac{1}{\pi} \text{Re} \left\{ \frac{z^2 + z(z_u + z_T) + z_u z_T + f_{uT}^2}{(z + z_h)\left[z + (\Gamma_s + i\omega_s) \right]\left[z + (\Gamma_s - i\omega_s) \right]} \right\} \tag{10.46a}$$

$$= \frac{1}{\pi} \text{Re} \left(\frac{A}{z + z_h} + \frac{B}{z + (\Gamma_s + i\omega_s)} + \frac{C}{z + (\Gamma_s - i\omega_s)} \right). \tag{10.46b}$$

We can determine the coefficients A, B and C with the help of residues. Set

$$D(z) = z^2 + z(z_u + z_T) + z_u z_T + f_{uT}^2, \tag{10.47a}$$

$$E(z) = \frac{z^2 + z(z_u + z_T) + z_u z_T + f_{uT}^2}{(z + z_h)\left[z + (\Gamma_s + i\omega_s) \right]\left[z + (\Gamma_s - i\omega_s) \right]}. \tag{10.47b}$$

Then A, B and C can be expressed as

$$A = \text{Res}\left[E(z), -z_h \right] = \frac{D(-z_h)}{\left[z_h - (\Gamma_s + i\omega_s) \right]\left[z_h - (\Gamma_s - i\omega_s) \right]}, \tag{10.47c}$$

$$B = \text{Res}\left[E(z), -(\Gamma_s + i\omega_s) \right] = \frac{D\left[-(\Gamma_s + i\omega_s) \right]}{\left[z_h - (\Gamma_s + i\omega_s) \right](-2i\omega_s)}, \tag{10.47d}$$

$$C = \text{Res}\left[E(z), -(\Gamma_s - i\omega_s) \right] = \frac{D\left[-(\Gamma_s - i\omega_s) \right]}{\left[z_h - (\Gamma_s - i\omega_s) \right](2i\omega_s)}. \tag{10.47e}$$

One can find that A is a real number and B, C are a pair of complex conjugate numbers. So we can rewrite Eq. (10.46b) as

$$\frac{S(Q, \omega)}{S(Q)} = \frac{1}{\pi} \text{Re} \left(\frac{A}{z + z_h} + \frac{B}{z + (\Gamma_s + i\omega_s)} + \frac{B^*}{z + (\Gamma_s - i\omega_s)} \right). \tag{10.48a}$$

Change z back to $i\omega$ and eliminate the imaginary part of the expression in the bracket, we can obtain

$$\frac{S(Q, \omega)}{S(Q)} = \frac{1}{\pi}\left[A_h\frac{z_h}{z_h^2 + \omega^2} + A_s\frac{\Gamma_s + b(\omega + \omega_s)}{\Gamma_s^2 + (\omega + \omega_s)^2} + A_s\frac{\Gamma_s - b(\omega - \omega_s)}{\Gamma_s^2 + (\omega - \omega_s)^2}\right],$$

(10.48b)

where

$$A_h = A, \quad A_s = \text{Re } B, \quad b = \text{Im } B/\text{Re } B, \tag{10.49a}$$

$$A_h = \frac{z_h^2 - z_u z_h - z_T z_h + z_T z_u + f_{uT}^2}{[z_h - (\Gamma_s + i\omega_s)][z_h - (\Gamma_s - i\omega_s)]}, \tag{10.49b}$$

$$A_s = \text{Re}B = \frac{1}{2}(B + B^*) = \frac{1}{2}(B + C). \tag{10.49c}$$

Now we need to calculate $B + C$:

$$\begin{aligned}
B + C &= -\frac{D[-(\Gamma_s + i\omega_s)]}{[z_h - (\Gamma_s + i\omega_s)](2i\omega_s)} + \frac{D[-(\Gamma_s - i\omega_s)]}{[z_h - (\Gamma_s - i\omega_s)](2i\omega_s)} \\
&= \frac{-D[-(\Gamma_s + i\omega_s)][z_h - (\Gamma_s - i\omega_s)] + D[-(\Gamma_s - i\omega_s)][z_h - (\Gamma_s + i\omega_s)]}{[z_h - (\Gamma_s + i\omega_s)][z_h - (\Gamma_s - i\omega_s)](2i\omega_s)} \\
&= \frac{-(2i\omega_s)(2\Gamma_s z_h - \Gamma_s^2 - z_u z_h - z_T z_h - \omega_s^2 + z_T z_u + f_{uT}^2)}{[z_h - (\Gamma_s + i\omega_s)][z_h - (\Gamma_s - i\omega_s)](2i\omega_s)} \\
&= \frac{-2\Gamma_s z_h + \Gamma_s^2 + z_u z_h + z_T z_h + \omega_s^2 - z_T z_u - f_{uT}^2}{[z_h - (\Gamma_s + i\omega_s)][z_h - (\Gamma_s - i\omega_s)]}.
\end{aligned} \tag{10.50a}$$

So that

$$A_s = \frac{1}{2}(B + C) = \frac{-2\Gamma_s z_h + \Gamma_s^2 + z_u z_h + z_T z_h + \omega_s^2 - z_T z_u - f_{uT}^2}{2[z_h - (\Gamma_s + i\omega_s)][z_h - (\Gamma_s - i\omega_s)]}. \tag{10.50b}$$

We can also obtain

$$b = \frac{(z_h - \Gamma_s)(\Gamma_s^2 - \omega_s^2 - \Gamma_s z_u - \Gamma_s z_T + z_u z_T + f_{uT}^2) - 2\Gamma_s \omega_s^2 + \omega_s^2(z_u + z_T)}{i\omega_s(2\Gamma_s z_h - \Gamma_s^2 - z_u z_h - z_T z_h - \omega_s^2 + z_T z_u + f_{uT}^2)}. \tag{10.51}$$

In addition, we can have a look at the value of $A_h + 2A_s$:

$$\begin{aligned}
A_h + 2A_s &= A + B + C \\
&= \frac{z_h^2 - z_u z_h - z_T z_h + z_T z_u + f_{uT}^2}{[z_h - (\Gamma_s + i\omega_s)][z_h - (\Gamma_s - i\omega_s)]} \\
&\quad + \frac{-2\Gamma_s z_h + \Gamma_s^2 + z_u z_h + z_T z_h + \omega_s^2 - z_T z_u - f_{uT}^2}{[z_h - (\Gamma_s + i\omega_s)][z_h - (\Gamma_s - i\omega_s)]}
\end{aligned}$$

$$= \frac{\Gamma_s{}^2 + z_h{}^2 - 2\Gamma_s z_h + \omega_s{}^2}{[z_h - (\Gamma_s + i\omega_s)][z_h - (\Gamma_s - i\omega_s)]} = 1. \tag{10.52}$$

The intermediate scattering function $F(Q, t)$ can be obtained by taking the inverse Fourier transformation of $S(Q, \omega)$. We use Eq. (10.48a) as the expression of $S(Q, \omega)$, take the inverse Fourier transformation and obtain

$$\frac{F(Q, t)}{S(Q)} = A \exp(-z_h t) + B \exp[(-\Gamma_s - i\omega_s)t] + B^* \exp[(-\Gamma_s + i\omega_s)t]$$

$$= A \exp(-z_h t) + \exp(-\Gamma_s t)\left[B \exp(-i\omega_s t) + B^* \exp(i\omega_s t)\right] \tag{10.53}$$

$$= A \exp(-z_h t) + 2 \exp(-\Gamma_s t)(\text{Re}B \cos \omega_s t + \text{Im}B \sin \omega_s t) \quad (t > 0).$$

Combining Eq. (10.49a) with Eq. (10.53), we can rewrite Eq. (10.54) finally as

$$\boxed{\frac{F(Q, t)}{S(Q)} = A_h \exp(-z_h t) + 2A_s \exp(-\Gamma_s t)(\cos \omega_s t + b \sin \omega_s t)}, \tag{10.54}$$

for $t > 0$.

10.2.3 Transformation into the continued fraction form and identify the memory function

From Eq. (10.43), one can also find that

$$\frac{S(Q, \omega)}{S(Q)} = \frac{1}{\pi}\text{Re}\left\{\frac{\left[z\overleftrightarrow{I} + \overleftrightarrow{H}(Q)\right]_{1,1}^{\text{adjoint}}}{\det\left[z\overleftrightarrow{I} + \overleftrightarrow{H}(Q)\right]}\right\}$$

$$= \frac{1}{\pi}\text{Re}\left\{\frac{z^2 + z(z_u + z_T) + z_u z_T + f_{uT}^2}{z[(z + z_u)(z + z_T) + f_{uT}^2] + f_{uT}^2(z + z_T)]}\right\}. \tag{10.55a}$$

After setting $z = i\omega$ and some algebra, Eq. (10.55a) can be rewritten into the so-called continued fraction form as

$$\boxed{\frac{S(Q, \omega)}{S(Q)} = \frac{1}{\pi}\text{Re}\left[i\omega + \frac{f_{un}^2(Q)}{i\omega + z_u(Q) + \frac{f_{uT}^2(Q)}{i\omega + z_T(Q)}}\right]^{-1}} \tag{10.55b}$$

Compare Eq. (10.55b) with the standard definition of the memory function $m_Q(\omega)$ in Balucani and Zoppi (1994) (see their Eq. (6.19)):

$$\frac{S(Q, \omega)}{S(Q)} = \frac{1}{\pi}\text{Re}\left[i\omega + \frac{\omega_0^2(Q)}{i\omega + m_Q(\omega)}\right]^{-1}. \tag{10.55c}$$

Setting

$$\omega_0(Q) = f_{un}(Q), \tag{10.56a}$$

then

$$m_Q(\omega) = z_u(Q) + \frac{f_{uT}^2(Q)}{i\omega + z_T(Q)} = m'_Q(\omega) - im''_Q(\omega), \tag{10.56b}$$

where

$$m'_Q(\omega) = z_u(Q) + \frac{\omega_0^2(Q)z_T(Q)}{z_T^2(Q) + \omega^2}. \tag{10.56c}$$

Then

$$\begin{aligned}
\frac{S(Q,\omega)}{S(Q)} &= \frac{1}{\pi}\mathrm{Re}\left[i\omega + \frac{\omega_0^2(Q)}{i\omega + m'_Q(\omega) - im''_Q(\omega)}\right]^{-1} \\
&= \frac{1}{\pi}\mathrm{Re}\left[\frac{m'_Q + i(\omega - m''_Q)}{\omega_0^2 + \omega m''_Q - \omega^2 + i\omega m'_Q}\right] \\
&= \frac{1}{\pi}\frac{\omega_0^2 m'_Q}{\left(\omega^2 - \omega_0^2 - \omega m''_Q\right)^2 + (\omega m'_Q)^2} \\
&= \frac{1}{\pi}\frac{\omega_0^2}{\omega}\mathrm{Im}\left[\frac{1}{(\omega^2 - \omega_0^2 - \omega m''_Q) - i\omega m'_Q}\right] \\
&= \frac{1}{\pi}\frac{\omega_0^2}{\omega}\mathrm{Im}[\omega^2 - \omega_0^2(Q) - i\omega m_Q(\omega)]^{-1}. \tag{10.57}
\end{aligned}$$

Monaco *et al.* (1999) chose the following form of the memory function in the time domain:

$$m_Q(t) = \omega_0^2(Q)[\gamma(Q) - 1]e^{-D_T(Q)Q^2 t} + 2\Gamma_\mu(Q)\delta(t) + \Delta^2(Q)e^{-t/\tau_\alpha(Q)}. \tag{10.58a}$$

Converting the above $m_Q(t)$ into the frequency domain function, we have

$$\begin{aligned}
m_Q(\omega) &= \int_0^{+\infty} m_Q(t)e^{-i\omega t}\,dt \\
&= \omega_0^2(Q)(\gamma(Q) - 1)\int_0^{+\infty} e^{-D_T(Q)Q^2 t}e^{-i\omega t}\,dt + 2\Gamma_\mu(Q) \\
&\quad + \Delta^2(Q)\int_0^{+\infty} e^{-t/\tau_\alpha(Q)}e^{-i\omega t}
\end{aligned}$$

$$= \frac{\omega_0^2(Q)(\gamma(Q) - 1)}{D_T(Q)Q^2 + i\omega} + 2\Gamma_\mu(Q) + \frac{\Delta^2(Q)}{1/\tau_\alpha(Q) + i\omega}$$

$$= 2\Gamma_\mu + \frac{D_T Q^2[\omega_0^2(\gamma - 1)]}{(D_T Q^2)^2 + \omega^2} + \frac{\Delta^2/\tau_\alpha}{1/\tau_\alpha^2 + \omega^2}$$

$$- i\omega\left[\frac{\omega_0^2(\gamma - 1)}{(D_T Q^2)^2 + \omega^2} + \frac{\Delta^2}{1/\tau_\alpha^2 + \omega^2}\right]. \qquad (10.58b)$$

Comparing Eq. (10.58b) with Eq. (10.56b), we finally arrive at the conclusion that the generalised three effective eigenmode (GTEE) theory (as shown in the following Eq. (10.59), corresponds to the case that

(i) $\gamma - 1 = 1$,
(ii) $2\Gamma_\mu = z_u$, $D_T Q^2 = z_T$,
(iii) $\Delta^2(Q) = 0$

$$m_Q(\omega) = \int_0^{+\infty} m_Q(t)\cos\omega t dt - i\int_0^{+\infty} m_Q(t)\sin\omega t dt$$

$$= m_Q'(\omega) - im_Q''(\omega). \qquad (10.59)$$

10.2.4 Extension of the TEE theory to polyatomic molecular simple atomic fluids[5]

The IXS spectrum of a simple atomic fluid is proportional to the dynamic structure factor of a single species of atom. In the case of a molecular fluid, however, the IXS spectrum is proportional to a weighted sum of the partial dynamic structure factors of pairs of atomic species. The weighting factors are products of the atomic form factors of the pairs. The dynamic structure factor in this case can be written as a weighted sum of the partial dynamic structure factor $S_{\alpha\beta}(k, \omega)$, namely,

$$S(k, \omega) = \sum_{\alpha,\beta} f_\alpha(k) f_\beta(k)\omega_\alpha\omega_\beta S_{\alpha\beta}(k, \omega) \qquad (10.60)$$

where $f_\alpha(k)$ is the atomic form factor of the alpha-species of atoms $\omega_\alpha = \sqrt{N_\alpha/N}$ and $S(k, \omega)$ is given by GTEE theory.

The structure factor measured by an X-ray diffraction experiment is thus shown as

$$S(k) = \sum_{\alpha,\beta} f_\alpha(k) f_\beta(k)\omega_\alpha\omega_\beta S_{\alpha\beta}(k), \qquad (10.61)$$

[5] Liao and Chen (2001).

where the partial structure factor $S_{\alpha\beta}(k)$ is given by

$$S_{\alpha\beta}(k) = \frac{1}{\sqrt{N_\alpha N_\beta}} \sum_{i\in\alpha,\, j\in\beta} \left\langle e^{i\vec{k}\cdot(\vec{r}_i - \vec{r}_j)} \right\rangle. \tag{10.62}$$

The second moment sum rule can then be shown to be

$$\langle\omega^2\rangle = k^2 v_0^2(k) = k^2 \sum_\alpha f_\alpha^2(k)\omega_\alpha^2 \frac{k_B T}{m_\alpha}. \tag{10.63}$$

In GTEE theory, the dynamic structure factor is then given as in Eq. (10.15),

$$S(k,\omega) = \frac{S(k)}{\pi} \mathrm{Re}\left\{ \frac{\overset{\leftrightarrow}{I}}{i\omega\overset{\leftrightarrow}{I} + \overset{\leftrightarrow}{H}(k)} \right\}_{1,1} \tag{10.64}$$

where $\overset{\leftrightarrow}{I}$ is the 3×3 identity matrix, and label 1,1 means the (1,1) element of the matrix. The generalised hydrodynamic matrix $\overset{\leftrightarrow}{H}(k)$ is given as in Eq. (10.16),

$$\overset{\leftrightarrow}{H}(k) = \begin{pmatrix} 0 & i f_{un}(k) & 0 \\ i f_{un}(k) & z_u(k) & i f_{uT}(k) \\ 0 & i f_{uT}(k) & z_T(k) \end{pmatrix}, \tag{10.65}$$

where the matrix element, $f_{un}(k)$ is determined by the second frequency moment of $S(k,\omega)$ to be $f_{un}(k) = kv_0(k)/[S(k)]^{1/2}$. The three independent parameters, $z_u(k)$, $f_{uT}(k)$, $z_T(k)$, are all real numbers and all have k dependence, which can be obtained by fitting IXS spectra.

As shown in Eq. (10.15) vs. Eq. (10.64), and Eq. (10.16) vs. Eq. (10.65), GTEE theory and TEE theory look formally the same although they are not. This is because the hydrodynamic matrix element $f_{un}(k)$ for GTEE theory is interpreted in a different way. In GTEE theory, the polyatomic nature of fluid is explicitly taken into account through the factor as given in Eq. (10.63). Thus, in an IXS experiment with a polyatomic fluid, it is important to measure explicitly so that one can calculate the numerical value of the element $f_{un}(k)$ rather than treating it as one of the four fitting parameters. By doing so, GTEE theory becomes a three-parameter model that enables one to obtain more unique fitting parameters.

10.3 Module – Finite Q collective modes in bio-macromolecular assemblies

10.3.1 Theory of the generalised dynamic structure factor of polyatomic macromolecules measured by inelastic X-ray scattering[6]

The study of atomic collective dynamics at large wave-vector transfers and energies in condensed matter is, traditionally, the domain of inelastic neutron scattering (INS) spectroscopy. Since the mid 1990s Sette *et al.* (1995) have developed a

---

[6] Chen *et al.* (2001); Liao and Chen (2001).

high-resolution inelastic X-ray scattering (IXS) technique that is complementary to INS. The IXS technique has the following advantages over the INS: IXS can investigate a much larger frequency and wave vector transfer ranges in the kinematic phase space at a constant scattering angle; namely, IXS is able to maintain a constant k (magnitude of the wave vector transfer) measurement at a constant scattering angle for the energy transfer up to the order of $k_B T$, which is impossible for INS; and the absence of multiple scattering processes allows direct measurement of the dynamics structure factor without having to use sophisticated procedures for the reduction of raw spectral data.

To interpret IXS data, one may directly extend the method for the analysis of INS data. This kind of analysis is straightforward if the scatterer consists of only a single species of atom, such as simple atomic fluids. Both INS and IXS measure the dynamic structure factor of the system. The only difference is that the prefactor is proportional to nuclear scattering length b of the atom for INS, but it is the product of the classical radius of electron r_0 and the electronic form factor $f(k)$ of the atom for IXS.

When the scatterer consists of multiple types of atoms, such as molecular fluids or supramolecular systems, the inelastic X-ray scattering intensity is not proportional to the dynamic structure factor of any single atom, but to a generalised dynamic structure factor defined as a weighted sum over partial dynamic structure factors of all pairs of atomic species. The weights are dependent on the form factor $f(k)$ and number density of atomic species of the molecule. The inclusion of the form factors into the generalised dynamic structure factor and its applications to analyse the IXS spectrum have been carried out by Liao and Chen (2001). They developed a generalised three effective eigenmode (GTEE) model to analyse IXS spectra of polyatomic molecular systems. Thus, the TEE model for simple fluids can be extended to the case for polyatomic molecular system in a rather straightforward way. As an example, they present the GTEE model analysis of IXS spectra of fully hydrated dilauroylphosphatidylcholine (DLPC) bilayers as described in the following.

The double differential cross section of the coherent X-ray scattering from a molecular system can be written as

$$\frac{d^2\sigma(E, \Omega)}{d\Omega dE} = N r_0^2 (\hat{\varepsilon}_i \cdot \hat{\varepsilon}_f)^2 \frac{k_f}{k_i} S(k, E). \tag{10.66}$$

The generalised intermediate scattering function, the inverse Fourier transform of the generalised dynamic structure factor $S(k, \omega)$, can be written as

$$F(k, t) = \frac{1}{N} \sum_{j,l} \left\langle f_j(k) f_l(k) \exp(i\vec{k} \cdot \vec{r}_l(t)) \exp(-i\vec{k} \cdot \vec{r}_j(0)) \right\rangle. \tag{10.67}$$

The generalised static structure factor $S(k)$ as measured by X-ray diffraction experiments can then given by

$$S(k) = \langle n^*(k)n(k) \rangle = \frac{1}{N} \sum_{j,l} \langle f_j(k) f_l(k) exp(i\vec{k} \cdot \vec{r}_l(0)) \exp(-i\vec{k} \cdot \vec{r}_j(0)) \rangle$$

$$= \sum_{\alpha,\beta} f_\alpha(k) f_\beta(k) \omega_\alpha \omega_\beta S_{\alpha\beta}(k), \tag{10.68}$$

where α and β denote the types of atom, $\omega_\alpha = \sqrt{N_\alpha/N}$, the square root of number fraction of atomic type α over the total number of atoms in the molecule, as already explained in Eq. (10.61).

One can define a generalised thermal velocity as the weighted average of thermal velocities of different types of atoms in the system,

$$v_0^2(k) = \sum_\alpha f_\alpha^2(k) \omega_\alpha^2 v_{0\alpha}^2, \tag{10.69}$$

where $v_{0\alpha} = \sqrt{k_B T/m_\alpha}$ is the thermal velocity of the atom type α.

The generalised dynamic structure factor is defined as the Fourier transform of the generalised intermediate scattering function Eq. (10.67). Then the generalised static structure factor $S(k)$ is given by Eq. (10.68) and the generalised thermal velocity $v_0(k)$ by Eq. (10.69). The normalised generalised dynamic structure factor is then given by the following equations according to GTEE theory, as in Eq. (10.64):

$$\frac{S(k, \omega)}{S(k)} = \frac{1}{\pi} \text{Re} \left(\frac{\overleftrightarrow{I}_{3\times3}}{i\omega \overleftrightarrow{I}_{3\times3} + \overleftrightarrow{H}(k)} \right)_{1,1} \tag{10.70}$$

where, as for Eq. (10.65),

$$\overleftrightarrow{H}(k) = \begin{pmatrix} 0 & i f_{un}(k) & 0 \\ i f_{un}(k) & z_u(k) & i f_{uT}(k) \\ 0 & i f_{uT}(k) & z_T(k) \end{pmatrix} \tag{10.71}$$

The $\overleftrightarrow{H}$ matrix in Eq. (10.71) explicitly depends on four k-dependent parameters: $f_{un}(k)$, $z_u(k)$, $z_T(k)$, $f_{uT}(k)$. Among them the first matrix element $f_{un}(k)$ can be expressed explicitly in terms of $S(k)$ and $v_0(k)$ as discussed in Section 10.2.4:

$$f_{un}(k) = k v_0(k)/[S(k)]^{1/2}. \tag{10.72}$$

10.3.2 Two-dimensional phonons propagating in the plane of lipid bilayers

Chen *et al.* (2001) carried out an IXS experiment at a very high-resolution IXS beamline (BL21-ID16) at ESRF. The undulator X-ray source was premonochromated by a Si(111) double crystal monochromator and then followed with a high energy resolution backscattering monochromator (temperature controlled and scanned), operating at the Si(11 11 11) Bragg reflection (X-ray energy, 21.748 KeV). A grooved spherical silicon crystal analyser operating at the same Bragg back reflections, and in Rowland geometry collected the scattered photons. The net energy resolution was measured by an elastic scattering of a plastic sample at its maximum of structure factor. This gives an energy resolution of 1.5 meV full width at half maximum (FWHM). The X-ray beam size at the sample was 0.15 mm × 0.35 mm.

The fully hydrated DLPC (dilaurylphosphatidylcholine) sample was in a state of partially oriented multi-lamellar vesicles. The sample was measured at two temperatures: $T = 269$ K and 294 K. Since the main transition temperature of hydrated DLPC is about 272 K, these two temperatures correspond to the gel phase ($L_{\beta'}$) and liquid crystal phase (L_{α}). Since a DLPC molecule has a large polar head group and double hydrophobic chains, one does not expect that the measured dynamic structure factor could be approximated by the centre of mass dynamic structure factor for DLPC. Therefore, one needs to use the GTEE model to fit the data. There are four fitting parameters: $f_{un}(k)$, $z_u(k)$, $z_T(k)$, $f_{uT}(k)$. The calculated $S(k, \omega)/S(k)$ was multiplied by a detailed balance factor and convoluted with the resolution function before comparing with the IXS spectra that have been normalised to have unit area. Figure 10.6 is two typical fits of the IXS spectra at $T = 269$ K and

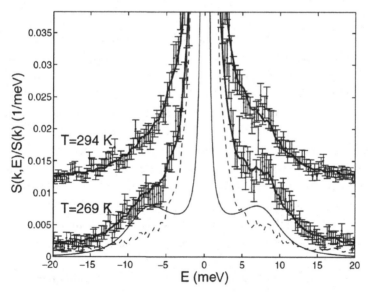

Figure 10.6 The IXS spectra of fully hydrated DLPC at $k = 5\,\text{nm}^{-1}$ at $T = 269$ K and 294 K. The thick solid lines are the fits by the GTEE theory in absolute scale. The thin solid line is the theoretical dynamic structure factor at $T = 269$ K, and the dashed line is the energy resolution function. The full height of the central peak of IXS spectra is about 0.35 in the ordinate scale (Chen *et al.*, 2001). Copyright 2001 by The American Physical Society.

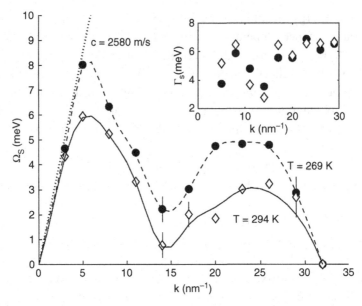

Figure 10.7 The high-frequency sound dispersion relation (symbols) and damping (inset) of fully hydrated DLPC at two different temperatures $T = 269$ K (solid circles) and $T = 294$ K (diamonds). The lines are drawn only to guide the eyes. The initial slopes of the dispersion relations give the high frequency sound speed indicated in the figure (Chen *et al.*, 2001). Copyright 2001 by The American Physical Society.

294 K, $k = 5$ nm^{-1}. The thick solid lines in the figure are the result of the fitting (χ^2 is of the order of unity).

Figure 10.7 shows that there is a significant difference in the high-frequency sound dispersion relations at the two temperatures studied. The sound frequencies in the gel phase ($T = 269$ K) are higher than those in the liquid crystalline phase ($T = 294$ K). This temperature dependence of the collective movements of hydrocarbon chains can also be observed directly from the IXS spectra at $T = 269$ K and $T = 294$ K (Figure 10.6). In both cases in the energy loss side, the sound modes can be observed as shifted peaks from the raw data. At the gel phase ($T = 269$ K), the hydrocarbon chains are ordered, and the dynamics is solid-like. The sound frequencies are higher and the sound mode is less damped, but in the liquid crystalline phase, the hydrocarbon chains are in a state of disordered liquid; the increase of viscosity increases the sound damping. This is the first experimental observation from collective dynamics that the hydrocarbon chains melt when temperature increases and crosses the main transition temperature.

One of the verifications of the GTEE analysis of the hydrated DLPC can be obtained by comparing the experimentally determined $S(k)$ with the calculated structure factor (Eq. (10.68)) from fitting parameters. Figure 10.8 shows such

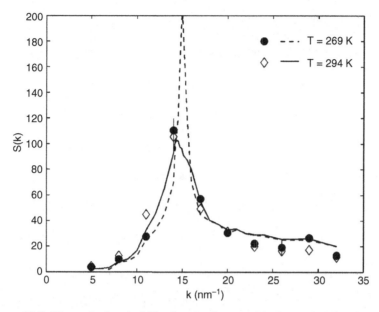

Figure 10.8 The calculated $S(k)$ (in absolute scale) from the fitted second moment of the dynamic structure factor $f_{un}^2(k)$ at $T = 269\,\text{K}$ (sold circles) and at $T = 294\,\text{K}$ (diamonds) are compared with the structure factors (in arbitrary scale) measured by X-ray diffraction at $T = 269\,\text{K}$ (dashed line) and at $T = 294\,\text{K}$ (solid line) (Chen *et al.*, 2001). Copyright 2001 by The American Physical Society.

a comparison. The calculated $S(k)$ is obtained by the equation: $S(k) = [kv_0(k)/f_{un}(k)]^2$, where the generalised thermal velocity $v_0(k)$ is calculated by Eq. (10.69), and the experimental $S(k)$ is obtained from X-ray scattering without energy analysis at the same IXS spectrometer. In order to compare them at the same scale, the experimental $S(k)$ were multiplied by one appropriate constant factor. The two curves (lines and symbols) obtained through different methods show over all good agreement. In this figure, the structure factor of the $L_{\beta'}$ gel phase at $T = 269\,\text{K}$ shows a sharp peak at $k = 15\,\text{nm}^{-1}$, corresponding to the ordered structure formed by hydrocarbon chains in the bilayers. However, measured points of the structure factor at 269 K did not include the $k = 15\,\text{nm}^{-1}$ point.

Figure 10.9 shows k dependence of the relaxation rate $\Gamma_h(k)$ of the central peak (the non-propagating mode) of the dynamic structure factor. There is no significant difference in relaxation rates in the two phases in the measured k range. The relaxation rate has a quadratic dependence on k with a constant thermal diffusivity of $7.5 \times 10^{-6}\,\text{cm}^2\text{s}^{-1}$.

We should point out here that, although GTEE theory is a very versatile and robust method for fitting IXS spectra of supramolecular liquids and for extracting

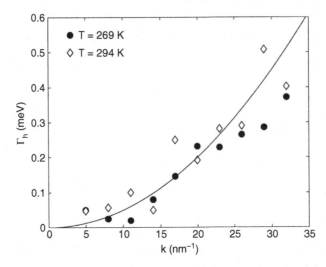

Figure 10.9 The k-dependent relaxation rate of the central peak of the dynamic structure factor at the two temperatures. The solid line is the relation αk^2, with $\alpha = 7.5 \times 10^{-6}\,\mathrm{cm^2 s^{-1}}$ (Chen *et al.*, 2001). Copyright 2001 by The American Physical Society.

experimental generalised dynamic structure factors from them, the model parameters so determined have no straightforward physical interpretations at this stage of development. Furthermore, the three effective eigenmodes the authors used in the analysis of IXS data may not be sufficiently *complete* to represent all the slow modes of the particular system studied. If this is the case, the generalised dynamic structure factors the authors extracted from the measured spectra may be of only limited accuracy. But in view of the fairly noisy data they obtained at that time with a limited integration time at a third generation synchrotron source, it was not necessary to further increase the number of modes in the model and thus also increasing the corresponding number of fitting parameters.

For the case of lipid bilayers at moderate temperatures that the three effective eigenmodes they used seems to have a sufficiently good approximation, judging from the results. The success of GTEE as a useful IXS data analysis tool could be assessed by applying it further to other systems such as liquid-crystalline DNA and hydrated globular proteins that we shall discuss in the following sections.

In summary, the high-resolution IXS technique has been used to study the collective dynamics of the fully hydrated DLPC phospholipid bilayer. The collective movements of hydrocarbon chains show significant temperature dependence due to hydrocarbon chain melting. The GTEE model analysis of IXS spectra allows Chen and coworkers to construct the dispersion relation of the high-frequency sound. They found a novel feature that the frequencies of the propagating density oscillations show a pronounced minimum at the wavelength corresponding to the two neighbouring inter-chain distances. The fact that this minimum frequency in the L_α phase is less than 1 meV may have theoretical implication on the possible mechanism for the passage of water molecules through a functioning bilayer. Although they did not have a theory for permeation of water molecules through the lipid bilayer, it was suggested that an appropriate theory would have to involve the large-amplitude, slow in-plane density fluctuations. In this regard the low-frequency inter-lipid collective oscillations along with the whole dispersion relation should play a key role in this theory. An immediate interest of this discovery is to see whether computer molecular dynamics simulation of hydrated DLPC bilayers can reproduce the measured dispersion relation given in Figure 10.7. This was indeed shown in another subsequent paper by Tarek *et al.* (2001).

10.3.3 One-dimensional phonons propagating along the axial direction of liquid crystalline DNA

Liu *et al.* (2005) used inelastic X-ray scattering (IXS) to study the phonon propagation and damping along the axial direction of films of aligned 40 wt% calf-thymus DNA rods.

The experiments used the high-resolution inelastic X-ray spectrometer, HERIX-3, at Beamline 3-ID, at the Advanced Photon Source (APS), Argonne National Lab in Chicago. The spectrometer was capable of measuring the dynamic structure factor $S(Q, E)$ with a Q resolution of about 0.7 nm^{-1}, and an energy resolution of about 2 meV. It had a Q range from 0 to 30 nm^{-1} and an energy transfer window from 0 to several hundreds of meV.

The experiments systematically studied the effects on the phonon dispersion relation and the phonon damping due to the presence of different types and valencies of counterions in the DNA solution. The counterions studied in this experiment were Zn^{2+}, Ca^{2+}, Cu^{2+}, and a trivalent counterion, spermidine. The ionic strength of aqueous electrolytic solution was 0.25. The DNA concentration was 40 wt% for all samples.

Typical IXS spectra of four DNA samples at $Q = 6.5$ nm^{-1} are shown in Figure 10.10 together with the fitting by the GTEE theory. The fitted curve with GTEE theory (thick solid line) agrees excellently with the experimental results (circles with error bars). The dashed line is the instrumental resolution function. The thin solid line is the extracted $S(Q, E)$ by the fitting.

Figure 10.11 shows the extracted $S(Q, E)$ at $Q = 4$ nm^{-1} and $Q = 18.7$ nm^{-1} for all four samples. At very low Q values, the phonon damping is very small. Therefore, all phonon peaks are all well defined compared to the extracted $S(Q, E)$ shown in Figure 10.10 at $Q = 6.5$ nm^{-1}. At $Q = 18.7$ nm^{-1}, it is at the

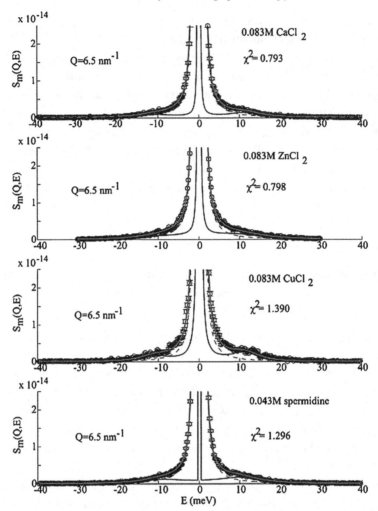

Figure 10.10 The typical fittings of IXS spectra with GTEE theory at $Q = 6.5\,\mathrm{nm}^{-1}$ are shown for four different samples with different counterions. Only the lower part of the measured spectra (circles with error bars) and their fitting curves (thick solid line) are shown to illustrate the feature of the collective excitations. The dashed line is the resolution function and the thin solid line is the extracted $S(Q, E)$ by the fitting. From top to bottom, the samples are 40 wt% calf-thymus Na-DNA with the different extra counterions. They are 0.083 M $CaCl_2$, 0.083 M $ZnCl_2$, 0.083 M $CuCl_2$, and 0.042 M spermidine, respectively (Liu *et al.*, 2005c). Reprinted with permission from Liu *et al.* (2005c). Copyright 2005, AIP Publishing LLC.

Brillouin-zone centre if one considers a DNA molecule as a one-dimensional crystal. There is no phonon feature at all in the extracted normalised dynamic structure factors at this Q value. The top panel of Figure 10.12 shows the phonon-dispersion relation of sample with Zn^{2+} counterions. At small Q, each spectrum contains a

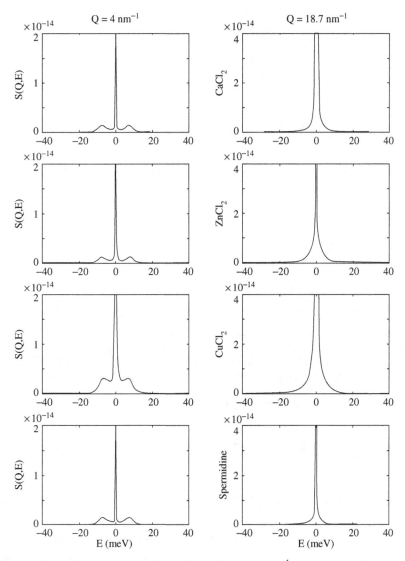

Figure 10.11 The extracted $S(Q, E)$ at $Q = 4\,\text{nm}^{-1}$ and $Q = 18.7\,\text{nm}^{-1}$ for four different samples is shown. From top to bottom, the samples are 40 wt% calf-thymus Na-DNA with different extra counterions, which are 0.083 M $CaCl_2$, 0.083 M $ZnCl_2$, 0.083 M $CuCl_2$ and 0.042 M spermidine, respectively (Liu *et al.*, 2005c). Reprinted with permission from Liu *et al.* (2005c). Copyright 2005, AIP Publishing LLC.

central peak with a shifted Brillouin doublet, from which the phonon energy is extracted with error bars. However, after $Q >\sim 12.5\,\text{nm}^{-1}$, the fitted result of a spectrum shows the three peaks all centring at $Q = 0$. Thus there is no phonon excitation energy to be extracted. This can be attributed to the effect that the phonon

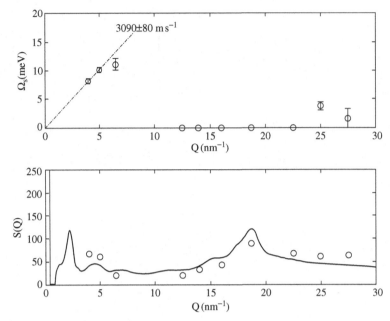

Figure 10.12 The longitudinal-acoustic-phonon-dispersion relation of 40 wt% calf-thymus Na-DNA in 0.083M $ZnCl_2$ is shown in the top panel. The extracted sound speed is about 3090±80 m s^{-1}. In the bottom panel, the approximate structure factor (circles) calculated through the fitted parameters is compared with the measured structure factor with a scaling factor (Liu *et al.*, 2005c) Reprinted with permission from Liu *et al.* (2005c). Copyright 2005, AIP Publishing LLC.

is over-damped. After $Q > 22.5\,\text{nm}^{-1}$, the fitted results show the phonon peaks again. The sound speed, of 3090 m s^{-1} is obtained by assuming that there is a linear relation between Ω_s and Q at $Q = 4\,\text{nm}^{-1}$.

The bottom panel of Figure 10.12 shows the calculated structure factor (circles) together with the measured structure factor (points with solid line) scaled with a constant. The equation $S(Q) = Q^2 v_0^2(Q)/f\,un^2(Q)$ is used to calculate $S(Q)$. The value of the form factor $f(Q)$ for different atoms can be found in the international tables for X-ray crystallography. As an approximation, it was assumed that phosphate has the dominating contribution to $v_0(Q)$. Since the measured structure factor by the instrument is in an arbitrary unit, the result is scaled by a constant in order to compare with the calculated result. The agreement is fair after considering the simplification of calculation of v_0. For the spermidine sample, the phonon-dispersion relation is shown in Figure 10.13. The ionic strength of this sample is kept the same as that for divalent counterion samples. Its dispersion-relation curve does not show much difference from the samples having divalent counterions. Nevertheless, it will be shown later that its damping is stronger than that added with divalent ions at the same ionic strength.

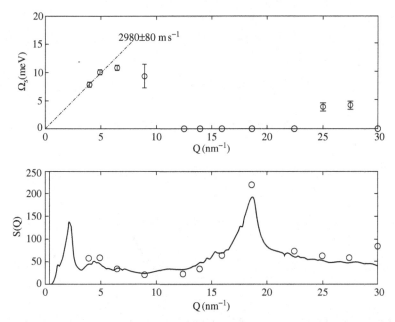

Figure 10.13 The longitudinal-acoustic-phonon-dispersion relation of 40 wt% calf-thymus Na-DNA in 0.042 M spermidine is shown in the top panel. The extracted sound speed is about 2980 ± 80 m s^{-1}. In the bottom panel, the approximate structure factor (circles) calculated through the fitted parameters is compared with the measured structure factor with a constant scaling factor (Liu *et al.*, 2005c). Reprinted with permission from Liu *et al.* (2005c). Copyright 2005, AIP Publishing LLC.

To compare the phonon-dispersion-relation curves with/without the addition of divalent counterions, the results from 40 wt% calf-thymus DNA molecules in H_2O (circles) and the Zn^{2+} sample (up triangles) together are shown in Figure 10.14. The sound speed of aligned 40 wt% DNA molecules in H_2O is about 3160 m s^{-1} estimated from the phonon excitation energy at $Q = 1.6$ nm^{-1}. It clearly shows the disappearance of phonon at the intermediate Q range due to stronger damping introduced by additional counterions. The curve is drawn to guide the eyes.

Together in this figure, the phonon-dispersion relation of DNA molecules from $Q = 16$ nm^{-1} to $Q = 21$ nm^{-1} measured by INS (stars and crosses) is also shown (Grimm *et al.*, 1987). The results from INS cover the Q range where no phonon peaks are detected in the IXS spectra.

The reason behind the failure to identify the phonon peaks in the IXS spectra is because the phonon excitation energy is too small, and the collective excitation features of the IXS spectrum are masked by the energy resolution function of the IXS instrument, so the excitation energy cannot be accurately measured. A direct comparison of the phonon-dispersion relation obtained with INS at the same Q

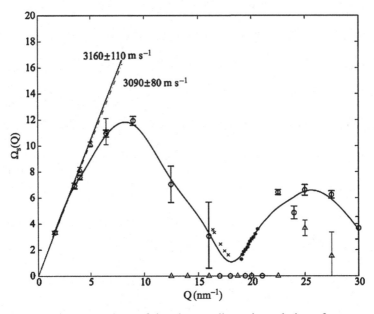

Figure 10.14 The comparison of the phonon-dispersion relation of two samples. The circles with error bars show the dispersion relation of the 40 wt% calf-thymus Na-DNA in H_2O. The solid curve is drawn to guide the eyes. The up triangles with error bars show the dispersion relation of 40 wt% calf-thymus Na-DNA in 0.083 M $ZnCl_2$. The added divalent counterions clearly overdamped phonons in the intermediate Q range (12.5–22.5 nm^{-1}). The star and cross symbols are the phonon-dispersion relations measured by INS (Grimm *et al.*, 1987). Reprinted with permission from Liu *et al.* (2005c). Copyright 2005, AIP Publishing LLC.

values requires a better energy resolution from the IXS instrument. However, the INS result seems to agree with the trend of the measured IXS spectra.

In Figure 10.15, one can see the dispersion-relation curves of 40 wt% calf-thymus DNA molecules in H_2O (circles) and the spermidine sample (stars) together to show the effect of the addition of trivalent counterion. It is instructive to compare the ratio between phonon damping and the phonon frequency from different samples, which is shown in Figure 10.16. The circles with error bars are the result of Na-DNA molecules in H_2O. One of the solid lines is drawn along the trend of the results of Na-DNA in H_2O, while another is drawn along that of the spermidine containing samples. These two lines are drawn to guide the eyes. The ratio of phonon damping and phonon frequency of samples with divalent counterions falls in between these two lines with the only exception for the case at $Q = 6.5$ nm^{-1} for the Zn^{2+} sample. At the beginning, the ratio is just a fraction of the phonon energy, which indicates that the IXS spectra contain well-defined phonon peaks. With increasing Q values, the ratio becomes larger than one. The damping is so strong that the phonon peaks become not well defined. At the hydrodynamic

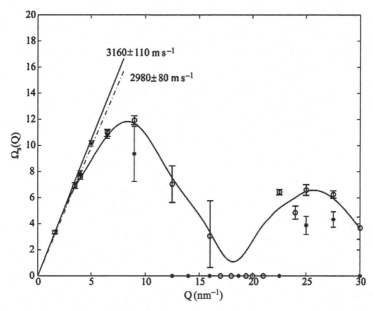

Figure 10.15 The comparison of the phonon-dispersion relation of two samples. The circles with error bars show the dispersion relation of the 40 wt% calf-thymus Na-DNA in H_2O. The solid curve is drawn to guide the eyes. The stars with error bars show the dispersion relation of 40 wt% calf-thymus Na-DNA in 0.042 M spermidine. The added trivalent counterions clearly overdamped phonons in the intermediate Q range (12.5–22.5 nm^{-1}) (Liu *et al.*, 2005c). Reprinted with permission from Liu *et al.* (2005c). Copyright 2005, AIP Publishing LLC.

limit, the ratio between the phonon damping and phonon energy should be linearly dependent on Q. From the results shown in Figure 10.16, the linear region is only up to about $Q \sim 6.5$ nm^{-1}, after which the ratio clearly deviates from linearity with Q.

In this example, Liu *et al.* (2005) have successfully applied the generalised three effective eigenmode theory (GTEE) to analyse the IXS spectra of aligned 40 wt% DNA samples with different counterions. They have constructed the phonon-dispersion relation along the DNA axial direction from the calculated eigenmodes. The sound speed obtained from the linear portion of the dispersion relations is about 3100 m s^{-1}, which is approximately twice the sound speed of a DNA sample with the similar hydration level observed in the hydrodynamic limit by Brillouin light scattering. Since it is known that the high-frequency sound speed of water in this Q range is also about 3000 m s^{-1}, they attribute this difference of the sound speed of aligned DNA rods to the strong coupling of DNA dynamics to that of the hydration water dynamics in the same Q range. It is observed that the addition of different counterions changes the sound speed at this Q range only slightly.

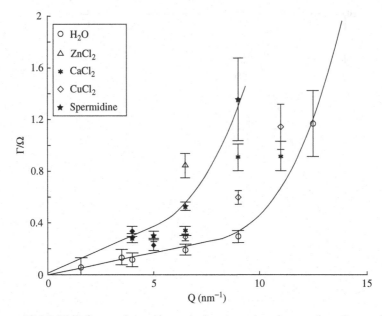

Figure 10.16 This figure shows the ratio between the phonon damping and the phonon frequency of DNA samples with different counterions. The circles with error bars are the results of 40 wt% calf-thymus Na-DNA in H_2O. The up triangle, star, and diamond symbols correspond to the results from the Zn^{2+}, Ca^{2+} and Cu^{2+} samples, respectively. The solid lines are drawn to guide the eyes. Reprinted with permission from Liu *et al.* (2005c). Copyright 2005, AIP Publishing LLC.

However, its effects on the phonon damping are very strong. The phonons at the range between ~ 12.5 and $\sim 22.5 \, nm^{-1}$ are over-damped by the added counterions according to their model analyses. The phonon damping becomes stronger with the addition of extra counterions. At the same ionic strength, the phonon damping due to addition of the trivalent counterion, spermidine, is stronger than those due to divalent counterions.

10.3.4 IXS studies of the collective vibrational motions in proteins

In order for a protein molecule to exhibit biological functions, it cannot be stationary. It is naturally moving because of the thermal fluctuations of all the constituent atoms in the protein molecule, generally known as protein dynamics (Frauenfelder *et al.*, 2009). Especially, intra-protein motions with characteristic time scale of less than 1 microsecond are of particular interest and have been extensively studied by using various types of probing methods, such as neutron scattering (Doster *et al.*, 1989; Chen *et al.*, 2006c) and Mössbauer spectroscopy (Lichtenegger *et al.*, 1999), in the last few decades. These investigations provided insights on how the atomic self-motion of constituent atoms, coupled with the protein structure and environ-ment, participates in the establishment of protein enzymatic function. However, the study of the short time collective motions of a protein molecule and its hydra-tion water is still limited, although significant efforts have been made on this topic (Tarek and Tobias, 2002; Born *et al.*, 2009; Orecchini *et al.*, 2009; He *et al.*, 2011; Paciaroni *et al.*, 2012).

Since 2008, with the development of the high-resolution inelastic X-ray scat-tering (IXS) technique, a powerful tool for the study of the protein collective motions on the sub-picosecond time scales has emerged (hereafter we will refer *sub-picosecond time scales* as *short time*) (Liu *et al.*, 2008b; Leu *et al.*, 2010; Yoshida *et al.*, 2010; Li *et al.*, 2011; Wang *et al.*, 2013). The results show that on such a short time scale, protein samples exhibit phonon-like excitations based on collective vibrations of constituent atoms. One interesting phenomenon is that, in certain Q (Q is the magnitude of the wave-vector transfer in scattering experiments) range, the energy of the phonon-like excitation, or equivalently, the frequency of the intra-protein collective vibration, will decrease significantly when the temperature increases from about 200 K to that of physiological temperature. Meanwhile, the phonon population increases also substantially (Liu *et al.*, 2008b; Li *et al.*, 2011; Wang *et al.*, 2013). In these papers, the former phenomenon is called *phonon energy softening* and the latter is called *phonon population enhancement*. It is important to understand these two phenomena in order to clarify the essence of the intra-protein short time collective vibration and its relation to the protein enzymatic function. Since the protein dynamics is closely related to structure and environment (Frauenfelder *et al.*, 2009), Wang *et al.* (2014c) are mainly exploring the following questions:

(i) How do the protein structures influence the phonon dispersion relation? In particular, whether a presence of tertiary and secondary structures, which are the unique features of a protein, is the necessary condition for the onset of the phonon energy softening and population enhancement?

(ii) What is the role of the hydration water in the short time intra-protein collective motion?

To answer these questions, Wang *et al.* (2014c) performed a series of IXS exper-iments at the 3-ID-XSD beamline as described in several papers (Sinn, 2001; Alatas *et al.*, 2011; Toellner *et al.*, 2011) at the Advanced Photon Source (APS), Argonne National Laboratory on five different powders, i.e. hydrated native lysozyme, dry lysozyme, hydrated denatured lysozyme, hydrated denatured α-Chymotrypsinogen A, and a dry mixture of amino acids. The hydration level (h, defined as $h = [\text{g H}_2\text{O}]/[\text{g dry protein}]$) for all the hydrated samples is 0.3. This value is well above $h = 0.2$–0.25, at which the polar groups on the protein surface are just completely saturated with water (Careri *et al.*, 1980). In this case the native lysozyme can gain enough mobility to function as enzyme. The IXS spectra can be analysed using the generalised three effective eigenmode (GTEE) theory (Liao and Chen, 2001), which is applicable to a polyatomic fluid such as protein molecules. GTEE theory has been successfully applied to the studies of several different biomaterials such as hydrated lipid bilayers (Chen *et al.*, 2001), liquid crystalline DNA (Liu *et al.*, 2004), and globular proteins (Wang *et al.*, 2013). The theoretical normalised dynamic structure factor (NDSF) from GTEE, multiplied by the detailed balance factor ($\exp(E/2k_BT)$) (Squires, 1978), and then convoluted with the energy resolution function of the IXS spectrometer, and then used to fit the measured IXS spectra. Figure 10.17 shows two examples of the fitting results for the hydrated denatured lysozyme sample at $T = 200$ K and $h = 0.3$. One can see that for both the Rayleigh peak and the two Brillouin peaks, the agreement between measured and calculated NDSFs is satisfactory.

The experimental results based on the GTEE analysis are discussed by Wang *et al.* (2014c) as follows. Figure 10.18(a) shows the dispersion relation of the

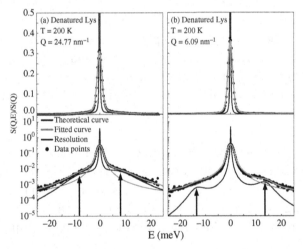

Figure 10.17 IXS spectra for the denatured lysozyme sample at $T = 200$ K and $h = 0.3$ for different Q values, (a) at 24.77 nm^{-1} and (b) at 6.09 nm^{-1}. For both figures, the upper panel shows data in linear scale, while the lower panel is in logarithmic scale. Thus one can check the fitting quality for both the Rayleigh peak and Brillouin peak. The black-circled points are the measured spectral data points. The darker line is the extracted NDSF (multiplied by the detailed balance factor). The light line is the fitted spectral intensity. For the lower panels, the lightest line shows the energy resolution function of the spectrometer. The arrows indicate the position of the Brillouin peaks (Wang *et al.*, 2014c). Reproduced by permission of The Royal Society of Chemistry.

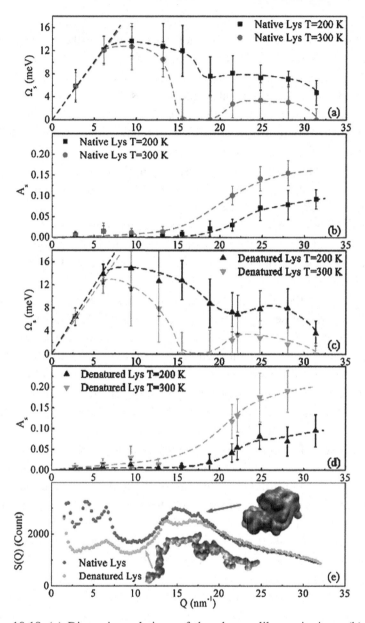

Figure 10.18 (a) Dispersion relations of the phonon-like excitations, (b) fractional area of the Brillouin peaks A_s at two temperatures for hydrated lysozyme at $h = 0.3$. Black squares and circles indicate $T = 200$ K and 300 K, respectively. (c) Dispersion relations of the phonon-like excitations, (d) the fractional area of the Brillouin peaks A_s at two temperatures for hydrated denatured lysozyme at $h = 0.3$. The black up triangles and light-coloured down triangles indicate $T = 200$ K and 300 K, respectively. (e) $S(Q)$ curves and 3D graphical representations for the above two samples. The dashed lines are drawn for easy reading (Wang *et al.*, 2014c). Reproduced by permission of The Royal Society of Chemistry.

phonon-like excitations for the native lysozyme sample with $h = 0.3$ at two temperatures, 200 K and 300 K. One notices that within the Q range of about 12–31 nm^{-1}, the phonon exhibits a substantial softening for the higher temperature (300 K) case indicated by the lower energy. Furthermore, the fractional area of the Brillouin peak A_s is significantly enhanced, as shown in Figure 10.8(b). Note that A_s is related to the phonon population n_p by $n_p(Q) \propto A_s(Q) \cdot S(Q)$, thus the increase of A_s directly indicates an increase of the phonon population. In previous papers of these authors, these phenomena were called as *phonon energy softening* and *phonon population enhancement*, respectively (Liu *et al.*, 2008b; Wang *et al.*, 2013). One notes that the Q range of 12–31 nm^{-1} corresponds to the length scale of about $l = 2$–5.2 Å ($l = 2\pi/Q$), which is similar to the typical length scale of the spatial order of protein secondary structures (β-sheet average distance and α-helix repeat and width), which is about 4–5 Å. Thus in a previous paper, Wang *et al.* (2013) suggested that the *phonon energy softening* was likely to be related to the protein secondary structure.

In order to study the relation between such *softening* and the protein structure, especially the protein secondary structure, Wang *et al.* (2014c) performed IXS measurements at $T = 200$ K and 300 K on a hydrated lysozyme sample denatured by heating in alkali environment (denoted as den-Lys in the following). For this sample, the amount of β-sheet increases, however, because the amount of α-helix is greatly reduced, the overall amount of secondary structure is still largely reduced compared to the native counterpart performed by Mamontov *et al.* (2010). This change of structure can also be reflected in the change of $S(Q)$, which is shown in Figure 10.18(e). Therefore, if the *softening* is strongly correlated to the protein tertiary structure or the secondary structure, the dispersion relation of the den-Lys is expected to be quite different from that of the hydrated native lysozyme. The dispersion relation and the fractional area of the Brillouin peaks A_s of den-Lys are shown in Figure 10.18(c) and (d) respectively. Remarkably, the den-Lys exhibits dispersion relations qualitatively similar to that of the native lysozyme, with an obvious phonon energy softening between 200 K and 300 K. The T–Q dependence of A_s is also similar to that of the native lysozyme. These results suggest that the native structure of protein is not the necessary condition for the *phonon energy softening* and *phonon population enhancement*. Furthermore, the *phonon energy softening* seems not directly related to the secondary structure either, even the maximum softening appears at the Q range of ~ 13–21 nm^{-1}, which corresponds to the length scale of ~ 3–5 Å, which is similar to the length scale of the spatial order of protein secondary structures (4–5 Å).

In order to further verify what was stated above, Wang *et al.* (2014c) performed another experiment on a denatured α-Chymotrypsinogen A sample at $h = 0.3$ (denoted as den-α-Ch). In this experiment, the denaturation is heat-induced in

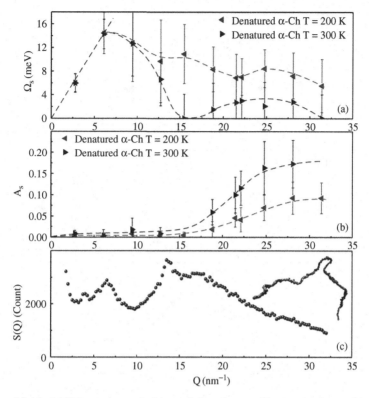

Figure 10.19 (a) Dispersion relations of the phonon-like excitations, (b) fractional area of the Brillouin peaks A_s at two temperatures for hydrated denatured α-Chymotrypsinogen A at $h = 0.3$. The left triangles and the right triangles indicate $T = 200\,\mathrm{K}$ and $300\,\mathrm{K}$ respectively. The dashed lines are drawn for easy reading of the results. (c) $S(Q)$ of the hydrated denatured α-Chymotrypsinogen A at $h = 0.3$ (Wang *et al.*, 2014c). Reproduced by permission of The Royal Society of Chemistry.

alkali environment. In this way, nearly all the tertiary structures and α-helices are removed, and only minor α-sheets are left (Chalikian *et al.*, 1997). Compared with den-Lys, it is found that den-α-Ch is even more unfolded. The dispersion relation of the phonon-like excitation, the fractional area of the Brillouin peaks A_s and $S(Q)$ of this sample are shown in Figure 10.19(a), (b) and (c), respectively. From Figure 10.19(a) and (b) one can find that the *phonon energy softening* and *phonon population enhancement* still exist, similar to the situations in hydrated native lysozyme and den-Lys. Thus, they can further confirm their conclusion that the *phonon energy softening* and *phonon population enhancement* are not directly related to the protein tertiary structure and secondary structure.

The weak dependence of the *phonon energy softening* on the tertiary and secondary structures indicates that this intriguing phenomenon may be closely

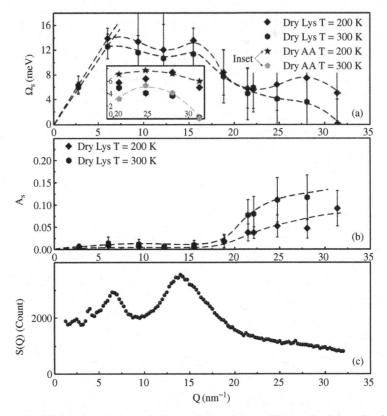

Figure 10.20 (a) Dispersion relations of the phonon-like excitations, (b) fractional area of the Brillouin peaks As at two temperatures for the dry lysozyme sample. The diamonds and hexagons indicate $T = 200\,K$ and $300\,K$, respectively. (c) $S(Q)$ of the dry lysozyme sample. Inset in (a) dispersion relations of the phonon-like excitation for the dry lysozyme sample and the dry mixture of amino acid molecules in the Q range of $20-35\,nm^{-1}$ at two temperatures. The stars and pentagons indicate $T = 200\,K$ and $300\,K$ for the dry mixture of amino acids, respectively. For clarity, we did not add error bars here. The dashed lines are drawn for easy reading (Wang *et al.*, 2014c). Reproduced by permission of The Royal Society of Chemistry.

related to the protein environment, specifically, the hydration water. To verify this, a dry lysozyme sample was measured. Its dispersion relation, the fractional area of the Brillouin peaks A_s and $S(Q)$ are shown in Figure 10.20(a), (b) and (c), respectively. From Figure 10.20(a) one can see that for lysozyme without sufficient amount of hydration water, the *phonon energy softening* is considerably suppressed within $Q < 24\,nm^{-1}$.

This indicates that the hydration water plays a key role in the onset of the *phonon energy softening*. Figure 10.20(b) shows that the *phonon population enhancement*

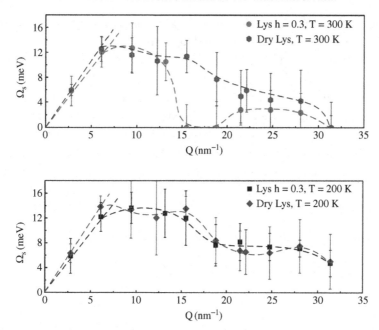

Figure 10.21 Upper panel shows that at $T = 300$ K a sufficient amount of hydration water ($h = 0.3$) causes the phonon energy softening of a lysozyme molecule while a dry molecule remains *hard*. Yet, as shown in the lower panel, both stay *hard* at $T = 200$ K (Wang *et al.*, 2014c). Reproduced by permission of The Royal Society of Chemistry.

still exists, since at high temperature it is easier to excite modes. In addition, for the same temperature, the phonon population of the dry sample is lower than that of the hydrated sample. This explanation can be supported by comparing Figure 10.18(b) with Figure 10.20(b), and is consistent with the previous result shown in (Wang *et al.*, 2013) that more hydration water (in a certain range) can lead to a larger phonon population. From Figure 10.20(a) it is found that for dry lysozyme, the significant phonon energy softening appears at $Q > 25$ nm^{-1}. This Q range corresponds to a length scale of $l < 2.5$ Å, which is highly local. Since it is even within the length scale of glycine, the smallest amino acid molecule, one may conjecture that the motions corresponding to this Q range are mostly the fast motions inside one amino acid residue. To give a preliminary test to this statement, a dry sample composed of amino acids (denoted as AA) in the Q range of 21–31 nm^{-1} was measured. The weight content of this mixture of amino acids is chosen to reproduce the lysozyme composition.

The measured dispersion relation for this sample is shown in the inset of Figure 10.20(a). One can find that as Q is larger than 25 nm^{-1}, the dispersion relation of dry AA becomes similar to that of the dry lysozyme at both temperatures of

$T = 200$ K and 300 K. Notice that in the dry AA, no peptide chain exists, and the motions are basically the highly localised motions within one amino acid. Thus the above-mentioned similarity may suggest that for the lysozyme sample, the motions corresponding to $Q > 25$ nm^{-1} are mostly the localised motions within one amino acid residue. Hydration water at $T = 300$ K causes the protein phonon excitation energy to become softer, as shown in the upper panel of Figure 10.21. But, this is not so for the hydration water at $T = 200$ K, as shown in the lower panel. This may be due to a substantially different values of the hydration water mobility at $T = 200$ K and $T = 300$ K. Hydration water is known to change from predominantly LDL form, at $T = 300$ K, to predominantly HDL form, at $T = 200$ K (Wang *et al.*, 2014c).

References

Adam, G., and Gibbs, J. H. 1965. On the temperature dependence of cooperative relaxation properties in glass-forming liquids. *J. Chem. Phys.*, **43**, 139.

Alatas, A., Leu, B. M., Zhao, J., Yavas, H., Toellner, T. S., and Alp, E. E. 2011. Improved focusing capacity for inelastic x-ray spectrometer at 3-ID of the APS: a combination of toroidal and Kirkpatrick–Baez (KB) mirror. *Nucl. Instrum. Methods Phys. Res. A*, **649**, 166–168.

Alley, W. E., and Alder, B. J. 1983. Generalized transport coefficients for hard spheres. *Phys.Rev. A*, **27**, 3158–3173.

Anderson, P. W. 1995. Through the glass lightly. *Science*, **267**, 1615–1616.

Anderson, V. J., and Lekkerkerker, H. N. V. 2002. Insights into phase transition kinetics from colloid science. *Nature*, **416**, 811–815.

Angell, C. A., and Kanno, H. 1976. Density maxima in high-pressure supercooled water and liquid silicon dioxide. *Science*, **193**, 1121–1122.

Azuah, R. T., Kneller, L. R., Qiu, Y., Tregenna-Piggott, P. L.W, Brown, C. M., Copley, J. R. D., and Dimeo, R. M. 2009. DAVE: a comprehensive software suite for the reduction, visualization, and analysis of low energy neutron spectroscopic data. *J. Res. Natl. Inst. Stan. Technol.*, **114**, 341.

Bafile, U., Verkerk, P., Barocchi, F., de Graaf, L. A., Suck, J.-B., and Mutka, H. 1990. Onset of depature from linearized hydrodynamic behavior in argon gas studied with neutron Brillouin scattering. *Phys. Rev. Lett.*, **65**, 2394–2397.

Ball, P. 2000. *Life's Matrix: A Biography of Water*. New York: Farrar, Straus and Giroux.

Ball, P. 2008. Water as an active constituent in cell biology. *Chem. Rev.*, **108**, 74–108.

Balucani, U., and Zoppi, M. 1994. *Dynamics of the Liquid State*. Oxford: Clarendon Press.

Bansil, R., Berger, T., Toukan, K., Ricci, M. A., and Chen, S.-H. 1986. A molecular dynamics study of the OH stretching vibrational spectrum of liquid water. *Chem. Phys. Lett.*, **132**, 165–172.

Bartsch, E., Eckert, T., Pies, C., and Sillescu, H. 2002. The effect of free polymer on the glass transition dynamics of microgel colloids. *J. Non-Cryst. Solids*, **307–310**, 802–811.

Baxter, R. J. 1967. Method of solution of the Percus–Yevick, hypernetted-chain, or similar equations. *Phys. Rev.*, **154**, 170–174.

Baxter, R. J. 1968a. Ornstein–Zernike relation for a disordered fluid. *Australian J. Phys.*, **21**, 563–570.

Baxter, R. J. 1968b. Percus–Yevick equation for hard spheres with surface adhesion. *J. Chem. Phys.*, **49**, 2770.

Bee, M. 1988. *Quasielastic Neutron Scattering.* Philadelphia, PA: Adam Hilger.

Bellissent-Funel, M.-C., Longeville, S., Zanotti, J. M., and Chen, S.-H. 2000. Experimental observation of the alpha relaxation in supercooled water. *Phys. Rev. Lett.*, **85**, 3644–3647.

Bellissent-Funel, M. C., Chen, S.-H., and Zanotti, J. M. 1995. Single-particle dynamics of water molecules in confined space. *Phys. Rev. E*, **51**, 4558–4569.

Belloni, L. 1986. Electrostatic interactions in colloidal solutions: comparison between primitive and one-component models. *J. Chem. Phys.*, **85**, 519.

Bengtzelius, U., Götze, W., and Sjölander, A. 1984. Dynamics of supercooled liquids and the glass transition. *J. Phys. C*, **17**, 5915.

Berendsen, J. C., Grigera, J. R., and Straatsma, T. P. 1987. The missing term in effective pair potentials. *J. Chem. Phys.*, **91**, 6269–6271.

Beresford-Smith, B., and Chan, D. Y. C. 1982. Highly asymmetric electrolytes: A model for strongly interacting colloidal systems. *Chem. Phys. Lett.*, **92**, 474–478.

Bergenholtz, J., and Fuchs, M. 1999a. Gel transitions in colloidal suspensions. *J. Phys. Condens. Matter*, **11**, 10171–10182.

Bergenholtz, J., and Fuchs, M. 1999b. Nonergodicity transitions in colloidal suspensions with attractive interactions. *Phys. Rev. E*, **59**, 5706–5715.

Berne, B. J., Pechukas, P., and Harp, G. D. 1968. Molecular reorientation in liquids and gases. *J. Chem. Phys.*, **49**, 3125.

Berthier, L., Biroli, G., Bouchaud, J.-P., Cipelletti, L., El Masri, D., L'Hote, D., Ladieu, F., and Pierno, M. 2005. Direct experimental evidence of a growing length scale accompanying the glass transition. *Science*, **310**, 1797–1800.

Bertrand, C. E., and Anisimov, M. A. 2011. Peculiar thermodynamics of the second critical point in supercooled water. *J. Phys. Chem.*, **B115**, 14099–14111.

Bertrand, C. E., Zhang, Y., and Chen, S.-H. 2013a. Deeply-cooled water under strong confinement: neutron scattering investigations and the liquid–liquid critical point hypothesis. *Phys. Chem. Chem. Phys.*, **15**, 721–745.

Bertrand, C. E., Liu, K.-H., Mamontov, E., and Chen, S.-H. 2013b. Hydration-dependent dynamics of deeply cooled water under strong confinement. *Phys. Rev.*, **E87**, 042312.

Bianchi, E., Largo, J., Tartaglia, P., Zaccarelli, E., and Sciortino, F. 2006. Phase diagram of patchy colloids: towards empty liquids. *Phys. Rev. Lett.*, **97**, 168301.

Bianchi, E., Tartaglia, P., La Nave, E., and Sciortino, F. 2007. Fully solvable equilibrium self-assembly process: fine-tuning the clusters size and the connectivity in patchy particle systems. *J. Phys. Chem. B*, **111**, 11765–11769.

Binder, K., and Kob, W. 2011. *Glassy Materials and Disordered Solids: An Introduction to Their Statistical Mechanics.* World Scientific.

Binder, K., Virnau, P., Wilms, D., and Winkler, A. 2011. Spurious character of singularities associated with phase transitions in cylindrical pores. *Eur. Phys. J. Special Topics*, **197**, 227–241.

Blum, L. 1980. Primitive electrolytes in the mean spherical approximation. In: *Theoretical Chemistry: Advances and Perspectives*, Vol. 5. New York: Academic Press, pp. 1–65.

Blum, L., and Høye, J. S. 1978. Solution of the Ornstein–Zernike equation with Yukawa closure for a mixture. *J. Stat. Phys.*, **19**, 319–324.

Boon, J. P., and Yip, S. 1980. *Molecular Hydrodynamics.* New York: McGraw-Hill.

Boonyaratanakornkit, B. B., Park, C. B., and Clark, D. S. 2002. Pressure effects on intra- and intermolecular interactions within proteins. *Biochimica et Biophysica Acta (BBA) – Protein Structure and Molecular Enzymology*, **1595**, 235–249.

Born, B., Weingartner, H., Brandermann, E., and Havenith, M. 2009. Observation of the onset of collective network moions. *J. Am. Chem. Soc.*, **131**, 3752–3755.

Boue, L., Hentschel, H. G. E., Ilyin, V., and Procaccia, L. I. 2011. Statistical mechanics of glass formation in molecular liquids with OTP as an example. *J. Phys. Chem. B*, **115**, 14301–14310.

Brambilla, G., El Masri, D., Pierno, M., Berthier, L., Cipelletti, L., Petekidi, G., and Schofield, A. B. 2009. Probing the equilibrium dynamics of colloidal hard spheres above the mode-coupling glass transition. *Phys. Rev. Lett.*, **102**, 085703.

Brovchenko, I., and Oleinikova, A. 2008. *Interfacial and Confined Water.* New York: Elsevier.

Burkel, E. 1999. *Inelastic Scattering of X-rays with Very High Energy Resolution.* Berlin: Springer-Verlag.

Burkel, E. 2000. Phonon spectroscopy by inelastic x-ray scattering. *Rep. Prog. Phys.*, **63**, 171–232.

Cabane, B., and Duplessix, R. 1982. Organization of surfactant micelles adsorbed on a polymer molecule in water: a neutron scattering study. *J. Phys.*, **43**, 1529–1542.

Caliskan, G., Kisliuk, A., and Sokolov, A. P. 2002. Dynamic transition in lysozyme: role of a solvent. *Journal of Non-Crystalline Solids*, **307**, 868–873.

Caliskan, G., Briber, R. M., Thirumalai, D., Garcia-Sakai, V., Woodson, S. A., and Sokolov, A. P. 2006. Dynamic transition in tRNA is solvent induced. *J. Am. Chem. Soc.*, **128**, 32–33.

Cametti, C., Codastefano, P., Tartaglia, P., Rouch, J., and Chen, S.-H. 1990. Theory and experiment of electrical conductivity and percolation locus in water-in-oil microemulsions. *Phys. Rev. Lett.*, **64**, 1461–1464.

Cardinaux, F., Gibaud, T., Stradner, A., and Schurtenberger, P. 2007. Interplay between spinodal decomposition and glass formation in proteins exhibiting short-range attractions. *Phys. Rev. Lett.*, **99**, 118301.

Careri, G., Gratton, E., Yang, P.-H., and Rupley, J. A. 1980. Correlation of IR spectroscopic, heat capacity, diamagnetic susceptibility and enzymatic measurements in lysozyme power. *Nature*, **284**, 572–573.

Carnahan, N. F., and Starling, K. E. 1969. Equation of state for nonattracting rigid spheres. *J. Chem. Phys.*, **51**, 635–636.

Carpineti, M., and Giglio, M. 1992. Spinodal-type dynamics in fractal aggregation of colloidal clusters. *Phys. Rev. Lett.*, **68**, 3327.

Cerveny, S., Colmenero, J., and Algeriá, A. 2006. Comment on: 'Pressure dependence of fragile-to-strong transition and a possible second critical point in supercooled confined water'. *Phys. Rev. Lett.*, **97**, 189802.

Chalikian, T. V., Voelker, J., Anafi, D., and Breslauer, K. J. 1997. The native and the heat-induced denatured states of τ-chymotrypsinogen A: thermodynamic and spectroscopic studies. *J. Mol. Biol.*, **274**, 237–252.

Chandler, D. 1987. *Introduction to Modern Statistical Mechanics.* Oxford: Oxford University Press.

Chandler, D. 2005. Interfaces and the driving force of hydrophobic assembly. *Nature*, **437**, 640–647.

Chandler, D., Weeks, J. D., and Andersen, H. C. 1983. Van der Waals picture of liquids, solids, and phase transformations. *Science*, **220**, 787–794.

Chang, J., and Sillescu, H. 1997. Heterogeneity at the glass transition: translational and rotational self-diffusion. *J. Phys. Chem. B*, **101**, 8794–8801.

Chen, S.-H. 1971. *Physical Chemistry: An Advanced Treatise*, Vol. 8a. New York: Academic Press.

Chen, S.-H. 1986. Small angle neutron scattering studies of the structure and interaction in micellar and microemulsion systems. *Ann. Rev. Phys. Chem.*, **37**, 351–399.

Chen, S.-H. 1991. Quasi-elastic and inelastic neutron scattering and molecular dynamics of water at supercooled temperature. In Dore, J. C., and Teixeira, J. (eds), *Hydrogen Bonded Liquids*, vol. 329. NATO ASI Series, pp. 289–332.

Chen, S.-H., and Lin, T. L. 1986. *Thermal Neutron Scattering*. New York: Academic Press.

Chen, S.-H., and Sheu, E. Y. 1990. *Micellar Solution and Microemulsions*. Berlin: Springer-Verlag.

Chen, S.-H., and Teixeira, J. 1986. Structure and fractal dimension of protein-detergent complexes. *Phys. Rev. Lett.*, **57**, 2583–2586.

Chen, S.-H., and Yip, S. 1976. Neutron molecular spectroscopy. *Phys. Today*, **29**, 32.

Chen, S.-H., Lefevre, Y., and Yip, S. 1973. Kinetic theory of collision line narrowing in pressurized hydrogen gas. *Phys. Rev. A*, **8**, 3163.

Chen, S.-H., Lai, C. C., Rouch, J., and Tartaglia, P. 1982. Critical phenomena in a binary mixture of n- hexane and nitrobenzene. Analysis of viscosity and light-scattering data. *Phys. Rev. A*, **27**, 1086.

Chen, S.-H., Toukan, K., Loong, C.-K., Price, D. L., and Teixeira, J. 1984. Hydrogen-bond spectroscopy of water by neutron scattering. *Phys. Rev. Lett.*, **53**, 1360–1363.

Chen, S.-H., Rouch, J., Sciortino, F., and Tartaglia, P. 1994. Static and dynamic properties of water-in-oil microemulsions near the critical and percolation points. *J. Phys. Condens. Matter*, **6**, 10855–10883.

Chen, S.-H., Liao, C., Sciortino, F., Gallo, P., and Tartaglia, P. 1999. Models for single-particle dynamics in supercooled water. *Phys. Rev. E*, **59**, 6708–6714.

Chen, S.-H., Liao, C.-Y., Huang, H.-W., Weiss, T. M., Bellisent-Funel, M. C., and Sette, F. 2001. collective dynamics in fully hydrated phospholipid bilayers studied by inelastic X-ray scattering. *Phys. Rev. Lett.*, **86**, 740–743.

Chen, S.-H., Chen, W. R., and Mallamace, F. 2003. The glass-to-glass transition and its end point in a copolymer micellar system. *Science*, **300**, 619.

Chen, S.-H., Liu, L., Chu, X.-Q., Zhang, Y., Fratini, E., Baglioni, P., Faraone, A., and Mamontov, E. 2006a. Experimental evidence of fragile-to-strong dynamic crossover in DNA hydration water. *J. Chem. Phys.*, **125**, 171103.

Chen, S.-H., Liu, L., Fratini, E., Baglioni, P., Faraone, A., and Mamontov, E. 2006b. Observation of fragile-to-strong dynamic crossover in protein hydration water. *Proc. Natl. Acad. Sci. USA*, **103**, 9012–9016.

Chen, S.-H., Liu, L., and Faraone, A. 2006c. Pressure dependence of fragile-to-strong transition and a possible second critical point in supercooled confined water [Reply to Comment]. *Phys. Rev. Lett.*, **97**, 189803.

Chen, S.-H., Mallamace, F., Mou, C.-Y., Broccio, M., Corsaro, C., Farone, A., and Liu, L. 2006d. The violation of Stokes–Einstein relation in supercooled water. *Proc. Natl. Acad. Sci. USA*, **103**, 12974–12978.

Chen, S.-H., Broccio, M., Liu, Y., Fratini, E., and Baglioni, P. 2007. The two-Yukawa model and its applications: the cases of charged proteins and copolymer micellar solutions. *J. Appl. Cryst.*, **40**, s321–s326.

Chen, S.-H., Mallamace, F., Liu, L., Liu, D., Chu, X.-Q., Zhang, Y., Kim, C., Faraone, A., Mou, C.-Y., Fratini, E., Baglioni, P., Kolesnikov, A, I., and Garcia-Sakai, V. 2008. Dynamic crossover phenomenon in confined supercooled water and its relation to the existence of a liquid-liquid critical point in water. *AIP Conference Proceedings*, **982**, 39–52.

Chen, S.-H., Chu, X.-Q., Lagi, M., Liu, D., Chu, X.-C., Zhang, Y., Kim, C., Farone, A., Mou, C.-Y., Fratini, E., Baglioni, P., Kolesnikov, A, I., and Garcia-Sakai, V. 2009a.

Dynamical coupling between a globular protein and its hydration water studied by neutron scattering and MD simulation. In: *WPI-AIMR-2009 Proceedings*.

Chen, S.-H., Zhang, Y., Lagi, M., S.-H., Chong, Baglioni, P., and Mallamace, F. 2009b. Evidence of dynamic crossover phenomena in water and other glass-forming liquids: experiments, MD simulations and theory. *J. Phys.: Condens. Matter*, **21**, 504102.

Chen, S.-H., Zhang, Y., Lagi, M., Chu, X.-Q., Liu, L., Faraone, A., Fratini, E., and Baglioni, P. 2010a. The dynamic response function $T(Q, t)$ of confined supercooled water and its relation to the dynamic crossover phenomenon. *Z. Phys. Chem.*, **224**, 109–131.

Chen, S.-H., Lagi, M., Chu, X.-Q., Zhang, Y., Kim, C., Faraone, A., Fratini, E., and Baglioni, P. 2010b. Dynamics of a globular protein and its hydration water studied by neutron scattering and MD simulations. *Spectroscopy: Biomedical Applications*, **24**, 1–24.

Chen, S.-H., Wang, Z., Kolesnikov, A. I., Zhang, Y., and Liu, K. H. 2013. Search for the first-order liquid-to-liquid phase transition in low-temperature confined water by neutron scattering. *AIP Conference Proc.*, **1518**, 77–85.

Chen, W.-R., Chen, S.-H., and Mallamace, F. 2002. Small-angle neutron scattering study of the temperature-dependent attractive interaction in dense L64 copolymer micellar solutions and its relation to kinetic glass transition. *Phys. Rev. E*, **66**, 021403.

Chiang, W.-S., Fratini, E., Baglioni, P., Liu, D., and Chen, S.-H. 2012. Microstructure determination of calcium-silicate-hydrate globules by small-angle neutron scattering. *J. Phys. Chem. C*, **116**, 5055–5061.

Chiew, Y. C., and Glandt, E. D. 1983. Percolation behaviour of permeable and of adhesive spheres. *J. Phys. A: Math. Gen.*, **16**, 2599.

Chong, S.-H. 2008. Connections of activated hopping processes with the breakdown of the Stokes–Einstein relation and with aspects of dynamical heterogeneities. *Phys. Rev. E*, **78**, 041501.

Chong, S.-H., Chen, S.-H., and Mallamace, F. 2009. A possible scenario for the fragile-to-strong dynamic crossover predicted by the extended mode-coupling theory for glass transition. *J. Phys. Condens. Matter*, **21**, 504101.

Chu, X.-Q., Kolesnikov, A. I., Moravsky, A. P., Garcia-Sakai, V., and Chen, S.-H. 2007. Observation of a dynamic crossover in water confined in double-wall carbon nanotubes. *Phys. Rev. E*, **76**, 021505.

Chu, X.-Q., Fratini, E., Baglioni, P., Faraone, A., and Chen, S.-H. 2008. Observation of a dynamic crossover in RNA hydration water which triggers a dynamic transition in the biopolymer. *Phys. Rev. E*, **77**, 011908.

Chu, X.-Q., Faraone, A., Kim, C., Fratini, E., Baglioni, P., Leao, J. B., and Chen, S.-H. 2009., Proteins remain soft at lower temperatures under pressure. *J. Phys. Chem.*, **B113**, 5001.

Chu, X.-Q., Liu, K.-H., Tyagi, M. S., Mou, C.-Y., and Chen, S.-H. 2010. Low-temperature dynamics of water confined in a hydrophobic mesoporous material. *Phys. Rev. E*, **82**, 020501.

Chu, X.-Q., Mamontov, E., O'Neill, H., and Zhang, Q. 2012. Apparent decoupling of the dynamics of a protein from the dynamics of its aqueous solvent. *J. Phys. Chem. Lett.*, **3**, 380–385.

Chumakov, A. I., Monaco, G., and Monaco, A. *et al.* 2011. Equivalence of the boson peak in glasses to the transverse acoustic van Hove singularity in crystals. *Phys. Rev. Lett.*, **106**, 225501.

Cohen, M. H., and Turnbull, D. 1959. Molecular transport in liquids and glasses. *J. Chem. Phys.*, **31**, 1164.

Cohen, E. D. G., and de Schepper, I. M. 1990. Effective eigenmode description of dynamical processes in dense classical fluids and fluid mixtures. *Nuovo Cimento*, **12**, 521.

Cohen, E. G. D., de Schepper, I. M., and Zuilhof, M. J. 1984. Kinetic theory of the eigenmodes of classical fluids and neutron scattering. *Physica B*, **C127**, 282–291.

Coniglio, A., and Klein, W. 1980. Clusters and Ising critical droplets: a renormalisation group approach. *J. Phys. A: Math. Gen.*, **13**, 2775.

Coniglio, A., De Angelis, U., and Forlani, A. 1977. Pair connectedness and cluster size. *J. Phys. A: Mat*, **10**, 1123.

Copley, J. R. D., and Cook, J. C. 2003. The disk chopper spectrometer at NIST: a new instrument for quasielastic neutron scattering studies. *Chem. Phys.*, **292**, 477.

Corezzi, S., De Michele, C., Zaccarelli, E., Fioretto, D., and Sciortino, F. 2008. A molecular dynamics study of chemical gelation in a patchy particle model. *Soft Matter*, **4**, 1173–1177.

Corezzi, S., De Michele, C., Zaccarelli, E., Tartaglia, P., and Sciortino, F. 2009. Connecting irreversible to reversible aggregation: time and temperature. *J. Phys. Chem. B*, **113**, 1233–1236.

Daniel, I., Oger, P., and Winter, R. 2006. Origins of life and biochemistry under high-pressure conditions. *Chem. Soc. Rev.*, **35**, 858–875.

Dawson, K. A., Foffi, G., Fuchs, M. *et al.* 2000. Higher-order glass-transition singularities in colloidal systems with attractive interactions. *Phys. Rev. E*, **63**, 011401–1–011401–17.

De Michele, C., Gabrielli, S., Tartaglia, P., and Sciortino, F. 2006a. Dynamics in the presence of attractive patchy interactions. *J. Phys. Chem. B*, **110**, 8064–8079.

De Michele, C., Tartaglia, P., and Sciortino, F. 2006b. Slow dynamics in a primitive tetrahedral network model. *J. Chem. Phys.*, **125**, 204710–1–204710–8.

de Schepper, I. C., Cohen, E. G. D., Bruin, C., van Rijs, J. C., Montfrooij, W., and de Graaf, L. A. 1988. Hydrodynamic time correlation functions for a Lennard-Jones fluid. *Phys.Rev. A*, **38**, 271–287.

Domb, C. 1996. *The Critical Point: A Historical Introduction to the Modern Theory of Critical Phenomena*. London: Taylor and Francis.

Dore, J. C., and Teixeira, J. 1991. *Hydrogen-bonded Liquids*. Dordrecht: Kluwer Academic.

Doster, W., Cusack, S., and Petry, W. 1989. Dynamical transition of myoglobin revealed by inelastic neutron scattering. *Nature*, **337**, 754–756.

Doster, W., Busch, S., Gaspar, A. M., Appavou, M.-S., Wuttke, J., and Scheer, H. 2010. Dynamical transition of protein-hydration water. *Phys. Rev. Lett.*, **104**, 098101.

Douglas, J. F., and Leporini, D. 1998. Obstruction model of the fractional Stokes–Einstein relation in glass-forming liquids. *J. Non-Cryst. Solids*, **235237**, 137–141.

Eckmann, J. P., and Procaccia, I. 2008. Ergodicity and slowing down in glass-forming systems withsoft potentials: No finite-temperature singularities. *Phys. Rev. E*, **78**, 011503.

Egelstaff, P. A. 1994. *An Introduction to the Liquid State*. Oxford: Clarendon Press.

Egelstaff, P. A. 1967. *Thermal Neutron Scattering*. New York: Academic Press.

Ehlers, G., Podlesnyak, A. A., Niedziela, J. L., Iverson, E. B., and Sokol, P. E. 2011. The new cold neutron chopper spectrometer at the Spallation Neutron Source: design and performance. *Rev. Sci. Instrum.*, **82**, 85108.

Eisenberger, P., and Platzman, P. M. 1970. Compton scattering of X-rays from bound electrons. *Phys. Rev. A*, **2**, 415.

Erko, M., Wallacher, D., Hoell, A., Hau, T., Zizak, I., and Paris, O. 2012. Density minimum of confined water at low temperatures: A combined study by small-angle scattering of X-rays and neutrons. *Phys. Chem. Chem. Phys.*, **14**, 3852–3858.

Evans, R. 1990. Fluids adsorbed in narrow pores – phase-equilibria and structure. *J. Phys. Conden. Matt.*, **2**, 8989–9007.

Fabbian, L., Götze, W., Sciortino, F., Tartaglia, P., and Thiery, F. 1999. Ideal glass–glass transitions and logarithmic decay of correlations in a simple system. *Phys. Rev. E*, **59**, R1347–R1350.

Fak, B., and Dorner, B. 1997. Damped harmonic oscillator fitting function, including Bose factor. *Physica B*, **234-236**, 1107.

Faraone, A., Chen, S.-H., Fratini, E., Baglioni, P., Liu, L., and Brown, C. 2002. Rotational dynamics of hydration water in dicalcium silicate by quasi-elastic neutron scattering. *Phys. Rev. E*, **65**, 040501.

Faraone, A., Liu, L., Mou, C.-Y., Shih, P.C., Brown, C., Copley, J.R.D., Dimeo, R.M., and Chen, S.-H. 2003a. Dynamics of supercooled water in mesoporous silica matrix MCM-48-S. *Eur. Phys. J.*, **E12**, S59–S62.

Faraone, A., Liu, L., Mou, C.-Y., Shih, P.-C., Copley, J.R.D., and Chen, S.-H. 2003b. Translational and rotational dynamics of water in mesoporous silica materials: MCM-41-S and MCM-48-S. *J. Chem. Phys.*, **119**, 3963–3971.

Faraone, A., Liu, L., Mou, C.-Y., Yen, C.-W., and Chen, S.-H. 2004. Fragile-to-strong liquid transition in deeply supercooled confined water. *J. Chem. Phys.*, **121**, 10843–10846.

Faraone, A., Zhang, Y., Liu, K.-H., Mou, C.-Y., and Chen, S.-H. 2009. Single particle dynamics of water confined in a hydrophobically modified MCM-41-S nanoporous matrix. *J. Chem. Phys.*, **130**, 134512.

Fayer, M.D., and Levinger, N.E. 2010. Analysis of water in confined geometries and at interfaces. *Annual Review of Analytical Chemistry*, **3**, 89–107.

Fenimore, P.W., Frauenfelder, H., McMahon, B.H., and Parak, F.G. 2002. Slaving: solvent fluctuations dominate protein dynamics and functions. *Proc. Natl. Acad. Sci. USA*, **99**, 16047.

Fenimore, P.W., Frauenfelder, H., McMahon, B.H., and Young, R.D. 2004. Bulk-solvent and hydration-shell fluctuations, similar to alpha- and beta-fluctuations in glasses, control protein motions and functions. *Proc. Natl. Acad. Sci. USA*, **101**, 14408.

Fenimore, P.W., Frauenfelder, H., Magazú, S. *et al.* 2013. Concepts and problems in protein dynamics. *Chem. Phys.*, **424**, 2.

Fernandez-Alonso, F., Bermejo, F.J., McLain, S.E. *et al.* 2007. Observation of fractional Stokes–Einstein behavior in the simplest hydrogen-bonded liquid. *Phys. Rev. Lett.*, **98**, 077801.

Fisher, M.E., and Barber, M.N. 1972. Scaling theory for finite-size effects in the critical region. *Phys. Rev. Lett.*, **28**, 1516.

Flory, P.J. 1953. *Principles of Polymer Chemistry*. Ithaca, NY: Cornell University Press.

Foffi, G., Zaccarelli, E., Sciortino, F., Tartaglia, P., and Dawson, K.A. 2000. Kinetic arrest originating in competition between attractive interaction and packing force. *J. Stat. Phys.*, **100**, 363.

Ford, M.H., Auerbach, S.M., and Monson, P.A. 2004. On the mechanical properties and phase behavior of silica: a simple model based on low coordination and strong association. *J. Chem. Phys.*, **121**, 8415.

Franks, F. 2000. *Water: A Matrix of Life*, 2nd edn. Cambridge: Royal Society of Chemistry.

Fratini, E., Chen, S.-H., Baglioni, P., and Bellissent-Funel, M.-C. 2001. Age-dependent dynamics of water in hydrated cement paste. *Phys. Rev. E*, **64**, 020201.

Frauenfelder, H., Chen, G., Berendzen, J. *et al.* 2009. A unified model of protein dynamics. *PNAS*, **106**, 5129.

Freidman, H. L. 1985. *A Course in Statistical Mechanics*. New York: Prentice-Hall.

Frick, B., and Richter, D. 1995. The microscopic basis of the glass transition in polymers from neutron scattering studies. *Science*, **267**, 1939.

Fuentevilla, D. A., and Anisimov, M. A. 2006. Scaled equation of state for supercooled water near the liquid-liquid critical point. *Phys. Rev. Lett.*, **97**, 195702.

Gallo, P., Sciortino, F., Tartaglia, P., and Chen, S.-H. 1996. Slow dynamics of water molecules in supercooled states. *Phys. Rev. Lett.*, **76**, 2730–2733.

Gallo, P., Rovere, M., and Spohr, E. 2000. Glass transition and layering effects in confined water: a computer simulation study. *J. Chem. Phys.*, **113**, 11324–11335.

Gallo, P., Rovere, M., and Chen, S.-H. 2010a. Anomalous dynamics of water confined in MCM-41 at different hydrations. *J. Phys.: Condens. Matter*, **22**, 284102.

Gallo, P., Rovere, M., and Chen, S.-H. 2010b. Dynamic crossover in supercooled confined water: Understanding bulk properties through confinement. *J. Phys. Chem. Lett.*, **1**, 729–733.

Gallo, P., Rovere, M., and Chen, S.-H. 2012. Water confined in MCM-41: a mode coupling theory analysis. *J. Phys.: Condens. Matter*, **24**, 064109.

Garrahan, J. P., and Chandler, D. 2003. Coarse-grained microscopic model of glass-formers. *PNAS*, **100**, 9710–9714.

Gelb, L. D., Gubbins, K. E., Radhakrishnan, R., and Sliwinska-Bartkowiak, M. 1999. Phase separation in confined systems. *Rep. Prog. Phys.*, **62**, 1573–1659.

Götze, W. 1991. *Liquids, Freezing and the Glass Transition*. Amsterdam: Elsevier.

Götze, W. 2009. *Complex Dynamics of Glass-Forming Liquids*. Oxford: Oxford University Press.

Götze, W., and Sjögren, L. 1987. The glass transition singularity. *Z. Phys. B*, **65**, 415–427.

Götze, W., and Sjögren, L. 1992. Relaxation processes in supercooled liquids. *Rep. Prog. Phys.*, **55**, 241–376.

Götze, W., and Sperl, M. 2004. Critical decay at higher-order glass-transition singularities. *J. Phys. Condens. Matt.*, **16**, S4807.

Götze, W., and Voigtmann, Th. 2003. Effect of composition changes on the structural relaxation of a binary mixture. *Phys. Rev. E*, **67**, 021502-1–021502-14.

Granroth, G. E., Kolesnikov, A. I., Sherline, T. E. *et al.* 2010. SEQUOIA: a newly operating chopper spectrometer at the SNS. *AIP Conference Proceedings*, **251**, 012058.

Grimm, G., Stiller, H., Majrzak, C. F., Rupprecht, A., and Dahlborg, U. 1987. Observation of acoustic umklapp-phonons in water-stabilized DNA by neutron scattering. *Phys. Rev. Lett.*, **59**, 1780–1783.

Guo, X. H., and Chen, S.-H. 1990. Reptation mechanism in protein-sodium-dodecylsulfate (SDS) polyacrylamide-gel electrophoresis. *Phys. Rev. Lett.*, **64**, 2579–2582.

Guo, X. H., Zhao, N. M., Chen, S. H., and Teixeira, J. 1990. Small-angle neutron scattering study of the structure of protein/detergent complexes. *Biopolymers*, **29**, 335–346.

Hansen, J.-P., and McDonald, I. R. 2006. *Theory of Simple Liquids*, 3rd edn. New York: Academic Press.

Hayter, J. B., and Penfold, J. 1981. An analytic structure factor for macroion solutions. *Mol. Phys.*, **42**, 109–118.

He, Y., Ku, P. I., Knab, J. R., Chen, J. Y., and Markelz, A. G. 2008. Protein dynamical transition does not require protein structure. *Phys. Rev. Lett.*, **101**, 178103.

He, Y., Chen, J.-Y., Knab, J. R., Zheng, W., and Markelz, A. G. 2011. Evidence of protein collective motions on the Picosecond timescale. *Biophys. J.*, **100**, 1058–1065.

Heremans, K., and Smeller, L. 1998. Protein structure and dynamics at high pressure. *Biochimica Et Biophysica Acta–Protein Structure and Molecular Enzymology*, **1386**, 353–370.

Higgins, J. S., Nicholson, L. K., and Hayter, J. B. 1981. Observation of single chain motion: a polymer melt. *Polymer*, **22**, 163–167.

Hildebrandt, G. 1979. X-ray linear absorption coefficients for silicon and germanium in the energy range 5 to 50 keV. *Acta. Cryst. A*, **35**, 696–697.

Hohenberg, P. C., and Halperin, B. I. 1977. Theory of dynamic critical phenomena. *Rev. Mod. Phys.*, **49**, 435–479.

Holten, V., Bertrand, C. E., Anisimov, M. A., and Sengers, J. V. 2012. Thermodynamics of supercooled water. *J. Chem. Phys.*, **136**, 094507.

Holz, M., and Chen, S.-H. 1978. Quasi-elastic light scattering from migrating chemo static bands of E. Coli. *Biophys. J.*, **23**, 15.

Horn, H. W., Swope, W. C., Pitera, J. W. *et al.* 2004. Development of an improved four-site water model for biomolecular simulations: TIP4P-Ew. *J. Chem. Phys.*, **120**, 9665–9678.

Hrubý, J., Vinš, V., Mareš, R., Hykl, J., and Kalová, J. 2014. Surface tension of supercooled water: no inflection point down to $\approx$ 25°C. *J. Phys. Chem. Lett.*, **5**, 425428.

Ito, K., Moynihan, C. T., and Angell, C. A. 1999. Thermodynamic determination of fragility in liquids and a fragile-to-strong liquid transition in water. *Nature*, **398**, 492–495.

Jennings, H. M. 2000. A model for the microstructure of calcium silicate hydrate in cement paste. *Cement and Concrete Res.*, **30**, 101–116.

Jennings, H. M. 2008. Refinements to colloid model of C-S-H in cement: CM-II. *Cement and Concrete Res.*, **38**, 275–289.

Jorgensen, W. L., and Tiradorives, J. 1988. The Opls potential functions for proteins – energy minimizations for crystals of cyclic-peptides and crambin. *J. Am. Chem. Soc.*, **110**, 1657–1666.

Jung, Y. J., Garrahan, J. P., and Chandler, D. 2004. Excitation lines and the breakdown of the Stokes–Einstein relation in supercooled liquids. *Phys. Rev. E*, **69**, 061205.

Kamgar-Parsi, B., Cohen, E. G. D., and de Schepper, I. M. 1987. Dynamical processes in hard-sphere fluids. *Phys. Rev. A*, **35**, 4781.

Kanno, H., Speedy, R. J., and Angell, C. A. 1975. Supercooling of water to −92°C under pressure. *Science*, **189**, 880–881.

Kautzmann, W. 1959. Some factors in the interpretation of protein denaturation. *Adv. Protein Chem.*, **14**, 1–63.

Kawasaki, K. 1986. Mode coupling and critical dynamics. In: Domb, C., and Green, M. S. (eds), *Phase Transitions and Critical Phenomena*, Vol. 5a. New York: Academic Press.

Kawasaki, K., and Lo, S. M. 1972. Nonlocal shear viscosity and order-parameter dynamics near the critical point of fluids. *Phys. Rev. Lett.*, **29**, 48.

Khodadadi, S., Pawlus, S., Roh, J. H., Sakai, V. G., and Mamontov, E. 2008. The origin of the dynamic transition in proteins. *J. Chem. Phys.*, **128**, 195101–195106.

Kim, C. 2008. *Simulation Studies of Slow Dynamics of Hydration Water in Lysozyme: Hydration Level Dependence and Comparison with Experiment using New Time Domain Analysis*. M.Phil. thesis, Department of Nuclear Science and Engineering, MIT.

Kittel, C. 1963. *Quantum Theory of Solids*. New York: John Wiley.

Koester, L., Rauch, H., Herkens, M., and Schrder, K. 1981. *Summary of Neutron Scattering Length. KFA-Report, Jl-1755.* Jlich, GmBH: Kernforschungsanlage.

Kolafa, J., and Nezbeda, I. 1987. Monte Carlo simulations on primitive models of water and methanol. *Mol. Phys.*, **61**, 161175.

Kolesnikov, A.I., Sinitsyn, V.V., Ponyatovsky, E.G., Natkaniec, I., and Smirnov, L.S. 1994. Neutron scattering studies of the vibrational spectrum of high-density amorphous ice in comparison with ice Ih and VI. *J. Phys.: Condens. Matter.*, **6**, 375–382.

Kolesnikov, A.I., Li, J., Dong, S. *et al.* 1997. Neutron scattering studies of vapor deposited amorphous ice. *Phys. Rev. Lett.*, **79**, 1869–1872.

Kolesnikov, A.I., Li, J., Parker, S.F., Eccleston, R.S., and Loong, C.-K. 1999. Vibrational dynamics of amorphous ice. *Phys. Rev., B*, **59**, 3569–3578.

Kotlarchyk, M., and Chen, S.-H. 1983. Analysis of small angle neutron scattering spectra from polydisperse interacting colloids. *J. Chem. Phys.*, **79**, 2461.

Krishnamurthy, S., Bansil, R., and Wiafe-Akenten. 1983. Low frequency Raman spectrum of supercooled water. *J. Chem. Phys.*, **79**, 5863.

Kumar, P. 2006. Breakdown of the Stokes–Einstein relation in supercooled water. *Proc. Natl. Acad. Sci. USA*, **103**, 12955–12956.

Kumar, P., Yan, Z., Xu, L., Mazza, M.G., Buldyrev, S.V., Chen, S.-H., Sastry, S., and Stanley, H.E. 2006. Glass transition in biomolecules and the liquid–liquid critical point of water. *Phys. Rev. Lett.*, **97**, 177802–177806.

Kumar, P., Wikfeldt, K.T., Schlesinger, D., Pettersson, L.G.M., and Stanley, H.E. 2013. The Boson peak in supercooled water. *Scientific Reports*, **3**, 1980.

Lagi, M., Chu, X.-Q., Kim, C., Mallamace, F., Baglioni, P., and Chen, S.-H. 2008. The low temperature dynamic crossover phenomenon in protein hydration water: simulations vs experiments. *J. Phys. Chem. B*, **112**, 1571–1575.

Lai, C.C. 1972. *Light Intensity Correlation Spectroscopy and its Application to Study of Critical Phenomena and Biological Problems.* Ph.D. thesis, MIT, Cambridge, MA.

Landau, L.D., and Lifshitz, E.M. 1960. *Electrodynamics of Continuous Media.* Reading, MA: Addison-Wesley.

Largo, J., Starr, F.W., and Sciortino, F. 2007. Self-assembling DNA dendrimers: a numerical study. *Langmuir*, **23**, 5896–5905.

Laughlin, W.T., and Uhlmann, D.R. 1972. Viscous flow in simple organic liquids. *J. Phys. Chem.*, **76**, 2317–2325.

Leone, N., Villari, V., and Micali, N. 2012. Modulated heterodyne light scattering set-up for measuring long relaxation time at small and wide angle. *Rev. Sci. Instrum.*, **83**, 083102.

Leu, B., Alatas, A., Sinn, H. *et al.* 2010. Protein elasticity probed with two synchrotron-based techniques. *J. Chem. Phys.*, **132**, 085103.

Li, J. 1996. Inelastic neutron scattering studies of hydrogen bonding in ices. *J. Chem. Phys.*, **105**, 6733–6755.

Li, J., and Kolesnikov, A.I. 2002. Neutron spectroscopic investigation of dynamics of water ice. *J. Mol. Liq.*, **100**, 1–39.

Li, M., Chu, X.-Q., Fratini, E., Baglioni, P., Alatas, A., Alp, E. and Chen, S.-H. 2011. Phonon-like excitation in secondary and tertiary structure of hydrated protein powders. *Soft Matter*, **7**, 9848.

Liao, C.-Y., and Chen, S.-H. 2001. Theory of the generalized dynamic structure factor of polyatomic molecular fluids measured by inelastic X-ray scattering. *Phys. Rev. E*, **64**, 021205.

Liao, C., Choi, S. M., Mallamace, F., and Chen, S.-H. 2000a. SANS study of the structure and interaction of L64 triblock copolymer micellar solution in the critical region. *J. Appl. Crystallogr.*, **33**, 677.

Liao, C.-Y., Chen, S.-H., and Sette, F. 2000b. Analysis of inelastic x-ray scattering spectra of low-temperature water. *Phys. Rev. E*, **61**, 1518–1526.

Lichtenegger, H., Doster, W., Kleinert, T., Birk, A., Sepiol, B., and Vogl, G. 1999. Heme-solvent coupling: A Mssbauer study of myoglobin in sucrose. *Biophys. J.*, **76**, 414–422.

Lide, D. R. 2007. *CRC Handbook of Chemistry and Physics*. Boca Raton, FL: Taylor and Francis.

Limmer, D. T., and Chandler, D. 2011. The putative liquid-liquid transition is a liquid-solid transition in atomistic models of water. *J. Chem. Phys.*, **135**, 134503.

Limmer, D. T., and Chandler, D. 2012. Phase diagram of supercooled water confined to hydrophilic nanopores. *J. Chem. Phys.*, **137**, 044509.

Lindahl, E., Hess, B., and van der Spoel, D. 2001. GROMACS 3.0: a package for molecular simulation and trajectory analysis. *J. Mol. Model.*, **7**, 306–317.

Liu, D., Zhang, Y., Chen, C.-C., Mou, C.-Y., Poole, P. H., and Chen, S.-H. 2007. Observation of the density minimum in deeply supercooled confined water. *Proc. Natl. Acad. Sci. USA*, **104**, 9570–9574.

Liu, D., Zhang, Y., Liu, Y., Wu, J. L., Chen, C. C., Mou, C. Y., and Chen, S.-H. 2008a. Density measurement of 1-d confined water by small angle neutron scattering method: pore size and hydration level dependences. *J. Phys. Chem. B*, **112**, 4309–4312.

Liu, D., Chu, X.-Q., Lagi, M., Zhang, Y., Fratini, E., Baglioni, P., Alatas, A., Said, A., Alp, E., and Chen, S.-H. 2008b. Studies of phononlike low-energy excitations of protein molecules by inelastic X-ray scattering. *Phys. Rev. Lett.*, **101**, 135501.

Liu, K.-H., Zhang, Y., Lee, J.-J., Chen, C.-C., Yen, Y.-Q., Chen, S.-H., and Mou, C.-Y. 2013. Density and anomalous thermal expansion of deeply cooled water confined in mesoporous silica investigated by synchrotron X-ray diffraction. *J. Chem. Phys.*, **139**, 064503.

Liu, L., Faraone, A., and Chen, S.-H. 2002. A model for the rotational contribution to quasi-elastic neutron scattering spectra from supercooled water. *Phys. Rev. E*, **65**, 041506.

Liu, L., Faraone, A., Mou, C.-Y., Shih, P. C., and Chen, S.-H. 2004a. Slow dynamics of supercooled water. *J. Phys.: Condens. Matter*, **16**, S5403–S5436.

Liu, L., Chen, S.-H., Faraone, A., Yen, C.-W., and Mou, C.-Y. 2005b. Pressure dependence of fragile-to-strong transition and a possible second critical point in supercooled confined water. *Phys. Rev. Lett.*, **95**, 117802.

Liu, L., Chen, S.-H., Faraone, A., Yen, C. W., Mou, C.-Y., Kolesnikov, A., Mamontov, E., and Leao, J. 2006. Quasielastic and inelastic neutron scattering investigation of fragile-to-strong transition in deeply supercooled water confined in nanoporous silica matrices. *J. Phys.: Condens. Matter*, **18**, S2261–S2284.

Liu, Y., Berti, D., Faraone, A., Chen, W.-R., Alatas, A., Sinn, H., Alp, E., Baglioni, P., and Chen, S.-H. 2004. Inelastic x-ray scattering studies of phonons in liquid crystalline DNA. *Phys. Chem. Chem. Phys.*, **6**, 1499.

Liu, Y., Chen, W.-R., and Chen, S.-H. 2005a. Cluster formation in two Yukawa fluids. *J. Chem. Phys.*, **122**, 044507.

Liu, Y., Chen, S.-H., Berti, D., Baglioni, P., Alatas, A. Sinn, H., Alp, E., and Said, A. 2005c. Effects of counterion valency on the damping of phonons propagating along the axial direction of liquid-crystalline DNA. *J. Chem. Phys.*, **123**, 214909.

Liu, Y., Palmer, J. C., Panagiotopoulos, A. Z., and Debenddetti, P. G. 2012. Liquid-liquid transition in ST2 water. *J. Chem. Phys.*, **137**, 214505.

Liu, Y.-C., Chen, S.-H., and Huang, J.-S. 1996. Relationship between the microstructure and rheology of micellar solutions formed by a triblock copolymer surfactant. *Phys. Rev. E*, **54**, 1698.

Lo, S. M., and Kawasaki, K. 1973. Frequency-dependence correction to the order-parameter decay rates near the critical point of fluids. *Phys. Rev. A*, **8**, 2176.

Lundahl, P. *et al.* 1986. A model for ionic and hydrophobic interactions and hydrogen-bonding in sodium dodecyl sulfate-protein complexes. *Biochem. Biophys. Acta*, **873**, 20–26.

Magazù, S., Migliardo, F., and Benedetto, A. 2011. Puzzle of protein dynamical transition. *J. Phys. Chem.*, **B115**, 7736–7743.

Makino, S. 1979. Interaction of proteins with amphiphilic substances. *Adv. Biophys*, **12**, 131–184.

Mallamace, F., Gambadauro, P., Micali, N., Tartaglia, P., Liao, C., and Chen, S.-H. 2000. Kinetic glass transition in a micellar system with short-range attractive interaction. *Phys. Rev. Lett.*, **84**, 5431–5434.

Mallamace, F., Broccio, M., Corsaro, C. *et al.* 2006. The fragile-to-strong dynamic crossover transition in confined water: NMR results. *J. Chem Phys.*, **124**, 161102.

Mallamace, F., Branca, C., Broccio, M., Corsaro, C., Mou, C.-Y., and Chen, S.-H. 2007a. The anomalous behaviour of the density of water in the range 30 K < T < 373 K. *Proc. Natl. Acad. Sci.*, **104**, 18387.

Mallamace, F., Broccio, M., Corsaro, C., Faraone, A., Majolino, D., Venuti, V., Liu, L., Mou, C.-Y., and Chen, S.-H. 2007b. Evidence of the existence of the low-density liquid phase in supercooled confined water. *Proc. Natl. Acad. Sci. USA*, **104**, 424–428.

Mallamace, F., Chen, S.-H., Broccio, M., Corsaro, C., Crupi, V., Majolino, D., Venuti, V., Baglioni, P., Fratini, E., Vannucci, C., and Stanley, H. E. 2007c. Role of the solvent in the dynamical transitions of proteins: the case of the lysozyme-water system. *J. Chem Phys.*, **127**, 045104.

Mallamace, F., Branca, C., Corsaro, C., Leone, N., Spooren, J., Chen, S.-H., and Stanley, H. E. 2010. Transport properties of glass-forming liquids suggest that dynamic crossover temperature is as important as the glass transition temperature. *Proc. Natl. Acad. Sci. USA*, **107**, 22457–22462.

Mallamace, F., Corsaro, C., Leone, N., Villari, V., Micali, N., and Chen, S.-H. 2014. On the ergodicity of supercooled molecular glass-forming liquids at the dynamical arrest: the o-terphenyl case. *Nature: Sci. Reports*, **4**, 3747.

Mamontov, E., and Chu, X.-Q. 2012. Water–protein dynamic coupling and new opportunities for probing it at low to physiological temperatures in aqueous solutions. *Phys. Chem. Chem. Phys.*, **14**, 11573.

Mamontov, E., and Herwig, K. W. 2011. A time-of-flight backscattering spectrometer at the Spallation Neutron Source, BASIS. *Rev. Sci. Instrum.*, **82**, 85109.

Mamontov, E., Burnham, C. J., Chen, S.-H., Moravsky, A. P., Loong, C. K., de Souza, N. R., and Kolesnikov, A. I. 2006. Dynamics of water confined in single- and double-wall carbon nanotubes. *J. Chem. Phys.*, **124**, 194703.

Mamontov, E., O'Neill, H., and Zhang, Q. 2010. Mean-squared atomic displacements in hydrated lysozyme, native and denatured. *J. Biol. Phys.*, **36**, 291–297.

Mancinelli, R., Bruni, F., and Ricci, M. A. 2010. Controversial evidence on the point of minimum density in deeply supercooled confined water. *J. Phys. Chem. Lett.*, **1**, 1277–1282.

Manoharan, V. N., Elsesser, M. T., and Pine, D. J. 2003. Dense packing and symmetry in small clusters of microspheres. *Science*, **301**, 483–487.

Mapes, M. K., Swallen, S. F., and Ediger, M. D. 2006. Self-diffusion of super-cooled o-terphenyl near the glass transition temperature. *J. Phys. Chem. B*, **110**, 507–511.

Masciovecchio, C., Bergmann, U., Krisch, M. H., Ruocco, G., Sette, F., and Verbeni, R. 1996a. A perfect crystal X-ray analyser with 1.5 meV energy resolution. *Nucl. Instrum. Methods*, **B117**, 339–340.

Masciovecchio, C., Bergmann, U., Krisch, M. H., Ruocco, G., Sette, F., and Verbeni, R. 1996b. A perfect crystal X-ray analyser with meV energy resolution. *Nucl. Instrum. Methods*, **B111**, 181–186.

Mézard, M., and Parisi, G. 1998. Thermodynamics of glasses: a first principles computation. *Phys. Rev. Lett.*, **82**, 747–750.

Mishima, O. 1994. Reversible first-order transition between two water amorphs at 0.2 GPa and 135 K. *J. Chem. Phys.*, **100**, 5910–5912.

Mishima, O. 2000. Liquid-liquid critical point in heavy water. *Phys. Rev. Lett.*, **85**, 334.

Mishima, O. 2005. Application of polyamorphism in water to spontaneous crystallization of emulsified $LiClH_2O$ solution. *J. Chem. Phys.*, **123**, 154506.

Mishima, O. 2007. Explanation of 'the mysteries of water' by a liquid-liquid critical point. *Rev. High Pressure Sci. Tech.*, **17**, 352–356.

Mishima, O., and Stanley, H. E. 1998a. Decompression-induced melting of ice IV and the liquid-liquid transition in water. *Nature*, **392**, 164–168.

Mishima, O., and Stanley, H. E. 1998b. The relationship between liquid, supercooled and glassy water. *Nature*, **396**, 329–335.

Monaco, G., Cunsolo, A., Ruocco, G., and Sette, F. 1999. Viscoelastic behavior of water in the terahertz-frequency range: an inelastic x-ray scattering study. *Phys. Rev. E*, **60**, 5505–5521.

Montfrooij, W., Svensson, E. C., de Schepper, I. M., and Cohen, E. G. D. 1996. Dynamics in He at 4 and 8 K from inelastic neutron scattering. *J. Low Temp. Phys.*, **105**, 149–183.

Mori, H. 1965. Transport, collective motion, and Brownian motion. *Prog. Theor. Phys.*, **33**, 423–455.

Morineau, D., Xia, Y., and Alba-Simionesco, C. 2002. Finite-size and surface effects on the glass transition of liquid toluene confined in cylindrical mesopores. *J. Chem. Phys.*, **117**, 8966.

Mozhaev, V. V., Heremans, K., Frank, J., Masson, P., and Balny, C. 1996. High pressure effects on protein structure and function. *Proteins–Structure Function and Genetics*, **24**, 81–91.

Nelson, C. A. 1971. The binding of detergents to proteins I. The maximum amount of dodecyl sulfate bound to proteins and the resistance to binding of several proteins. *J. Biol. Chem.*, **246**, 3895–3901.

Novikov, V. N., and Sokolov, A. P. 2003. Universality of the dynamic crossover in glassforming liquids: a magic relaxation time? *Phys. Rev. E*, **67**, 031507.

Nozieres, P. 1964. *Theory of Interacting Fermi Systems*. New York: W. A. Benjamin.

Orecchini, A., Paciaorni, A., Bizzarri, A. R., and Cannistraro, S. 2001. Low-frequency vibrational anomalies in -lactoglobulin: contribution of different hydrogen classes revealed by inelastic neutron scattering. *J. Phys. Chem. B*, **105**, 12150.

Orecchini, A., Paciaroni, A., De Francesco, A., Petrillo, C., and Sacchetti, F. 2009. Collective dynamics of protein hydration water by Brillouin neutron spectroscopy. *J. Am. Chem. Soc.*, **131**, 4664–4669.

Ornstein, L. S., and Zernike, F. 1914. Accidental deviations of density and opalescence at the critical point of a single substance. *Akad. Sci. (Amsterdam)*, **17**, 793.

Oxtoby, D. W., and Gelbart, W. M. 1974. Shear viscosity and order parameter dynamics of fluids near the critical point. *J. Chem. Phys.*, **61**, 2957.

Paciaroni, A., Cinelli, S., and Onori, G. 2002. Effect of the environment on the protein dynamical transitions: a neutron scattering study. *Biophys. J.*, **83**, 1157.

Paciaroni, A., Orecchini, A, Haertlein, M. *et al.* 2012. Vibrational collective dynamics of dry proteins in the terahertz region. *J. Phys. Chem. B*, **116**, 3861–3865.

Parisi, G. 1980. A sequence of approximated solutions to the S-K model for spin glasse. *J. Phys. A: Math. Gen.*, **13**, L115–121.

Paschek, D. 2005. How the liquid-liquid transition affects hydrophobic hydration in deeply supercooled water. *Phy. Rev. Lett.*, **94**, 217802.

Pawlus, S., Khodadadi, S., and Sokolov, A. P. 2008. Conductivity in hydrated proteins: no signs of the fragile-to-strong crossover. *Phys. Rev. Lett.*, **100**, 108103.

Perl, R., and Ferrel, R. A. 1972. Critical viscosity and diffusion in the binary-liquid phase transition. *Phys. Rev. Lett.*, **29**, 51.

Pham, K. N. *et al.* 2002. Multiple glassy states in a simple model system. *Science*, **296**, 104–106.

Poole, P. H., Sciortino, F., Essmann, U., and Stanley, H. E. 1992. Phase-Behavior of Metastable Water. *Nature*, **360**, 324–328.

Poole, P. H., Sciortino, F., Essmann, U., and Stanley, H. E. 1993. Spinodal of liquid water. *Phys. Rev. E*, **48**, 3799–3817.

Poole, P. H., Saika-Voivod, I., and Sciortino, F. 2005. Density minimum and liquid–liquid phase transition. *J. Phys.: Condens. Matt.*, **17**, L431–L437.

Power, E. A., and Thirunamachandran, T. 1978. On the nature of the Hamiltonian for the interaction of radiation with atoms and molecules. *Am. J. Phys.*, **46**, 370.

Pusey, P. N., and van Megen, W. 1986. Phase behavior of concentrated suspensions of nearly hard colloidal spheres. *Nature*, **320**, 340–342.

Pusey, P. N., and van Megen, W. 1989. Dynamic light scattering by non-ergodic media. *Physica (Amsterdam)*, **157A**, 705–741.

Rasmussen, B. F., Stock, A. M., Ringe, D., and Petsko, G. A. 1992. Crystalline ribonuclease-a loses function below the dynamic transition at 220-K. *Nature*, **357**, 423–424.

Rauch, H., and Petrascheck, D. 1978. *Neutron Diffraction*. Berlin: Springer-Verlag.

Reynolds, J. A., and Tanford, C. 1970. Binding of dodecyl sulfate to proteins at high binding ratios. Possible implications for the state of proteins in biological membranes. *Proc. Natl. Acad. Sci. USA*, **66**, 1002–1007.

Ricci, M. A., and Chen, S.-H. 1986. Chemical-bond spectroscopy with neutrons. *Phys. Rev. A*, **34**, 1714–1719.

Ricci, M. A., Chen, S.-H., Price, D. L., Loong, C.-K., Toukan, K., and Teixeira, J. 1986. Observations and interpretation of H-bond breaking by neutron scattering. *Physica B&C*, **136**, 190.

Ridi, F., Fratini, E., and Baglioni, P. 2011. Cement: a two thousand year old nano-colloid. *J. Colloid Interface Sci.*, **357**, 255–264.

Roh, J. H., Curtis, J. E., Azzam, S. *et al.* 2006. Influence of hydration on the dynamics of lysozyme. *Biophys. J.*, **91**, 2573.

Romano, F., Tartaglia, P., and Sciortino, F. 2007. Gas-liquid phase coexistence in a tetrahedral patchy particle model. *J. Phys.: Condens. Matter*, **19**, 322101/1–322101/11.

Röntgen, W. C. 1892. Über die Constitution des Flussig–Wassers. *Ann. Phys. (Leipzig)*, **281**, 91–97.

Rossky, P. J., and Dale, W. D. T. 1980. Generalized recursive solutions to Ornstein Zernike integral equations. *J. Chem. Phys.*, **73**, 2457.

Rouch, J., Tartaglia, P., and Chen, S.-H. 1993. Experimental evidence of nonexponential relaxation near the critical point of a supramolecular liquid mixture. *Phys. Rev. Lett.*, **71**, 1947–1950.

Rovere, M., Ricci, M. A., Vellati, D., and Bruni, F. 1998. A molecular dynamics simulation of water confined in a cylindrical SiO_2 pore. *J. Chem. Phys.*, **108**, 9859–9867.

Rubinstein, M., and Colby, R. H. 2003. *Polymer Physics*. New York: Oxford University Press.

Ruocco, G., and Sette, F. 1999. The high-frequency dynamics of liquid water. *J. Phys.: Condens. Matter*, **11**, R259.

Rupley, J. A., and Careri, G. 1991. Protein hydration and function. *Adv. Protein Chem.*, **41**, 37.

Rupley, J. A., Yang, P. H., and Tollin, G. 1980. Thermodynamic and related studies of water interacting with proteins. In *Water in Polymers. ACS Symp. Series*, **127**, 111–132.

Russel, W. B. 1996. *Colloidal Dispersions*. London: Cambridge University Press.

Sachs, R. G. 1953. *Neutron Diffraction*. Reading, MA: Addison-Wesley.

Sastry, S., Debenedetti, P. G., Sciortino, F., and Stanley, H. E. 1996. Singularity-free interpretation of the thermodynamics of supercooled water. *Phys. Rev. E*, **53**, 6144–6154.

Schiró, G., Caronna, C., Natali, F., Koza, M. M., and Cupane, A. 2011. The protein dynamical transition does not require the protein polypeptide chain. *J. Phys. Chem. Lett.*, **2**, 2275–2279.

Schiró, G., Natali, F., and Cupane, A. 2012. Physical origin of anharmonic dynamics in proteins: new insights from resolutiondependent neutron scattering on homomeric polypeptides. *Phys Rev. Lett.*, **109**, 128102.

Sciortino, F. 2002. One liquid, two glasses. *Nature Mat.*, **1**, 145–146.

Sciortino, F. 2008. Gel forming patchy colloids and network glass formers: thermodynamic and dynamic analogies. *Eur. Phys. J.*, **B64**, 505–509.

Sciortino, F., and Tartaglia, P. 1995. Structure factor scaling during irreversible cluster-cluster aggregation. *Phys. Rev. Lett.*, **74**, 282–285.

Sciortino, F., and Tartaglia, P. 2005. Glassy colloidal systems. *Adv. Phys.*, **54**, 471–524.

Sciortino, F., Belloni, A., and Tartaglia, P. 1995. Irreversible diffusion-limited cluster aggregation: the behavior of the scattered intensity. *Phys. Rev. E*, **52**, 4068–4079.

Sciortino, F., Gallo, P., Tartaglia, P., and Chen, S.-H. 1996. Supercooled water and the kinetic glass transition. *Phys. Rev. E*, **54**, 6331–6343.

Sciortino, F., Fabbian, L., Chen, S.-H., and Tartaglia, P. 1997. Supercooled water and the kinetic glass transition II: collective dynamics. *Phys. Rev. E*, **56**, 5397–5404.

Sciortino, F., Tartaglia, P., and Zaccarelli, E. 2003. Evidence of a higher-order singularity in dense short-ranged attractive colloids. *Phy. Rev. Lett.*, **91**, 268301.

Sciortino, F., Mossa, S., Zaccarelli, E., and Tartaglia, P. 2004. Equilibrium cluster phases and low-density arrested disordered states: the role of short-range attraction and long-range repulsion. *Phys. Rev. Lett.*, **93**, 055701–1–055701–4.

Sciortino, F., Bianchi, E., Douglas, J. F., and Tartaglia, P. 2007. Self-assembly of patchy particles into polymer chains: a parameter-free comparison between Wertheim theory and Monte Carlo simulation. *J. Chem. Phys.*, **126**, 194903.

Sciortino, F., Saika-Voivod, I., and Poole, P. H. 2011. Study of the ST2 model of water close to the liquid-liquid critical point. *Phys. Chem. Chem. Phys.*, **13**, 19759–19764.

Sears, V. F. 1967. Cold neutron scattering by molecular liquids: III. Methane. *Can. J. Phys.*, **45**, 237–254.

Sears, V. F. 1978. Dynamic theory of neutron diffraction. *Can. J. Phys.*, **56**, 1262.

Sears, V. F. 1989. *Neutron Optics*. Oxford: Oxford University Press.

Segré, P. N., Prasad, V., Schofield, A. B., and Weitz, D. A. 2001. Glasslike kinetic arrest at the colloidal–gelation transition. *Phys. Rev. Lett.*, **86**, 6042–6045.

Sette, F., Ruocco, G., Krisch, M. *et al.* 1995. Collective dynamics in water by high energy resolution inelastic X-ray scattering. *Phys. Rev. Lett.*, **75**, 850–853.

Sette, F., Krisch, M. H., Masciovecchio, C., Ruocco, G., and Monaco, G. 1998. Dynamics of glasses and glass-forming liquids studied by inelastic X-ray scattering. *Science*, **280**, 1550.

Sheu, E. Y. 1987. *Theoretical Development and Experimental Verification of a Primitive Model for the Inter-Micellar Interactions*. Ph.D. thesis, MIT, Cambridge, MA.

Shintani, H., and Tanaka, H. 2008. Universal link between the boson peak and transverse phonons in glass. *Nature Materials*, **7**, 870.

Shirahama, K. *et al.* 1974. Free-boundary electrophoresis of sodium dodecyl sulfate-protein polypeptide complexes with special reference to SDS-polyacrylamide gel electrophoresis. *J. Biochem. (Tokyo)*, **75**, 309–319.

Shvyd'ko, Y., Stoupin, S., Shu, D., and Khachatryan, R. 2011. Using angular dispersion and anomalous transmission to shape ultra-monochromatic x-rays. *Phys. Rev. A*, **84**, 053823.

Shvyd'ko, Y. *et al.* 2013. High contrast inelastic x-ray scattering with sub-meV resolution for nano- and mesoscale science. Preprint.

Silvestre-Albero, A., Jardim, E. O., Bruijn, E., Meynen, V., and Cool, P. 2008. Is there any microporosity in ordered mesoporous silicas? *Langmuir: ACS j. surf. colloids*, **25**, 939–943.

Sinha, S. K., Freltoft, T., and Kjems, J. 1984. *Kinetics of Aggregation and Gelation*. Amsterdam: Elsevier Press.

Sinn, H. 2001. Spectroscopy with meV energy resolution. *J. Phys.: Condens. Matter*, **13**, 7525.

Sjöström, J., Swenson, J., Bergman, R., and Shigeharu, K. 2008. Investigating hydration dependence of dynamics of confined water: Monolayer, hydration water and Maxwell–Wagner processes. *J. Chem. Phys.*, **128**, 154503.

Sokolov, A. P., Roh, J. H., Mamontov, E., and Sakai, V. G. 2008. Role of hydration water in dynamics of biological macromolecules. *Chem. Phys.*, **345**, 212–218.

Soper, A. K., and Ricci, M. A. 2000. Structures of high-density and low-density water. *Phys. Rev. Lett.*, **84**, 2881–2884.

Speedy, R. J. 1982. Stability-limit conjecture: an interpretation of the properties of water. *J. Phys. Chem.*, **86**, 982–991.

Speedy, R. J., and Angell, C. A. 1976. Isothermal compressibility of supercooled water and evidence for a thermodynamic singularity at 45°C. *J. Chem. Phys.*, **65**, 851–858.

Squires, G. L. 1978. *Introduction to the Theory of Thermal Neutron Scattering*. New York: Dover.

Stauffer, D., and Aharony, A. 1992. *Introduction to Percolation Theory*. London: Taylor and Francis.

Stillinger, F. H., and Rahman, A. 1974. Improved simulation of liquid water by molecular dynamics. *J. Chem. Phys.*, **60**, 1545.

Stillinger, F. H., and Weber, T. A. 1984. Packing structures and transitions in liquids and solids. *Science*, **225**, 983–989.

Stockmayer, W. H. 1943. Theory of molecular size distribution and gel formation in branched-chain polymers. *J. Chem. Phys.*, **11**, 45–55.

Swallen, S. F., Bonvallet, P. A., McMahon, R. J., and Ediger, M. D. 2003. Self-diffusion of trisnaphthylbenzene near the glass transition temperature. *Phys. Rev. Lett.*, **90**, 015901.

Swenson, J. 2006. Comment on 'Pressure dependence of fragile-to-strong transition and a possible second critical point in supercooled confined water'. *Phys. Rev. Lett.*, **97**, 189801.

Swenson, J., Jansson, H., and Bergman, R. 2006. Relaxation processes in supercooled confined water and implications for protein dynamics. *Phys. Rev. Lett.*, **96**, 247802.

Takahara, S., Nakano, M., Kittaka, S. *et al.* 1999. Neutron scattering study on dynamics of water molecules in MCM-41. *J. Phys. Chem. B*, **103**, 5814–5819.

Tanaka, H. 1998. Simple physical explanation of the unusual thermodynamic behavior of liquid water. *Phys. Rev. Lett.*, **80**, 5750–5753.

Tanford, C. 1968. Protein denaturation. *Adv. Protein Chem.*, **23**, 121–282.

Tarek, M., and Tobias, D. J. 2000. The dynamics of protein hydration water: a quantitative comparison of molecular dynamics simulations and neutron-scattering experiments. *Biophys. J.*, **79**, 3244–3257.

Tarek, M., and Tobias, D. J. 2002. Single-particle and collective dynamics of protein hydration water: A molecular dynamic study. *Phys. Rev. Lett.*, **89**, 275501.

Tarek, M., Tobias, D. J., Chen, S.-H., and Klein, M. L. 2001. Shortwavelength collective dynamics in phospholipid bilayers: a molecular dynamics study. *Phys. Rev. Lett.*, **87**, 238101–1.

Tartaglia, P. 2007. Models of gel-forming colloids. *AIP Conf. Proc.*, **982**, 295–303.

ten Wolde, P. R., and Frenkel, D. 1997. Enhancement of protein crystal nucleation by critical density fluctuations. *Science*, **277**, 1975–1977.

Thiele, E. 1963. Equation of state for hard spheres. *J. Chem. Phys.*, **39**, 474.

Toellner, T. S., Alatas, A., and Said, A. H. 2011. Six-reflection meV-monochromator for synchrotron radiation. *J. Synchrotron Radiat.*, **18**, 605–611.

Toukan, K., and Rahman, A. 1985. Molecular-dynamics study of atomic motions in water. *Phy. Rev. B*, **31**, 2643.

Toukan, K., Ricci, M. A., Chen, S.-H., Loong, C.-K., Price, D. L., and Teixeira, J. 1988. Neutron-scattering measurements of wave-vector-dependent hydrogen density of states in liquid water. *Phys. Rev. A*, **37**, 2580–2589.

Tsujii, K., and Takagi, T. 1975. Proton magnetic resonance studies of the binding of an anionic surfactant with a benzene ring to a protein polypeptide with special reference to SDS-polyacrylamide gel electrophoresis. *J. Biochem. (Tokyo)*, **77**, 511–519.

Turnbull, D., and Cohen, M. H. 1961. Free-volume model of the amorphous phase: glass transition. *J. Chem. Phys.*, **34**, 120.

Turnbull, D., and Cohen, M. H. 1970. On the free-volume model of the liquid-glass transition. *J. Chem. Phys.*, **52**, 3038.

Uhlenbech, G. A., and Ornstein, L. S. 1930. On the theory of Brownian motion. *Phys. Rev.*, **36**, 823.

van Beest, B. W. H., Kramer, G. J, and van Santen, R. A. 1990. Force fields for silicas and aluminophosphates based on ab initio calculations. *Phys. Rev. Lett.*, **64**, 1955–1958.

van Dongen, P., and Ernst, M. 1984. Kinetics of reversible polymerization. *J. Stat. Phys*, **37**, 301–324.

Verwey, E. J., and Overbeek, J. Th. G. 1948. *Theory of the Stability of Lyophobic Colloids.* New York: Elsevier.

Vogel, M. 2008. Origins of apparent fragile-to-strong transitions of protein hydration waters. *Phys. Rev. Lett.*, **101**, 225701.

Vural, D., and Glyde, H. R. 2012. Intrinsic mean-square displacements in proteins. *Phys. Rev. E*, **88**, 011926.

Waisman, E. 1973. The radial distribution function for a fluid of hard spheres at high densities: Mean spherical integral equation approach. *Mol. Phys.*, **25**, 45–48.

Walrafen, G. E. 1964. Raman spectral studies of water structure. *J. Chem. Phys.*, **40**, 3249.

Walrafen, G. E. 1972. In Franks, F. (ed.), *Water: A Comprehensive Treatise*, vol. 1. New York: Plenum Press, p. 151.

Wang, J. S. 1989. Clusters in the three-dimensional Ising model with a magnetic field. *Physica A*, **161**, 249.

Wang, Z., Bertrand, C. E., Chiang, W.-S., Fratini, E., Baglioni, P., Alatas, A., Alp. E. E., and Chen, S.-H. 2013. Inelastic X-ray scattering studies of the short-time collective vibrational motions in hydrated lysozyme powders and their possible relation to enzymatic function. *J. Phys. Chem. B*, **117**, 1186–1195.

Wang, Z., Liu, K.-H., Harriger, L., Leao, J. B., and Chen, S.-H. 2014a. Evidence of the existence of the high-density and low-density phases in deeply-cooled confined heavy water under high pressures. *J. Chem. Phys.*, **141**, 014501.

Wang, Z., Fratini, E., Li, M., Le, P., Mamontov, E., Baglioni, P., and Chen, S.-H. 2014b. Hydration-dependent dynamic crossover phenomenon in protein hydration water. *Phys. Rev. E*, **90**, 042705.

Wang, Z., Chiang, W.-S., Le, P., Fratini, E., Li, M., Alatas, A., Baglioni, P., and Chen, S.-H. 2014c. One role of hydration water in protein: key to the softening of short time intraprotein collective vibrations of specific length scale. *Soft Matter*, **10**, 4298–4303.

Wang, Z., Liu, K.-H., Le, P., Li, M., Chiang, W.-S., Leao, J., Tyagi, M., Copley, J. R. D., Podlsnyak, A., Kolesnikov, A. I., Mou, C.-Y., and Chen, S.-H. 2014d. Boson peak in deeply cooled confined water: a possible way to explore the existence of the liquid-to-liquid transition in water. *Phys. Rev. Lett.*, **112**, 237802.

Wang, Z., Chen, S.-H. *et al.* 2014e. Unpublished.

Wang, Z., Ito, K., Chen, S.-H. *et al.* 2015. To be published.

Warren, B. E. 1969. *X-ray Diffraction*. New York: Addison-Wesley.

Weber, K., and Osborn, M. 1969. The reliability of molecular weight determinations by dodecyl sulfate-polyacrylamide gel electrophoresis. *J. Biol. Chem.*, **241**, 4406–4412.

Wertheim, M. 1984. Fluids with highly directional attractive forces. I. Statistical thermodynamics. *J.Stat.Phys.*, **35**, 19–34.

Wertheim, M. S. 1963. Exact solution of the Percus–Yevick integral equation for hard spheres. *Phys. Rev. Lett.*, **10**, 321.

Wertheim, M. S. 1964. Analytic solution of the Percus–Yevick equation. *J. Math. Phys.*, **5**, 643.

Williams, S. R., and Evans, D. J. 2007. Statistical mechanics of time independent nondissipative nonequilibrium states. *J. Chem. Phys.*, **127**, 184101.

Williams, S. R., and van Megen, W. 2001. Motions in binary mixtures of hard colloidal spheres: melting of the glass. *Phys. Rev. E*, **64**, 041502–1–041502–9.

Wu, C. F., and Chen, S.-H. 1987. SANS studies of concentrated protein solutions: Determination of the charge, hydration, and H/D exchange in cytochrome. *J. Chem. Phys.*, **87**, 6199–6200.

Xia, Y., Dosseh, G., Morineau, D., and Alba-Simionesco, C. 2006. Phase diagram and glass transition of confined benzene. *J. Phys. Chem. B*, **110**, 19735–19744.

Xu, L., Mallamace, F. Z., Yan., Starr, F. W., Buldyrev, S. V., and Stanley, H. E. 2009. Appearance of a fractional Stokes–Einstein relation in water and a structural interpretation of its onset. *Nature Phys.*, **5**, 565569.

Xu, L.-M., Kumar, P., Buldyrev, S. V., Chen, S.-H., Poole, P. H., Sciortino, F., and Stanley, H. E. 2005. Relation between the Widom line and the dynamic crossover in systems with a liquid–liquid phase transition. *Proc. Natl. Acad. Sci. USA*, **102**, 16558–16562.

Yoshida, K., Yamaguchi, T., Kittaka, S., Bellissent-Funel, M.-C., and Fouquet, P. 2008. Thermodynamic, structural, and dynamic properties of supercooled water confined in mesoporous MCM-41 studied with calorimetric, neutron diffraction, and neutron spin echo measurements. *J. Chem. Phys.*, **129**, 054702.

Yoshida, K., Hosokawa, S., Baron, A., and Yamaguchi, T. 2010. Collective dynamics of hydrated lactogloblin by inelastic x-ray scattering. *J. Chem. Phys.*, **133**, 134501.

Yoshida, K., Yamaguchi, T., Kittaka, S., Bellissent-Funel, M.-C., and Fouquet, P. 2012. Neutron spin echo measurements of monolayer and capillary condensed water in MCM-41 at low temperatures. *J. Phys.: Condens. Matter*, **24**, 064101.

Zaccarelli, E. 2007. Colloidal gels: equilibrium and non-equilibrium routes. *J. Phys.: Condens. Matter*, **19**, 323101.

Zaccarelli, E., Foffi, G., Dawson, K. A., Sciortino, F., and Tartaglia, P. 2001. Mechanical properties of a model of attractive colloidal solutions. *Phys. Rev. E.*, **63**, 031501-1–031501-11.

Zaccarelli, E., Foffi, G., Sciortino, F., and Tartaglia, P. 2003. Activated bond-breaking processes preempt the observation of a sharp glass–glass transition in dense short-ranged attractive colloids. *Phys. Rev. Lett.*, **91**, 108301.

Zaccarelli, E., Buldyrev, S., La Nave, E. *et al.* 2005. Model for reversible colloidal gelation. *Phys. Rev. Lett.*, **94**, 218301.

Zaccarelli, E., Sciortino, F., and Tartaglia, P. 2007. A spherical model with directional interactions. I. Static properties. *J. Chem. Phys.*, **127**, 174501-1–174501-10.

Zanotti, J.-M., Bellissent-Funel, M.-C., and Chen, S.-H. 1999. Relaxational dynamics of supercooled water in porous glass. *Phys. Rev. E*, **59**, 3084–3093.

Zhang, Y., Lagi, M., Ridi, F., Fratini, E., Baglioni, P., Mamontov, E., and Chen, S.-H. 2008. Observation of dynamic crossover and dynamic heterogeneity in hydration water confined in aged cement paste. *J. Phys.: Condens. Matter*, **20**, 502101.

Zhang, Y., Liu, K.-H., Lagi, M., Liu, D., Littrell, K., Mou, C.-Y., and Chen, S.-H. 2009a. Absence of the density minimum of supercooled water in hydrophobic confinement. *J. Phys. Chem. B*, **113**, 5007–5010.

Zhang, Y., Lagi, M., Fratini, E., Baglioni, P., Mamontov, E., and Chen, S.-H. 2009b. Dynamic susceptibility of supercooled water and its relation to the dynamic crossover phenomenon. *Phys. Rev. E*, **79**, 040201 (R).

Zhang, Y., Faraone, A., Kamitakahara, W. A., Liu, K.-H., Mou, C.-Y., Leao, J. B., Chang, S., and Chen, S.-H. 2011. Density hysteresis of heavy water confined in a nanoporous silica matrix. *Proc. Natl. Acad. Sci. USA*, **108**, 12206–12211.

Zwanzig, R. 1961. *Lectures in Theoretical Physics*, Vol. 3. New York: Wiley-Interscience.

Index

Printed in the United States
By Bookmasters